Association for Women in Mathematics Series

Volume 7

Series editor

Kristin Lauter, Redmond, WA, USA

Focusing on the groundbreaking work of women in mathematics past, present, and future, Springer's Association for Women in Mathematics Series presents the latest research and proceedings of conferences worldwide organized by the Association for Women in Mathematics (AWM). All works are peer-reviewed to meet the highest standards of scientific literature, while presenting topics at the cutting edge of pure and applied mathematics, as well as in the areas of mathematical education and history. Since its inception in 1971, The Association for Women in Mathematics has been a non-profit organization designed to help encourage women and girls to study and pursue active careers in mathematics and the mathematical sciences and to promote equal opportunity and equal treatment of women and girls in the mathematical sciences. Currently, the organization represents more than 3000 members and 200 institutions constituting a broad spectrum of the mathematical community in the United States and around the world.

More information about this series at http://www.springer.com/series/13764

Jacqueline Dewar • Pao-sheng Hsu
Harriet Pollatsek
Editors

Mathematics Education

A Spectrum of Work in Mathematical Sciences Departments

Springer

Editors
Jacqueline Dewar
Department of Mathematics
Loyola Marymount University
Los Angeles, CA, USA

Pao-sheng Hsu
Independent
Columbia Falls, ME, USA

Harriet Pollatsek
Department of Mathematics and Statistics
Mount Holyoke College
South Hadley, MA, USA

ISSN 2364-5733 ISSN 2364-5741 (electronic)
Association for Women in Mathematics Series
ISBN 978-3-319-44949-4 ISBN 978-3-319-44950-0 (eBook)
DOI 10.1007/978-3-319-44950-0

Library of Congress Control Number: 2016948741

Printed on acid-free paper

This Springer imprint is published by Springer Nature
The registered company is Springer International Publishing AG
The registered company address is: Gewerbestrasse 11, 6330 Cham, Switzerland

Foreword

I am delighted to introduce the first volume devoted to Mathematics Education in our budding Association for Women in Mathematics (AWM) Series with Springer. The idea and the philosophy of the series is to highlight important work by women in the mathematical sciences as reflected in the activities supported by the AWM. Ensuring the mathematics education of the next generation of humans is surely one of the most important roles of our profession. Thus I am very proud of all of the work in mathematics education done by AWM members in mathematical sciences departments as well as the ongoing work of the AWM Education Committee.

This volume was inspired by the panel at the 2016 Joint Mathematics Meetings on "Work in Mathematics Education in Departments of Mathematical Sciences," co-sponsored by the AWM Education Committee and the American Mathematical Society Committee on Education, and co-organized by two of the editors of this volume. The editors sought out contributors from across the mathematical community.

The table of contents reveals the broad scope of the work discussed in the 25 chapters, and the introductory chapter provides further context for the volume. Topics covered reflect ongoing work on mentoring; outreach; policy change; development of faculty, content, and pedagogy; and mathematics education research. It spans work affecting students and teachers of mathematics at all levels. I have high hopes that this volume will advance the discussion of *the value* of this work in mathematics education to our community and to society.

Redmond, WA, USA

Kristin Lauter
AWM President (2015-2017)

Organizers, panelists, and moderator of the 2016 Joint Mathematics Meetings panel, "Work in Mathematics Education in Departments of Mathematical Sciences," co-sponsored by the AWM Education Committee and the AMS Committee on Education

Acknowledgements

We thank the members of the AWM Committee on Education whose thoughtful discussions inspired the 2016 JMM panel. We are grateful to Maura Mast for proposing the idea of a volume on mathematics education in the Springer AWM Series, to Kristin Lauter for enthusiastically endorsing the proposal, and to both of them for their support. Forty-one mathematicians and mathematics educators as well as a social scientist served as reviewers for the chapters in this volume. We appreciate their care and their insight.

Contents

Part I
Benefitting the Readers of this Volume

Chapter 1
Opening Lines: An Introduction to the Volume

Jacqueline Dewar, Pao-sheng Hsu, and Harriet Pollatsek

Abstract In this opening chapter, the editors set the stage for the wide-ranging description and discussion of work in mathematics education awaiting readers of this volume. They define how the phrase "work in mathematics education" is to be understood for this volume and explain how the 25 chapters are grouped according to intended beneficiaries of the work. The editors describe the genesis of the book: how the idea arose in June 2015 and how it was intended to be an extension of the conversation that would take place at the 2016 Joint Mathematics Meetings panel on "Work in Mathematics Education in Departments of Mathematical Sciences," co-sponsored by the Association for Women in Mathematics (AWM) Education Committee and the American Mathematical Society Committee on Education. To entice the reader to explore the volume, the editors highlight some of the contents and note common themes and connections among the chapters. This chapter also summarizes the multi-stage process that brought the idea for this book to fruition so that the reader may understand the selection and peer review process. As many of the chapters do, this one closes with a final reflection by its authors on their involvement in this project.

Keywords Work in mathematics education • Mathematical sciences departments • AWM Education Committee

---

MSC Code
97Axx

J. Dewar (✉)
Department of Mathematics, Loyola Marymount University,
Los Angeles, CA 90045, USA
e-mail: jdewar@lmu.edu

P.-s. Hsu
Independent, Columbia Falls, ME 04623, USA
e-mail: hsupao@maine.edu

H. Pollatsek
Department of Mathematics and Statistics, Mount Holyoke College,
South Hadley, MA 01075, USA
e-mail: hpollats@mtholyoke.edu

J. Dewar et al. (eds.), *Mathematics Education*, Association for Women in Mathematics Series 7, DOI 10.1007/978-3-319-44950-0_1

1.1 Introduction

Many members of the mathematics community in the United States are involved in mathematics education in various capacities. Indeed, through its professional societies and many of their committees, the mathematics community has been working for many decades on improving mathematics education at all levels (See Sect. 25.4.2). Government agencies, private foundations, and the professional societies themselves have funded a great many projects with this goal. Many of these projects involved the efforts and contributions of members of departments of mathematical sciences.

This volume focuses at the level of the people doing the work, often collaboratively, in mathematics education. The contributors tell how their work has been informed by research findings and educational theories. They describe impacts that go well beyond their own classrooms; some have published articles in professional journals about their work. Some authors discuss how their work might be adapted for use elsewhere or direct the interested reader to additional resources. This volume does not contain research articles; instead the authors narrate their efforts and successes (supported in many cases with data collected locally). The volume seeks to initiate a conversation in the mathematical community about difficult issues of how work in mathematics education is perceived and valued.

1.2 Our Definition of Work in Mathematics Education

This volume in Springer's Association for Women in Mathematics Series, *Mathematics Education: A Spectrum of Work in Mathematical Sciences Departments*, offers a sampling of the work in mathematics education undertaken by members of departments of mathematical sciences.[1] For the purposes of this volume, we will take the phrase "work in mathematics education" to mean:

> *endeavors concerning the teaching or learning of mathematics, done by mathematical scientists or mathematics educators in their professional capacity.*

Examples of work encompassed by our definition (and appearing in this volume) include:

- Mathematical outreach,
- Mentoring of those learning or doing mathematics,
- Work with pre-service and in-service teachers of mathematics,
- Development or dissemination of instructional content, materials, activities or teaching practices in mathematics,

[1]Throughout the volume, the word "mathematics" is often used as shorthand for "mathematical sciences."

- Efforts aimed at effecting departmental or disciplinary change relative to the teaching and learning of mathematics,
- Scholarly study (whether considered scholarship of teaching and learning or mathematics education research) of any of the above.

Each chapter illustrates one or more of these to varying degrees.

1.3 The Organization and Goal of the Volume

The participants in and the intended beneficiaries of any work in mathematics education are an important consideration. Collectively, the work described in this volume involves students at all levels from kindergarten through graduate school, K-12 teachers, college and university faculty and administrators, and in some cases the general public. To emphasize this, we have organized the book into five parts according to the primary beneficiaries of the work:

- The readers of this volume (Part I),
- Pre-service and in-service teachers and graduate student instructors (Part II),
- STEM majors (Part III),
- Students in general education courses (Part IV), and
- The general public and the mathematical community at large (Part V).

The writing style is expository, not technical, and should be accessible to and inform a diverse audience of faculty, administrators, and graduate students. Contributors were asked to describe their work, its impact, and how it has been perceived and valued. Some have been willing to be quite candid about the last of these. The overarching goal for publishing this volume is to inform the readership of the breadth of this work and to encourage discussion of its value to the mathematical community and beyond to society at large.

1.4 The Genesis of this Volume

In early June 2015, Kristin Lauter, then President of the Association for Women in Mathematics (AWM), emailed two of the editors, Jacqueline Dewar and Pao-sheng Hsu, in their capacity as co-chairs of the AWM Education Committee. She wrote:

> Maura [Mast] and I met with Springer at the AWM Symposium and we discussed ideas for new volumes [in the Springer AWM Series]. Maura suggested the idea of a volume on math education, and it would be natural for you to lead this effort, and perhaps tie it to the panel you are organizing in January and get contributions from the speakers on your panel. You could also solicit other contributions from people in the community (personal communication, June 9, 2015).

So from the very beginning, this volume was envisioned as an extension of the conversation that would take place at the 2016 Joint Mathematics Meetings[2] (JMM) panel, "Work in Mathematics Education in Departments of Mathematical Sciences." Dewar and Hsu agreed to undertake the task of putting together such a volume and invited Harriet Pollatsek, a member of the AWM Education Committee, to join them in this effort.

1.4.1 The Panel that Inspired this Volume

Discussions within the AWM Education Committee during 2014–2015 prompted and shaped the proposal for the panel. The panel, which took place on January 7, 2016, in Seattle, WA, was co-sponsored with the American Mathematical Society's Committee on Education. Beth Burroughs, Professor, Montana State University, a member of the AWM Education Committee and a contributor to this volume, moderated the panel. Four panelists discussed their work in mathematics education and reflected on its impact and how it has been received in their respective departments:

- Curtis Bennett, Professor and former Associate Dean for Faculty Development and Graduate Studies, Loyola Marymount University,
- Brigitte Lahme, Professor and Department Chair, California State University, Sonoma,
- Yvonne Lai, Assistant Professor, University of Nebraska, Lincoln,
- Kristin Umland, then Associate Professor, University of New Mexico.

Three of the panelists (Bennett, Lai, and Umland) contributed to this volume. Other commitments prevented the fourth panelist from doing so, but she provided other support. A summary of the panelists' remarks can be found in Dewar and Hsu (2016). At the end of the panel a lively discussion with the audience of approximately 60 people ensued.

1.5 The Process that Resulted in this Volume

Prior to this, the volumes in the Springer AWM Series grew out of research conferences or symposia and are collections of research papers. This one, inspired by the JMM Panel, is the first book in the series on mathematics education and is

[2] The Joint Mathematics Meetings conference is jointly sponsored by two major professional societies: the American Mathematical Society and the Mathematical Association of America. It also hosts sessions by other associations, such as the Association for Symbolic Logic, the Association for Women in Mathematics, the National Association for Mathematicians, and the Society for Industrial and Applied Mathematics. Approximately 6000 have attended each year from 2014 to 2016.

expository. In order to present a broad spectrum of work in mathematics education, we recruited beyond the original panel participants. Throughout the process we sought to represent a wide diversity in terms of the type of work in mathematics education, the career stage (early, mid, or late) of the contributor, the institutional type of the contributor (liberal arts, comprehensive and research-intensive institutions, and several secondary schools), as well as gender and ethnicity. The three editors, all mathematicians who have had long careers in mathematics and collegiate education, drew upon many networks of colleagues and scoured abstracts of papers presented at national meetings to develop a list of potential contributors. Thirty-four invitations were extended to submit a 500–1000 word proposal for an expository contribution about their work in mathematics education including how it is received by and affects its intended audience, how the work has affected the proposer's career, and how it has been received by the proposer's colleagues, department, and institution.

The three editors reviewed and discussed each proposal and gave feedback for expanding the proposal into a full chapter draft. Meanwhile, we recruited 41 mathematical scientists and a social scientist as reviewers for the chapters that would be submitted. We aimed to enlist reviewers who had expertise in the type of work in mathematics education that would appear in the volume, and also reviewers who would, in essence, be "general readers." Each submitted chapter was then subjected to a single-blind review by at least three individuals—one expert reviewer, one general reviewer, and at least one editor. In addition, each editor read all of the submissions. The editors discussed the reviews and returned all the formal review material along with a joint editorial report and advice for revising the chapter. The revised submissions were again read by all three editors, and some further editing was done or requested. The result of a nearly year-long intensive process is this volume.

1.6 Reflections on the Volume

With any work in mathematics education, mathematics and its related sciences should be a central feature. Equally important are the participants involved: students, faculty, and sometimes the general public. This volume represents a selection of work in mathematics education by members in departments of mathematical sciences.

For some authors, the work focuses on courses or topics in the core undergraduate mathematics curriculum, including those for the mathematics ***majors***[3] and non-majors: ***calculus*** (Cohen et al., Tomlinson), statistics (Johnson, Williams and Martonosi), linear algebra (Bremser, Wawro), differential equations (Sumner, Tomlinson), group theory (Maycock, Yackel), number theory (Bremser), non-Euclidean geometry for teachers (Burroughs and Burke), introduction to mathemat-

[3] The words in bolded italic in the next few paragraphs are the 11 items listed as aspects of a department's work by the AMS Task Force on Excellence (Ewing 1999, p. 12).

ical modeling (Sumner), complex variables (Tomlinson), and history of mathematics (Sumner). Also included are first-year seminars (Bremser, Catepillán, Fung, Sumner) and capstone courses (Bennett, Cohen et al., Williams and Martonosi).

Teacher preparation is an important mission of a department and plays a critical role in the health of the discipline. Several chapters (Bennett, Bremser, Burroughs and Burke) document different aspects of this work within the department, including one (Lai et al.) that describes the preparation of graduate teaching assistants to be future mathematics faculty. Bremser, Karakok et al., Seshaiyer and Kappmeyer, and Umland and Black work with K-12 teachers outside of the physical space of a department.

Indeed, ***outreach*** takes different forms: in addition to Math Circles for teachers and Math Circles for students (Karakok et al.), there are talks with the public at the National Museum of Mathematics (Greenwald) and traveling workshops for teachers and college faculty (von Renesse).

Several authors include designs of a ***graduate*** course for teachers: Bremser, Sumner, and Wawro.

For the large number of students who need a course that is mathematically before the ***precalculus*** level, there is a discussion about teaching college algebra and intermediate algebra (Lai et al.). For ***general education*** students, there are two versions of a quantitative reasoning course, a class that serves many in place of college algebra (Lopez et al.) and an interdisciplinary seminar (Fung). There are also a course for liberal arts students using dance movement (von Renesse) and a course in ethnomathematics (Catepillán) on mathematics in non-Western cultures.

Several authors (Catepillán, Fung, von Renesse) describe ***interdisciplinary*** courses they created. Sometimes the first-year seminar is the venue for these courses.

In terms of teaching methods, many authors discuss their preference for inquiry-based methods (Bremser, von Renesse), several want students to discover the mathematics they are learning (Maycock, von Renesse, Yackel), several use "tactile" techniques (Karakok et al., Tomlinson, Yackel), and one employs a flipped or blended approach (Tomlinson). Many use student projects and research (Bennett, Bremser, Catepillán, Cohen et al., Johnson, Sumner, Williams and Martonosi). Several chapters in the volume (Chaps. 11, 12, 20, 22, and 23)[4] focus on the use of writing. Another format in the form of a "Clinic" is discussed in the chapter by Williams and Martonosi (Chap. 12) where students produce "deliverables" for real clients. Greenwald describes some mathematical activities she and a colleague have developed from animated sitcoms, bringing popular culture into the classroom.

We asked our authors to provide any information on assessments of what they have done. Quantitative methods were used in two chapters (Chaps. 17 and 22) and many others employed qualitative methods to assess some aspects of the work.

One kind of work that this volume does not contain is a research paper, although some authors (Bennett, Burroughs and Burke, Johnson, Wawro) report on the ***research*** they did. All use research or professional guidelines to support and inform

[4]The reader will find both "write to learn" and "write-to-learn" appearing in a chapter, as they do in many texts in the literature in writing across the curriculum.

their work. As editors, we made no attempt to distinguish what is from what is not "research" or "scholarly" work in mathematics education. Instead, there is a chapter on language use among different communities (Chap. 2, Hauk and Toney). As a research mathematician, Bennett (Chap. 4) gives a glimpse of his struggle with the language in mathematics education.

We want the reader to evaluate each piece of work on its merits. Two mathematicians (Cordero and Mast) who have moved to administration provide their perspectives as academic leaders on the value of the kind of work described in the volume. The chapter by Umland and Black delineates several categories of work that they label "scholarly" while noting that "traditionally [these would not be] considered research" (Chap. 9, p. 127). The authors then detail specific ways to evaluate each type of work based on the tangible product it produces.

External funding does make a difference in much of the work. In fact, over half of our chapters acknowledge that the work was supported by outside funding. One entire chapter is devoted to a description of the Carleton College Summer Mathematics Program, a funded program that has built a community of women becoming mathematicians (Richardson).

Several authors also connect their work with a "social justice" theme in paying special attention to students in groups underrepresented in mathematics: ethnic minorities such as Native American, Hispanic, African American and those with economic hardship. Also included are first-generation college and university students, as well as students who work or are considered "non-traditional" (Bremser, Catepillán, Lopez et al., Cohen et al., Johnson, von Renesse). Catepillán's ethnomathematics course qualifies as a ***diversity*** course at her university. Some programs are specifically aimed at underrepresented groups (Seshaiyer and Kappmeyer).

The word "change" used to describe an institutional transformation appears explicitly in two chapters in the volume. In one, Cohen et al. describe how their department managed a change in departmental culture: faculty collaborated, shared ideas and results, and provided mutual support. In the other, Holm discusses efforts toward achieving systemic change in the teaching of undergraduate mathematics. Our authors are from different types of institutions that vary in governance, mission, and culture. From the descriptions of their work, we also get a glimpse of the complexities in the enterprise we call mathematics education.

Collaboration is a key word in this volume. Even in chapters with one author, many describe the work they do as a collaborative effort. Support from their institutions, colleagues, and students is also crucial for the work that these authors do. From their reports, we see that the authors have different backgrounds, with a majority on a more or less straight-forward career path, some with a small twist (Bennett, Bremser). Black was and Kappmeyer is a K-12 teacher. Some have changed their careers: Kappmeyer was a civil engineer; Johnson worked as a statistician in medical and in marketing research; Craviotto left a university position to work in a school district; more recently, Umland has moved from academia to a non-profit organization working on K-12 curricular materials.

While some of the courses and work described in this volume are not preparing students for the content of a next mathematics course per se, they will shape students' views of mathematics and their habits of learning mathematics. These views and habits are important for all students whether or not they continue with a course of studies in or using mathematics. All of them will carry experiences from the courses into their lives as parents, members of the work force, citizens who vote, or decision-makers in society.

1.7 Reflection on Our Involvement

From the start, our primary goal has been to draw attention to the breadth and variety of work in mathematics education done in departments of mathematical sciences and to encourage discussion of its value. We will be very satisfied if the volume creates opportunities for those discussions. But, we also hope that the many examples contained in this volume will not just inform, but inspire, readers.

Through our involvement in this project we have learned about a great deal of notable work in mathematics education. We have been impressed by the imagination and dedication, not just of our contributors, but also of all those involved in the work that is described in this volume. Our original belief in the value of this work to the mathematical community, the academy, and society has been further strengthened through the examples presented here. We offer this volume to our readers for their consideration.

References

Dewar, J., & Hsu, P.-s. (2016). AWM-AMS mathematics education panel. *AWM Newsletter, 46*(2), 20–21.

Ewing, J. (Ed.). (1999). *Towards excellence*. Washington, DC: American Mathematical Society. Retrieved June 13, 2016, from http://www.ams.org/profession/leaders/workshops/part_1.pdf.

Chapter 2
Communication, Culture, and Work in Mathematics Education in Departments of Mathematical Sciences

Shandy Hauk and Allison F. Toney

Abstract Communication is much more than words—written, spoken, or unspoken. It is also in how a person participates in or orchestrates discussion (in a hallway or in a meeting). Conversation is shaped by what a person knows or anticipates about colleagues' previous experiences and how to attend to that in the context of the goals of a given professional interaction. This chapter builds a foundation of ideas from discourse theory and intercultural competence development as aspects of communication. The presentation is grounded in two vignettes and several small examples of discourse about work in mathematics education. The ideas and vignettes provide touchstones for noticing and understanding what happens when people communicate across professional cultures within departments of mathematics.

Keywords Professional cultures • Post-secondary mathematics education • Intercultural orientation • Discourse

2.1 Introduction to Noticing *This* and *That*

The human capacity to reason includes a reliance on comparison, on noticing difference: *this* is, or is not, like *that*. Grouping makes for comparison of *these* and *those*, for *us* and *them*. When we compare, we discern similarity and difference. With

---

MSC Code
97B40

S. Hauk (✉)
Science, Technology, Engineering, & Mathematics Program, WestEd,
400 Seaport Court, Ste 222, Redwood City, CA 94063, USA
e-mail: shauk@wested.org

A.F. Toney
Department of Mathematics & Statistics, University of North Carolina Wilmington,
601 S College Road, Wilmington, NC 28403, USA
e-mail: Toneyaf@uncw.edu

J. Dewar et al. (eds.), *Mathematics Education*, Association for Women in Mathematics Series 7, DOI 10.1007/978-3-319-44950-0_2

practice, more fine-grained noticing happens. In mathematics, the noticing happens about elements (propositions) that are fairly stable. A theorem, once proved in a particular axiomatic system, pretty much stays proved.

In education, the noticing happens about elements (people) that are quite dynamic. Any lesson learned from work in mathematics education is subject to revision, debate, reframing, and change.

The chapter is about becoming aware of nuance in the observation of this and that. Yet, the path to awareness is fraught with pitfalls. A unifying feature of these pitfalls is over-reliance on the polarizing of *this* and *that* into *this* VERSUS *that*. In fact, dissimilar perspectives on what constitutes work in mathematics education—even among collaborators on a single project—can result in uncertainty that becomes confusion, turmoil, or conflict. The journey begins with a question for the reader: Would everyone in your department agree that the communication about work in mathematics education in the department is effective, appropriate, inclusive, and respectful?

2.2 Noticing Difference

Successful professional communication involves interacting with the multiplicity of discourse styles that colleagues, curriculum, and department history bring to a conversation. Some faculty work in largely monocultural departments in the sense that most colleagues share experience of a common set of personal and professional norms and practices. However, in the US, departments may have a dozen different foci of professional work. It means faculty, staff, and students are destined to have regular opportunities for cross-cultural experience that, for many, may be fraught with unavoidable uncertainty.

2.2.1 A Note on "Cases"

We ground our discussion of uncertainty in two vignettes, real examples of communication in departments of mathematical sciences (all names have been changed). These are gleaned from the authors' own work in mathematics education. It is our hope to offer windows (and possibly mirrors) on the experiences of those navigating the challenges of communicating across different sub-cultures in mathematics departments.

A vignette-based case is not just a short story. A case combines a vignette that is a context-rich description of a dilemma, challenge, or epitome with an analysis of the vignette. A worthwhile case will give rise to discomfort for the reader. An effective case generates dissonance between what case users thought they knew to be true and what they experience in the vignette and analysis. Such cognitive dissonance is the basis on which new understanding is constructed.

2.2.2 Top Tier Journals: *Noticing Across Two Professional Sub-cultures*

As academics, we have within-professional-group standards for communication about our work. Standards can be seen, for example, in the ways faculty generate and disseminate the various publications they create. Yet, norms vary across different sub-communities within a department (e.g., researchers in undergraduate mathematics education, mathematicians, statistics education researchers, statisticians, teacher educators, etc.). Getting a paper into a particular peer-reviewed journal involves different activities for the author than publishing a book, contributing to a grant proposal, or conducting and reporting on a program change. What they all share, however, is the scholarly standard of peer review. The tricky bit is who is a "peer" and who decides the standards for review? Uncertainty in this aspect of interaction across professional sub-cultures and how some might handle it are illustrated in the first vignette, *Top Tier Journals*.

> ***Top Tier Journals***
> A tenure-track colleague of mine was preparing for her third-year review. Because the department chair was not familiar with her research area, he told her to put together "a list of the top tier journals in the field of math education."
> The colleague immediately sought advice from her peers. She asked questions of 20 faculty members across the US who worked in mathematics education: "What is on the top 10 list for sharing research work, the top 10 list for sharing applied and program-level work (like the report of how we redesigned our sequence of courses for pre-service elementary teachers), and the top 10 list for sharing course-level work (like particular lesson materials or advice on how to use certain approaches in teaching such as inquiry-based learning (IBL) Learning (primary) inquiry-based (secondary)?"
> This group of 20 people agreed on a list of 30 dissemination outlets, though not necessarily on the ordering within a list. Then my colleague came to me. She described what she had done, and said, "Would you go over these lists and let me know what you think? Is there anything obvious that is left out or something you would move from one list to another?"
> My first hint this was going to be an unusual conversation *should have been* noticing that she had taken the chair's instructions and made a task of not one list, but three—one for research, one for applied program work, and one for materials development work. But no, I only noticed that in passing, thinking, "Well the first list is what she was asked for, the other two are useless." Then, reading the first list, I was stunned to see that the *Journal of Mathematics Teacher Education* (*JMTE*), what I would consider—what *my* peers would consider—the top tier journal in our field, was absent from the list.
> At first I was very angry. I thought to myself, "Oh, this is a typical demonstration of the narrowness of the fields and the ignorance of some of my colleagues and the fact that they don't pay attent..."—then I stopped myself.
> I realized, "Wait a minute: She came to me and *asked* me." She recognized there might be something she doesn't know. She is saying it would be worthwhile for her to understand my values. She asked me for help.
> So, while she and I were both surprised she didn't know about *JMTE*, I ended up being ashamed (quietly, to myself) when I reflected on my first response to the other two lists as "useless." In reviewing them, I realized there was a lot of sharing going on out there through open-source resources and conferences and organizations like the Mathematical Association of America (MAA) and the American Mathematical Association of Two-Year Colleges (AMATYC) about which I was completely ignorant. I had trouble coming up with outlets I could add to the last two lists and, to mitigate my shame, I am proud to say it occurred to

me to say, "Let's go talk with Pat and Xie. I remember them talking about IBL. I don't know much about it, but I wonder if the outlets are on the lists."
In the end, it actually turned out to be a positive experience. In part, this was because I was careful not to go off into a rant (except in my head, perhaps). It was an opportunity for us to unpack the subtle and not-so-subtle differences between our work worlds, the way scholarship is valued and the locations in which work in mathematics education is valued.

The first part of the vignette highlights the ways different sub-communities exist within departments—specifically, within the field of research in mathematics education. For both the narrator and her colleague, what was valued depended on what respected peers saw as valuable. Also, note that the colleague was aware of and valued other forms of dissemination, beyond research products, in a way the narrator did not. In the second part of the vignette, the narrator noticed, reflected, and then acted on the difference between what she valued and what the colleague asserted as valuable.

Top Tier Journals highlights the fact that meaning is *situated*. Consider how to interpret each of these statements: "The coffee spilled, get a mop" and "The coffee spilled, get a broom" (Gee 1999, p. 48). In each case, context-based storylines that may or may not be consciously considered are connected to the word "coffee." In the first statement, the cue of "mop" is likely to trigger a situated meaning for coffee as a liquid while, depending on one's experience and available storylines, "broom" may be more likely to bring to mind dried beans (perhaps whole, or perhaps ground up). Meaning also is situated in larger conversations of current and historical social experiences and cultural practices. Situated meanings are dynamic in that they are assembled on the spot, based on past and present experience, "customized in, to, and for context, used always against a rich store of cultural knowledge (cultural models) that are themselves 'activated' in, for, and by contexts." (Gee 1999, p. 63).

2.2.3 Department Dynamics: *Noticing About Department Norms*

In each department a variety of norms exist for how we talk with each other about teaching. A department's norms for respectful communication about other work may be quite different. Consider the uncertainty of the narrator in *Departmental Dynamics*, in noticing the habits sanctioned by her department's norms.

Departmental Dynamics
I was so totally caught by surprise when two colleagues made snarky comments about our colleague Bea's recent work to include attention to social justice in her liberal arts math class. Partly my surprise came from the fact that earlier the same day, in a department meeting, they had spoken up in favor of her efforts to put together summer support for graduate students to be research assistants on various department projects. But a few hours later in the hallway, they were snide and disrespectful.
I had to ask myself: *Why did these people feel comfortable making offensive statements in front of me in the first place? Are they really that free-of-clue?*
Instead of doing or saying anything, I froze – not knowing what to say, what to do, how to respond.
Then I thought about my freezing up. I felt like a bystander at a robbery. I asked myself: *Have I been clear about* my *values?*

> And I answered: *Um, no.*
> Why not? What am I afraid of? What about this department and how communication happens is pumping "frozen in the headlights" juice through my veins? And then I realized I didn't know whom I could talk with about it.
> Who could I turn to and have a reasonable expectation for a productive conversation about examining and possibly modifying communication in the department? We have norms for feedback on research, on teaching, and on service. But what are the department norms for constructive feedback on communication about our work within the department – or even the university? Who decides? How are the norms changed?

Unexamined customs can encourage unexamined habits. Being informed is the first step in challenging a habit. As obvious as this is, it conflicts with one common conversational practice in departments: to speculate about what others think based on conclusions drawn from a few interactions. In scholarly work, such incomplete data gathering would be considered intellectually sloppy.

How might the narrator in *Departmental Dynamics* learn about the habits on which the observed norm rests? What are the (unspoken) assumptions about how people view and discuss teaching? A first step might be to gather more information. She might have conversations with one or two colleagues at a time, as a fact-finding mission, driven by questions like: "What makes teaching worth talking about? What is good teaching? How do you know it when you see someone else do it?" The onus would be on the narrator to avoid evaluating or judging the answers she gets—the purpose is to discover how others think, not to persuade them to think like she does. How people answer can help make explicit some assumptions and provide information for shaping subsequent change-oriented discussions.

This section gave two examples of communication about the contexts in which the work of mathematics education is conducted. The next four sections address ways of being aware of nuance within such interactions.

2.3 Discourse (Big D) and discourse (Little d)

Interactions with other people are shaped by our orientation to noticing and engaging with difference. In the present case, interactions are situated in the tensions among types of work in a mathematics department. Professional awareness includes noticing what a colleague says, and also is present in how a person participates in or orchestrates conversation and discussion (in a hallway or in a meeting). Effective, professionally aware, conversation is molded by what a person knows or anticipates about colleagues' previous experiences and how to attend to that in the context of the goals of a given interaction. For example, knowing how to launch a discussion and negotiate the conflicts that can emerge from a department's norms about each variety of work in mathematics education can require well-developed awareness of multiple professional cultures.

Gee (1996) distinguished between "little d" discourse and "big D" Discourse. "Little d" discourse is about written and spoken language-in-use. It is what we say and what we write. In post-secondary mathematics and mathematics education, this may include connected stretches of utterances, symbolic statements, and mathematical diagrams.

In *Top Tier Journals*, discourse (little d) between the narrator and colleague, what each person said, is absent. Instead, it is summarized by the narrator. Similarly, in *Departmental Dynamics*, the discourse in the narrator's witnessing of what was said by colleagues in two different contexts is summarized. In both cases, the nature of the interaction involved more than the words spoken.

Discourse (big D) describes *situated discourse*. Written with the capital D, Discourse indicates language *and* the norms influencing its use *and* the processes for perpetuating or changing both, in context. Little d discourse is a subset of big D Discourse.

In *Top Tier Journals*, the Discourse included the ways the narrator's interaction with her junior colleague challenged her existing notions about what was valuable in reporting on work in mathematics education. The result was twofold. First was the expansion of the narrator's awareness, noticing and acknowledging the value of types of work other than her own. Second was the willingness to seek advice from others, just as the junior colleague sought her advice. Big D Discourse appears in *Departmental Dynamics* in that the narrator reflected on her desire to contribute to the norms for professional communication in her department. Her inner dialogue examined the kinds of conversation she thought might be needed with her colleagues. The vignette highlights her awareness of herself as a part of the Discourse, rather than a non-participant observer of discourse. As a result, at the end of the vignette she formulated questions whose answers she needed to move forward. In each case, the narrator in the vignette sought ways to use language *and* ways of thinking and valuing that were associated with a group in which the narrator saw herself participating. As Gee described it:

> A Discourse is a socially accepted association among ways of using language, other symbolic expressions, and 'artifacts', of thinking, feeling, believing, valuing, and acting that can be used to identify oneself as a member of a socially meaningful group or 'social network', or to signal (that one is playing) a socially meaningful 'role' (Gee 1996, p. 131).

As in any culture, a department culture has a set of values, beliefs, behaviors, and norms in use by a group that can be reshaped and handed along to others (e.g., existing and new faculty, graduate students, and administrative staff can contribute to the reshaping and handing along). Not everyone in a department may describe or experience the culture in the same way. As evidenced by *Top Tier Journals*, Discourses may differ from person to person or group to group within a department. The narrator in *Departmental Dynamics* thought there was something to navigate, reflected on what needed navigating, but did not yet know how to do the navigation. The Discourse in *Departmental Dynamics* included aspects of the departmental cultural context.

2.4 Framework for Intercultural Awareness and Competence

The ways we are aware of and respond to Discourse is a consequence of our *intercultural orientation*. This is not a reference to our beliefs about culture or about the doing, teaching, or learning of mathematics. Rather, intercultural orientation is the perspective about *difference* each person brings to interacting with other people, in context. For faculty, it includes perceptions about the differences between their own views and values around various types of work in mathematics education, and the views of their colleagues.

To build skill at establishing and maintaining relationships in, and exercising judgment relative to, cross-cultural situation requires the development of intercultural sensitivity (Bennett 2004). The developmental continuum for intercultural sensitivity has five milestone orientations to noticing and making sense of difference: *denial*, *polarization*, *minimization*, *acceptance*, and *adaptation*.

With mindful experience a person can develop from ethno-centric ignoring or *denial* of differences, moving through an equally ethno-centric *polarization* orientation that views the world through an us-versus-them mindset. With growing awareness of commonality, a person enters the less ethno-centric orientation of *minimization* of difference, which may over-generalize sameness and commonalities. From there, development leads to an ethno-relative *acceptance* of the existence of intra- and intercultural differences. Further development aims at a highly ethno-relative *adaptation* orientation in which differences are anticipated and responses to them readily come to mind.

2.4.1 Denial

As noted earlier, a central part of awareness is to observe. In the context of a conversation with colleagues, the *denial* orientation might take the form: "I know the math and the math ed discourse I use, I don't really notice any other discourse." Such an orientation is not denial in the sense of "I'm going to say it is not there" but denial as in "I can't even see it." The view is "we're all members of the department and we all do our work" without attention to what "our work" might mean to others.

2.4.2 Polarization

The *polarization* orientation towards orchestrating conversation might be characterized as: "There's a RIGHT way to talk about things and there's a WRONG way to talk about things. And we're going to make sure we use the right way." For example, depending on the experience and values of the conversant, the "right" way to talk about work in mathematics education may or may not include education discourse or the language of assessment, curriculum, program, or teacher development. Nonetheless, enacting a polarized orientation in talking about work in mathematics education would mean seeing, for instance, that a practice is happening or noticing a norm being developed.

Perhaps, when a faculty member strongly identifies with a particular sub-culture, like research in computational proof, scholarship of teaching and learning (SoTL), or assessment development, that person is loyal to it. And, when focused on right ways and wrong ways of talking, a person may not attend to what is done by people in another group: "What they say they are doing in mathematics education is not worthy of my time or energy." In transitioning from polarization to a minimization of difference, a person may come to a new, still polarized, sense of things: "What you do in math education is so different from what I do, I can't possibly understand, review, or evaluate it."

2.4.3 Minimization

From a *minimization* orientation, in minimizing differences and paying attention to similarities, colleagues may also be very true to their own version of professional culture and valued ways of communicating. For someone mathematically trained, this might be characterized as, "Look how this stuff called math ed is LIKE mathematics teaching. It has a lot in common with teaching, even if the way it is said is a little different. Let's talk about how it is similar. Let's leverage the fact that we have seen this before." From this perspective, any work in mathematics education is similar to all other work in mathematics education—whether one is reflecting on teaching a mathematics class, writing a textbook, engaging in SoTL, leading professional development workshops for in-service teachers, or is researching how students learn to validate proofs.

Consider a basic example in the representation of effective teaching. Suppose the standard in the department is that teaching is successful when numbers from a student evaluation are high. Yet some faculty members, who are also familiar with educational theories, say that teaching is effective when students demonstrate learning in some directly measurable way, such as on a common final exam. It may be characteristic of a minimization orientation to consider both representations once and then note "But these are basically the same, so we'll use the one I know, the one commonly used in the department, the student evaluations."

2.4.4 Acceptance

In developing an *acceptance* orientation, it might be more characteristic to notice and accept either representation of "effective teaching" and suggest faculty use whichever makes most sense for them. A well-developed acceptance orientation might be evidenced when a faculty member alternated between using student evaluations and direct measures of student learning when talking with a colleague. Additionally, she might encourage peers to accept and understand the difference in the two ways of thinking about teaching effectiveness.

More generally, an acceptance orientation might be characterized by statements like: "I'm a mathematician, but am accepting the fact that not all of my colleagues are going to be mathematicians" or "I'm a researcher in mathematics education, but am accepting the fact that not all of my colleagues are going to be interested in that approach" and "I'm accepting the fact that there may be other ways, teacher ed, assessment, or math ed research ways, of talking about the idea of effectiveness in teaching that are valuable and may be even more valuable to my colleagues than my way of talking about it. I can accept that those various ways will come out in the conversation in the department." But a general intention of accepting the different ways may not provide guidance about how to make decisions about which Discourse(s) are useful in a given context (e.g., solving problems in teaching pre-service teachers may not be facilitated by a research mathematics vocabulary, and vice versa).

2.4.5 Adaptation

A further developmental orientation is *adaptation*. Now, not only does one accept that there are these differences, adaptation-oriented people seek for themselves, and find ways to give colleagues, opportunities in noticing, articulating, and responding to those differences. This might be characterized by statements such as, "I am looking for ways to work with colleagues to pursue the opportunities that arise from variety in approach or strategy. I don't have to assert or defend many, or even one method. Effective teaching is a relative thing. My goals are for teaching and learning of rigorous math and those goals include the standard math language and representations. How my colleagues and I connect ideas and access, organize, or value ideas is not necessarily strictly limited to the ways valued by my perspective." In adaptation, a person can converse well with people of differing mindsets, understanding and appropriately using Discourse familiar to conversational partners.

2.4.6 Integration

Though not yet fully tested by researchers, the theory of intercultural competence development also hypothesizes something called an *integration* orientation. This is something that is likely to be very rare. This perspective might be characterized by a statement like: "Okay, that particular approach to this problem of what effective teaching is, that is a whole other way of looking at the world. It's internally consistent, which I value. So, it's okay. And I'm going to integrate what I can while remaining true to mathematics and to my own work in mathematics education. I'm going to be myself as a professional, in that environment." We suspect such a view might be analogous to the ultimate mission of the scholarship of theology: studying a variety of belief systems, without disagreement or approval of the system, while remaining authentic in one's own beliefs. In the research about intercultural competence development, examples of how an integration orientation might be realized come in the shape of expert and effective negotiators in high stakes endeavors (e.g., diplomat, hostage negotiator).

2.5 Being Intentional in Noticing Professional Differences

In a recently concluded project, we spent time and attention on dealing with the realities of navigating the multiple cross-cultural relationships in creating and running graduate courses for secondary mathematics teacher professional development (Hauk et al. 2011, 2014, 2015). Project participants included university staff (26 faculty members and graduate students) in three departments of mathematical sciences whose work included research mathematics, research in mathematics education and teacher education, curriculum development for undergraduate and graduate mathematics, and professional development of in- and pre-service secondary mathematics teachers. Some of the university staff developed and taught courses for teachers and teacher leaders (71 teachers, 23 leaders) while others conducted research on the teaching and learning in those courses.

2.5.1 Example of Difference in Orientation Across Professional Groups

All staff, teachers, and teacher leaders completed a valid and reliable measure of intercultural sensitivity (Hammer 2009). In Fig. 2.1 are the distributions of intercultural orientation for the university staff on the project (faculty members and graduate students). As a group, their orientations were largely in minimization.

In Fig. 2.2, the distribution for university staff is situated in the larger view of intercultural orientations for all of the participants in the project. Notice that the orientations of teachers were more evenly distributed between polarization and minimization while the distribution for teacher leaders was more like that of university staff.

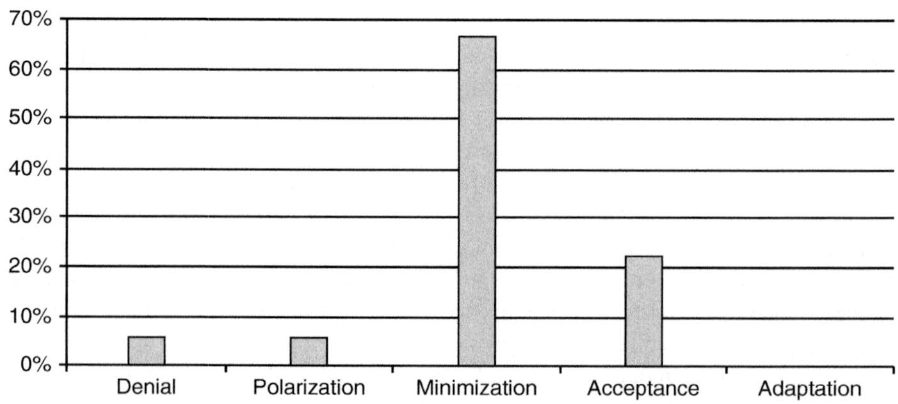

Fig. 2.1 Distribution of university faculty and graduate student intercultural orientations

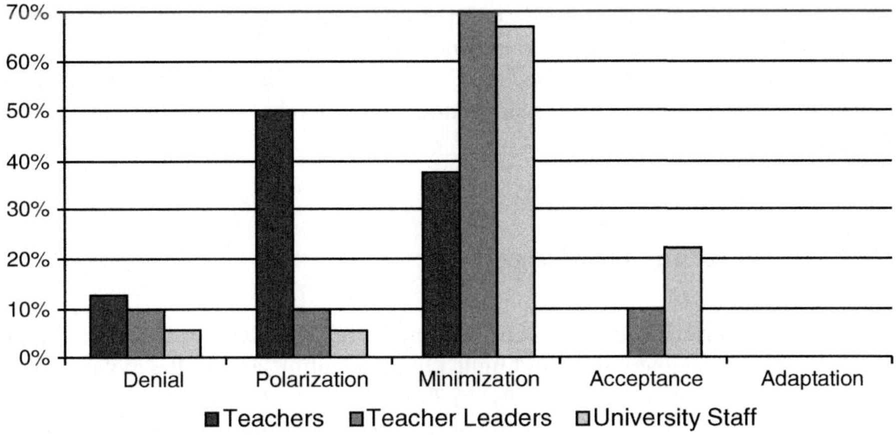

Fig. 2.2 Distributions of all three groups' intercultural orientations

As part of the project, we conducted a debriefing session with each group. The session explained the framework and the five milestone orientations for intercultural sensitivity. In each case, the group saw the distribution of their orientations and that of the other two groups. Each group discussed in their session what knowing this information could contribute to knowledge about themselves and about working with the other two groups.

In particular, university faculty members and graduate students said they felt a challenge in getting teachers to see the connections, the similarities, among ideas. The large proportion of teachers with a polarization orientation meant teacher-participants were willing and able to notice difference. University staff (who were mostly minimizers seeking common ground) often found themselves uncomfortable with this attention to difference. They were stymied about how to negotiate conversations with teachers whose Discourse was framed to highlight difference using right-wrong, strong-weak, good-bad polarization. In the debriefing session, university staff learned that noticing differences within and among things that may appear to be similar is a hallmark of acceptance. The opportunity existed to encourage more detailed exploration of difference and similarity in ways that would support intercultural development for polarizers and minimizers.

With knowledge of the intercultural developmental continuum, and their mostly minimization orientation, the group of university staff also explored the assumption that equality and equity are the same. One approach to teasing apart the two ideas is to think about the distinctions between "fairness" and equality. Consider the following example.

One university faculty member had broken a leg skiing and was using a small cart under one knee when walking. If each program faculty member was expected to give teacher-participants a 40-min walking tour of some part of the university, then the cart-bound faculty member was unfairly burdened. An alternate way to fulfill the responsibility was needed. An unequal but fair solution: the colleague would sit with participants during their first lunch in the dining hall. Not only would this be an excellent addition to the "tour" of the campus, it would give participants a chance to talk informally with a program faculty member (an opportunity absent in the previous plan).

2.5.2 Connecting to the Vignettes

Given these experiences in the recent project, for this chapter we selected material for the two case vignettes to highlight communication across the polarization-minimization-acceptance orientations. In *Top Tier Journals*, the narrator was challenged in a way that might be seen as moving her from polarization towards minimization, while the colleague generating the lists had a minimization orientation, perhaps moving towards acceptance—she was seeking to understand the large and small differences across some types of work in mathematics education. In *Departmental Dynamics*, the acceptance orientation of the narrator might be seen in that she noticed difference and wanted to learn how to negotiate the difference—these are earmarks of early adaptation.

What is more, the vignettes were designed to keep other aspects of Discourse in the background, such as gender. While a deep discussion of the role of gender in communication is beyond the scope of this chapter, communication about work in mathematics education in a department of mathematical sciences may be gender connected in several ways.

2.6 Gender, Discourse, and Professional Culture

By one estimate, two-thirds of the mathematics department faculty who do professional work in mathematics education are women (Reys 2008). This has consequences for how the work is communicated, perceived, and valued. The Discourse resources of women are often different from those of men. In fact, "there are two abiding truths on which the general public and research scholars find themselves in uneasy agreement: (a) men and women speak the same language, and (b) men and women speak that language differently" (Mulac 1998, p. 127). And, we would add, (c) not all women "speak that language differently" in the same way!

2.6.1 Women Speak Differently in Different Ways

International and national variation means factors of ethnic, racial, and other types of group and institutional enculturation and socialization are involved in same-gender professional intergroup communication. For example, one comparison of

African American and European American women found a direct communication style to be more common among African American women than the indirect framing most used by their European American peers. Both groups of women had a goal of reducing potential conflict in the workplace (or, largely in the case of the European American participants, conflict avoidance), but their methods for how to articulate and achieve it were different (Shuter and Turner 1997).

From a gender-as-culture perspective, communication habits emerge from a childhood and adolescence filled with same-sex conversational partners and a lifetime of social expectation (Maltz and Borker 1982). Review of the literature on studies of language and gender has found that women may have access to power (and more acceptance) in a majority culture context when using indirect language, uncertainty, and hedges in relatively long sentences: "Well, I was wondering if...," "Perhaps we might...," "It's kind of...," while men fulfill expectations by referencing quantity or judgments in direct statements: "An evaluation of 3.8...," "It's good...," "Write it down." (Mulac et al. 2001, p. 125).

The fact that interaction in most universities occurs in the context of historically male Discourses makes every interaction between the sexes a *doing* of gender in some way (Uchida 1992). Consequently, gendered communication structures can be (dis)empowering depending on context. For example, one "ironic consequence" for women who adopt a more direct communication style is that they "are rated as less warm and likeable, and evaluators indicate less willingness to comply with their requests" (von Hippel et al. 2011, p. 1312).

Additionally, those whose work focuses on teaching tend to value a pragmatic approach and may seek career rewards based on personal motivation rather than external distinction (Wang et al. 2015). Some have written about the importance of women seeking to participate in the career reward structures and other status quo value systems in the academy (Nicholson and de Waal-Andrews 2005; Olsen et al. 1995). However, embracing the status quo without also attempting to change it has the danger of derailing progress in the intellectual and professional work of mathematics education.

2.6.2 Views of Work in Mathematics Education

What does work in mathematics education in a department of mathematical sciences look like from the various intercultural perspectives, taking gender as an aspect of the Discourse? From a polarized orientation, the situation regarding work in a department may seem to be one of unending conflict, of the male-dominated status quo (them) versus women (us).

From a minimization view, the situation would seem mutable, if slowly, towards a goal of commonality. The more equivocal each type of language use becomes, the more that women use male language features and vice versa, the closer the department comes to an equality in talk. The problem in this over-reliance on commonality is that equality in discourse style is not equity in Discourse. As Marilyn Cochran-Smith and colleagues have recently described it, "With the former, the valence of the terms is primarily about sameness (equality) or difference (inequality), while with the latter, the valence of the terms has primarily to do with fairness and justice (equity) or unfairness and injustice (inequity)" (Cochran-Smith et al. 2016, p. 69).

From an acceptance orientation, gender-as-culture and gender-as-power are overlapping ways of seeing the world and the goal might be a hazy one of "better communication" (though it would be difficult to know what steps to take to move towards the goal). Additionally, in the acceptance view, noticing of differences in language usage would be a tool to understanding the intentions and perceptions of colleagues, with such understanding seen as contributing to "better communication."

Building on this noticing of difference in communication, the adaptation orientation would attend to creating infrastructure that validates and leverages the subtleties of difference and uses variety in Discourses to mitigate marginalization. Here is a very small example: in preparation for every run-of-the-mill department meeting, the chair might provide faculty with the agenda a few days in advance and have each person email her back with a short written summary statement (25–100 words) about one agenda item, perhaps addressing "The things I am wondering about topic *X*" or "Where I'd like to see the department in two years regarding topic *Y*." Creating the norm of considering one's perspective and how to communicate it as preparation for a meeting becomes profoundly useful when the department faces a meeting where a highly charged or high stakes topic will be discussed. It can position the meeting as a place to air ideas and to collaborate on solving a community problem (rather than a place to air grievances).

2.7 Conclusion

Central to effective communication across multiple professional cultures is the strategy of information gathering. We cannot notice nuances in difference until we have enough information to see difference. Tackling the ideas of equity, diversity, and inclusion are current challenges in U.S. schools, colleges, and universities (Darling-Hammond 2015). In the latter-half of the twentieth century, "equality" was the watchword—a minimization orientation concept. In the twenty-first century, more people are developing an acceptance orientation, in which gradations of commonality and difference are noticed. This has brought attention to fairness and equity. Further progress along the continuum foreshadows a need, in the not too distant future, to have conversational resources that allow adaptation to the diversity of Discourses we encounter daily.

In providing information about the intercultural orientation continuum in this chapter, we have offered language and perspective for examining professional interactions. Keep in mind, the continuum is *developmental*. This means a person can take intentional and mindful action to move along the continuum towards adaptive intercultural competence. What is more, such personal growth can support greater effectiveness as an agent of change in a department.

As noted at the start, humans compare, including comparison of themselves to others. In fact, this book is an effort in that direction. Readers get to see some of *this* and some of *that* without being put in the position of having to pit *this* and *that* against each other.

Acknowledgements This material is based upon work supported by the National Science Foundation (NSF) under Grant Nos. DUE 0832026 and DUE 1504551. Any opinions, findings and conclusions or recommendations expressed are those of the authors and do not necessarily reflect the views of the NSF.

References

Bennett, M. J. (2004). Becoming interculturally competent. In J. Wurzel (Ed.), *Towards multiculturalism: A reader in multicultural education* (2nd ed., pp. 62–77). Newton, MA: Intercultural Resource Corporation.

Cochran-Smith, M., Ell, F., Grudnoff, L., Haigh, M., Hill, M., & Ludlow, L. (2016). Initial teacher education: What does it take to put equity at the center? *Teaching and Teacher Education, 57*, 67–78. doi:10.1016/j.tate.2016.03.006.

Darling-Hammond, L. (2015). *The flat world and education: How America's commitment to equity will determine our future*. New York: Teachers College Press.

Gee, J. P. (1996). *Social linguistics and literacies: Ideology in discourses* (2nd ed.). London: Taylor & Francis.

Gee, J. P. (1999). *An introduction to discourse analysis: Theory and method.* London: Routledge.

Hammer, M. R. (2009). The intercultural development inventory. In M. A. Moodian (Ed.), *Contemporary leadership and intercultural competence: Exploring the cross-cultural dynamics within organizations* (pp. 203–217). Thousand Oaks, CA: Sage.

Hauk, S., Toney, A. F., Jackson, B., Nair, R., & Tsay, J.-J. (2014). Developing a model of pedagogical content knowledge for secondary and post-secondary mathematics instruction. *Dialogic Pedagogy: An International Online Journal, 2*, A16–A40. Retrieved March 22, 2016, from dpj.pitt.edu/ojs/index.php/dpj1/article/download/40/50.

Hauk, S., Toney, A. F., Nair, R., Yestness, N. R., & Troudt, M. (2015). Discourse in pedagogical content knowledge. In T. Fukakawa-Connelly, N. E. Infante, K. Keene, & M. Zandieh (Eds.), *Proceedings of the 18th Conference on Research in Undergraduate Mathematics Education,* February 19–21, 2015 in Pittsburg, PA (pp. 170–184). Retrieved June 24, 2016, from http://sigmaa.maa.org/rume/RUME18-final.pdf

Hauk, S., Yestness, N., & Novak, J. (2011). Transitioning from cultural diversity to cultural competence in mathematics instruction. In S. Brown, S. Larsen, K. Marrongelle, & M. Oehrtman (Eds.), *Proceedings of the 14th Conference on Research in Undergraduate Mathematics Education,* February 24–27, 2011 in Portland, OR (pp. 128–131). Retrieved June 24, 2016, from http://sigmaa.maa.org/rume/RUME_XIV_Proceedings_Volume_1.pdf

Maltz, D. J., & Borker, R. A. (1982). A cultural approach to male-female miscommunication. In J. J. Gumpertz (Ed.), *Language and social identity* (pp. 196–216). Cambridge, UK: Cambridge University Press.

Mulac, A. (1998). The gender-linked language effect: Do language differences really make a difference? In D. J. Canary & K. Dindia (Eds.), *Sex differences and similarities in communication: Critical essays and empirical investigations of sex and gender in interaction* (pp. 127–155). Mahwah, NJ: Erlbaum.

Mulac, A., Bradac, J. J., & Gibbons, P. (2001). Empirical support for the gender as culture hypothesis. *Human Communication Research, 27*(1), 121–152.

Nicholson, N., & de Waal-Andrews, W. (2005). Playing to win: Biological imperatives, self-regulation, and trade-offs in the game of career success. *Journal of Organizational Behavior, 26*(2), 137–154.

Olsen, D., Maple, S. A., & Stage, F. K. (1995). Women and minority faculty job satisfaction: Professional role interests, professional satisfactions, and institutional fit. *The Journal of Higher Education, 66*(3), 267–293.

Reys, R. E. (2008). Jobs in mathematics education in institutions of higher education in the United States. *Notices of the American Mathematical Society, 55*(6), 676–680.

Shuter, R., & Turner, L. H. (1997). African American and European American women in the workplace. *Management Communication Quarterly, 11*(1), 74–96.

Uchida, A. (1992). When "difference" is "dominance": A critique of the "anti-power-based" cultural approach to sex differences. *Language in Society, 21*(4), 547–568.

von Hippel, C., Wiryakusuma, C., Bowden, J., & Shochet, M. (2011). Stereotype threat and female communication styles. *Personality and Social Psychology Bulletin, 37*, 1312–1324. doi:10.1177/0146167211410439.

Wang, H., Hall, N. C., & Rahimi, S. (2015). Self-efficacy and causal attributions in teachers: Effects on burnout, job satisfaction, illness, and quitting intentions. *Teaching and Teacher Education, 47*, 120–130.

Chapter 3
Valuing and Supporting Work in Mathematics Education: An Administrative Perspective

Minerva Cordero and Maura B. Mast

Abstract In this chapter we reflect on the roles and responsibilities of academic leaders in encouraging faculty in mathematics departments to value contributions to mathematics teaching and learning. We discuss how academic leaders can and should use their perspective, position and influence to: encourage productive dialogue between practitioners of mathematics and mathematics education; use assessment of student learning as an opportunity to further this dialogue; and value and reward work in mathematics teaching and learning in the hiring, evaluation, tenure, promotion, and merit processes.

Keywords Assessment • Mathematics education • Scholarship of teaching and learning

3.1 Introduction

Faculty at US colleges and universities are responsible for teaching, research, and service, or what the American Association of University Professors (AAUP) describes as "student-centered work," "disciplinary- or professional-centered work," and "community-centered work" (AAUP n.d., p. 1). Academic leaders in today's colleges and universities, especially deans, are responsible for supporting this tripartite work of the faculty with the overall goal of promoting excellence in their

---

MSC Code
97B40

M. Cordero
Department of Mathematics, The University of Texas at Arlington,
Box 19047, 501S. Nedderman Drive, Arlington, TX 76019-0047, USA

M.B. Mast (✉)
Office of the Dean, Fordham College at Rose Hill, Fordham University,
Keating Hall, Room 201, 441 East Fordham Road, Bronx, NY 10458, USA
e-mail: mmast@fordham.edu

J. Dewar et al. (eds.), *Mathematics Education*, Association for Women in Mathematics Series 7, DOI 10.1007/978-3-319-44950-0_3

institutions and advancing institutional mission. As such, an academic leader must have a future-oriented perspective; take a wide, cross-campus view; and prioritize the support and nurturing of activities that contribute to the institution's broad and strategic goals.

In this chapter, we discuss how academic leaders can use their perspective, position, and influence to encourage and value the work of mathematics faculty in mathematics education. In keeping with the approach taken in this volume, we regard the definition of mathematics education and the associated contributions to be intentionally broad and to encompass work in pedagogy, curricula, and outreach, as well as research in mathematics education. We focus primarily on approaches to supporting the professional-centered work of the mathematician in mathematics education, as this connects more closely with our backgrounds and experiences. We also provide examples to illustrate what academic leaders actually do and what results they achieve.

An important note: for convenience, we will often refer to the academic leader in this chapter as the dean. In reality, the academic leader could be a department chair, a program director, a division head, an associate dean, an associate or vice president for academic affairs, a provost, or even a president. What matters here is that the individual is a respected leader, has some financial discretion and some influence, and possesses a viewpoint that can encompass both local issues (at the department level) and global issues (at the campus or community level).

In Sect. 3.2 we consider the role of academic leaders in facilitating productive interactions and discuss their contributions to this area. In Sect. 3.3 we address the increasing emphasis on assessment of student learning in higher education and the relationship between assessment and disciplinary-centered work. In Sect. 3.4 we discuss contributions to the teaching and learning of mathematics in the context of a faculty member's professional and career development, with particular attention to how these contributions may be evaluated. Throughout the chapter, we highlight the important role of academic leaders in supporting and valuing all forms of contributions to mathematics education.

3.2 Mathematics and Mathematics Education: Facilitating Productive Interactions

3.2.1 Building Productive Interactions: Difficult but Essential Work

Our fundamental conviction is that practitioners in mathematics and mathematics education have much to learn from one another; in fact, these practitioners could and should be extended to include in-service teachers, education faculty, psychology faculty, and policy makers. We acknowledge the difficulties of past interactions as summarized, for example, by Ralston (2004): "...instead of cooperation, we have had for the past decade... the Math Wars, which pit (mainly) research mathematicians against (mainly) college and university mathematics educators and school mathematics teachers" (p. 403). But we agree with Ralston and others that cooperation is essential for real progress to be made in K-12 and, therefore, post-secondary mathematics education in the US.

A significant challenge in this context is that research mathematicians have not always demonstrated an understanding of, or appreciation for, the nature of work in mathematics education. Ball and Forzani (2007) addressed this challenge and noted:

> One impediment is that solving educational problems is not thought to demand special expertise. Despite persistent problems of quality, equity, and scale, many Americans seem to believe that work in education requires common sense more than it does the sort of disciplined knowledge and skill that enable work in other fields. Few people would think they could treat a cancer patient, design a safer automobile, or repair a bridge, for these obviously require special skill and expertise (p. 529).

In mathematics departments, this misunderstanding frequently results in an undervaluing of the work performed by mathematics faculty whose focus has shifted to mathematics education. As McCallum diplomatically put it, "Collaborative efforts between mathematicians and mathematics educators are sometimes hampered by a general lack of mutual respect between the two fields" (2003, p. 1097).

Hyman Bass (2005) articulated the challenges further, arguing that there are two common myths regarding research mathematicians becoming involved in mathematics education. Mathematicians promulgate the first myth, sharing "... a common belief ... that attention to education is a kind of pasturage for mathematicians in scientific decline." Educators are responsible for the second myth, with doubts about "... the relevance of contributions made by research mathematicians, whose experience and knowledge is so remote from the concerns and realities of school mathematics education" (p. 418). Bass acknowledged that mathematics and mathematics education are not the same, but that "productive interactions" between these fields can (and do) exist (p. 430).

Mathematics departments should bear the primary responsibility for supporting these collaborations and "productive interactions." Sometimes, however, they need help in initiating or sustaining these efforts. The silo-like nature of today's higher education makes this a challenge. The prevalent organizational structure in US higher education is one of departments within colleges or schools. This somewhat vertical structure supports discipline-based teaching and research, but isolates departments and inhibits collaboration. As a result, faculty may not see opportunities to work with colleagues in other departments (sometimes even within their own department!) or in other areas of the institution.

3.2.2 The Role of the Academic Leader in Initiating and Sustaining Collaborations

Our experience is that an academic leader who resides outside the department and who has an understanding of the need for cooperation across different areas of the university can bring faculty together around projects that may lead to deeper collaborations. We have seen this at several different institutions, including our own. In each of the following examples, an academic leader identified, encouraged, or brought together faculty members from across the institution to work on a common project. In several cases, the results went beyond the immediate project to include the deepening of interdisciplinary understandings, the implementation of curricular change, or the advancement of new research partnerships.

- Education faculty wanted to develop a graduate degree in education for in-service high school mathematics teachers. Because such a degree needed to have a significant amount of mathematics content (both as a good practice and as a requirement for advanced certification), the input of mathematics faculty was vital. The resulting collaboration led to the development of a joint graduate degree.
- In a different institution, faculty in the mathematics department sought out a collaboration with faculty in education to design a subject-based master's degree in mathematics education for in- or pre-service teachers. The resulting discussions brought mathematics and education faculty together in new ways, leading to other joint projects that included grant proposals and curriculum development.
- With support from a federal grant, mathematics and education faculty at a 4-year institution and a community college met over the period of a year to compare syllabi for first- and second-year mathematics courses, discuss student success concerns, and review transfer policies.
- Mathematics and mathematics education faculty served together on a state-wide committee charged with evaluating mathematics placement testing and the role of developmental mathematics in public higher education in that state.
- A mathematics department chair initiated a collaboration with education faculty and K-12 teachers, supported by National Science Foundation funding, to vertically bridge the school curriculum to research-level mathematics. The innovative partnership benefitted graduate students and faculty at the institution, as well as teachers and students in local K-12 schools, and provided a model for other institutions.

How can deans help? The nature of their role is that they take a cross-institutional perspective. This perspective gives them insights into connections and opportunities, as well as synergies, across a college or university. With this perspective, deans bring together groups of faculty to initiate new programs or build out areas of potential strength. In this context, collaborations that connect mathematics faculty and mathematics education faculty, or that support mathematics faculty with research interests in teaching and learning, should be connected to specific programs (such as the graduate education programs mentioned earlier) or focused to support assessment of student learning (more on that below), student retention goals, or revisions of academic support services.

Bennett's chapter in this volume describes work with education faculty who had concerns about a required course for a mathematics education program. The request to replace this course with a "mathematically rigorous capstone course for secondary mathematics teachers that would make explicit connections between college and high school mathematics" (Chap. 4, p. 45) gave Bennett the opportunity to not only think carefully about what these students should learn, but to begin a research project in the scholarship of teaching and learning.

Sultan and Artzt (2005) presented an example of how faculty from two different departments, with two different worldviews, can collaborate. What began as an initial discussion about the preparation of secondary mathematics teachers led to National Science Foundation funding for a project to recruit high school seniors into a mathematics teacher preparation program. As they noted, "If there is one thing that we have learned it is that collaboration is a complex process. We have to be willing to learn from each other, we have to respect each other, and we have to be willing to change" (p. 53).

Holm's chapter in this volume echoes this sentiment. She wrote, "I believe strongly that the mathematical sciences community must maintain the bridges between researchers in mathematics education and practitioners of mathematics education, particularly at the post-secondary level. Moving forward we need to improve our communication and collaboration" (Chap. 25, p. 377). (In this context, a practitioner of mathematics education is someone who strives to use mathematics education research in his or her teaching.)

3.3 Using Assessment of Student Learning to Build Productive Interactions

3.3.1 The Pressure of Assessment

The call for assessment of student learning in higher education has grown across the United States over the last decade. A recent survey by the Association of American Colleges and Universities (AAC&U) reported that "[t]he proportion of AAC&U member institutions assessing learning outcomes both in general education and more broadly at the institutional level has increased from 6 years ago" (AAC&U 2016, p. 2). The use of business terminology in this context, such as "return on investment" and "value-added," is increasingly common at the administrative level and this usage suggests that non-academic models and approaches are being imposed on higher education. Colleges and universities are under tremendous pressure to not only articulate student learning outcomes in general education and in the major, but to assess progress toward these learning outcomes and to demonstrate that this progress is a direct result of the educational experience. These pressures come from regional accreditors, federal and state agencies including the United States Department of Education, the media and the public (see, for example, the U. S. Department of Education's report (2006) by the Spellings Commission and Stratford's *Inside Higher Ed* article (2015)).

At the departmental level, regular program reviews may include an expectation for assessment of the curriculum, of student learning, and of other areas such as tutoring and other academic support. While the immediate pressure for assessment and departmental reviews may be external, or from the higher administration, there are potential benefits to a department that engages seriously in this work. The guidelines prepared by the Mathematical Association of America's (MAA 2010) Committee on Departmental Reviews frames a self-study as an opportunity for renewal, suggesting that, "…it provides an opportunity for members of a department to move forward together with a shared understanding, a shared set of goals, and a shared commitment" (p. 5).

3.3.2 Finding Value in the Assessment of Student Learning: The Role of Faculty

Faculty may rightly argue about the emphasis on assessment in today's higher education world, raising questions about time and effort along with the concern that "the entire premise of 'assessment to improve instruction'—especially if it is offered by outsiders—is that there is something wrong with instruction to begin with" (Hutchings 2010, p. 4). But when faculty have significant roles in assessment, the results can be meaningful. Discussions about assessment should become conversations about student learning, which then should become transformative conversations about curricula and teaching. The MAA has long supported assessment of learning in undergraduate mathematics with workshops, publications (Gold et al. 1999; Steen 2006), and other resources. As Steen noted in his introduction to Gold et al., "Assessment not only places value on things, but also identifies the things we value" (p. 1). The value of assessment itself is described by Pat Hutchings of the Carnegie Foundation for the Advancement of Teaching, as follows: "… the real promise of assessment—and the area in which faculty involvement matters first and most—lies precisely in the questions that faculty, both individually and collectively, must ask about their students' learning in their regular instructional work: what purposes and goals are most important, whether those goals are met, and how to do better" (Hutchings 2010, p. 7).

3.3.3 Assessment as an Opportunity for Productive Interactions

Deans should use assessment expectations (or mandates) as opportunities to bring faculty together to have discussions to identify what is valued, to ask questions about students learning within the context of their discipline, and to think deeply about course and curriculum goals. The assessment discussions also present an opportunity for deans to demonstrate to faculty involved in mathematics education that their work is valued.

The chapter by Sumner in this volume provides one example of how an understanding of student learning outcomes can and does inform teaching and curriculum. She described several examples of aspects of integrating writing and speaking into courses to address student learning outcomes. Sumner noted that her success in introducing these alternative forms of assessment led to changes in how she taught and assessed expectations in mathematics major courses such as differential equations (Chap. 11). In their chapter in this volume, Lopez et al. discussed how three faculty in the department used research literature and the experiences of other institutions to design a quantitative reasoning course that would "strengthen the quantitative and financial literacy of the students" (Chap. 17, p. 249). This group not only focused on the student learning outcomes in designing the course, but thought carefully about the pedagogy and the learning environment. Catepillán's chapter in this volume outlined a different challenge: how to use her background in ethnomathematics to develop a mathematics course that also met the student learning outcomes to qualify as a diversity course (Chap. 19).

Deans play a role in supporting these discussions and this work in multiple ways. The most fundamental should be the encouragement, if not the expectation, for broad faculty involvement and leadership in the development and assessment of student learning outcomes. In the context of mathematics, this assessment work presents a very natural opportunity for mathematicians with interests in and experience with mathematics education to take meaningful and significant roles. Deans should provide financial support to begin and sustain these activities. This can take several forms beyond offering stipends or course releases for assessment work; the results often last well beyond the initial activity. We have seen faculty who receive funding to attend conferences and workshops on the assessment of student learning and the scholarship of teaching and learning return energized and excited. This enthusiasm can be infectious, leading to real change in teaching and learning at the institution. In some cases this engagement revitalizes a faculty member's research interests, leading to new work in the scholarship of teaching and learning. We are most familiar with the positive impact resulting from participation in activities such as Project Kaleidoscope and AAC&U conferences with a focus on science and mathematics, MAA's Project NExT (New Experiences in Teaching), and the workshops and resources offered by Science Education for New Civic Engagements and Responsibilities (SENCER), but this is not an exhaustive list.

Deans should use grant or institutional funding to bring in respected speakers to work with faculty on engaging in meaningful assessment. As Hutchings noted:

> Clearly there are productive bridge-building possibilities here, as the scholarship of teaching and learning and assessment share overlapping agendas, practices, and institutional constituencies and as growing faculty involvement in the former shifts understandings of the latter to more clearly align assessment with what faculty actually do as teachers (Hutchings 2010, p. 11).

Deans should also lead conversations about recognition and rewards for faculty work in assessment (see below for a broader discussion of this). This can take many forms: providing guidance to hiring committees as they write position descriptions; supporting graduate program directors in the incorporation of assessment into the preparation of graduate students (see Chap. 6 in this volume by Lai et al. for more information about professional development for mathematics graduate teaching assistants); working with faculty to incorporate student self-assessment (in the form of an e-portfolio, for example) into the curriculum; and leading campus-wide discussions about assessment as scholarship and as part of the departmental and institutional reward system. These discussions may fit naturally within a larger campus discussion about scholarship. In 1990, Boyer called for a rethinking of how higher education institutions prioritize the "activities of the professoriate" (p. *xi*). He argued for a new, shared vision of scholarship, one that defines "the work of the faculty in ways that reflect more realistically the full range of academic and civic mandates" (p. 16). These discussions are as "vital" now (in Boyer's terminology) as they were more than 25 years ago.

3.4 Faculty Professional and Career Development

3.4.1 The Arc of a Faculty Member's Career

Many faculty (ourselves included) tend to view faculty careers in a traditional progression: first finish graduate school, then perhaps have a post-doctoral position or other short-term experience, then move to a tenure-track position as an assistant professor. From there, the steps are tenure, promotion to associate professor, and then promotion to full professor. (And as we can attest, a move to the administration may be part of a faculty member's academic journey.) Of course there are many versions of this progression, and alternate career paths could include an appointment as a full-time lecturer or instructor (perhaps a "professor of the practice," as the teaching-intensive positions are commonly called), working in industry or in government before taking a position as a faculty member, or laboring as an adjunct at several different institutions. Regardless of the path, an individual faculty member's professional development should never stop. Department chairs and deans should not only recognize this, but should work actively with faculty at all stages of their career to help them reflect on their own professional development and to set short- and long-term goals.

Bremser's chapter in this volume provides a compelling reflection of how faculty may naturally change focus and direction as their careers develop. Educated as a research mathematician, Bremser described how she "...began to direct more intellectual energy toward educational issues, with the explicit goal of finding constructive ways to get involved" (Chap. 23, p. 336). This led to participation in a workshop on social justice, which in turn led to scholarly engagement in a number of new and unexpected ways.

The chapter by Karakok et al. provides another example of how careers develop and are shaped in unanticipated ways by joint work between mathematics and mathematics education faculty (Chap. 7). The authors outlined their collaboration to develop a Math Teachers' Circle. The program grew out of an identified need to improve the mathematical and pedagogical content knowledge of local middle school mathematics teachers. The approach meant doing something new: designing an interactive and in-depth experience of mathematics for these teachers using the Math Circle model. As the authors noted, the impact of this work went beyond the audience of middle school teachers. All three authors benefitted from the experience and saw positive impacts on their careers. It also benefitted the department, since the authors brought their reflections about the program back to the department through discussions at department meetings.

Bass noted that "... the knowledge, practices, and habits of mind of research mathematicians are not only relevant to school mathematics education, but ... this mathematical sensibility and perspective is essential for maintaining the mathematical balance and integrity of the educational process—in curriculum development, teacher education, assessment, etc." (Bass 2005, p. 418). We know many mathematicians who have become engaged in very natural ways with K-12 education, often with support and encouragement from senior leaders on campus. These have ranged from informal presentations in local schools to structured programs like the ones described by Karakok et al. and Seshaiyer and Kappmeyer (Chaps. 7 and 8). Deans should support this type of work for multiple reasons: it represents a valuable faculty contribution; it is a form of community engagement, something that is often a key piece of an institution's mission; and this type of work has great impact on today's K-12 students, who are tomorrow's college and university students.

3.4.2 Valuing and Rewarding Work in Mathematics Education Done by Mathematics Faculty

Mathematics faculty face inherent challenges when undertaking work in mathematics teaching and learning. A primary concern is that neither mathematics faculty nor education faculty will perceive this work as scholarship. Because faculty evaluation of these contributions heavily influences how the higher administration appraises them, the result is that the institution as a whole may not value this work. Furthermore, departments used to evaluating mathematics research as part of a portfolio for tenure or promotion may not feel equipped to adequately assess work in mathematics education. A faculty member may view his or her work in mathematics education as contributing to the department's mission (and perhaps the larger institutional mission), but other members of the department may not share this perception or agree that this should be a priority. Faculty may feel uncomfortable with a colleague's focus on undergraduate teaching, concerned that the outcome may be an increase in their workload due to changes in curriculum or pedagogy, or a criticism of their own teaching. Faculty hired under one set of expectations may move in a different direction as their career progresses, leading to concerns about teaching responsibilities and research coherence within the department. Finally, different members of the same department may have very different perspectives and understandings of words such as scholarship and research, effective teaching, and outreach (see Chap. 2 by Hauk and Toney in this volume).

Mathematics faculty members need to think carefully about work in mathematics teaching and learning and how this activity aligns with both their professional development and with departmental expectations. This is a particularly important concern for pre-tenure faculty. Bremser's advice in this regard is so good that we repeat it here: "…start with a careful assessment of your own department and institution, as well as your tolerance for risk" (Chap. 23, p. 344). Holm sounds a similar caution, from the perspective of a faculty member at an Ivy League institution: "I count myself lucky to be a member of a supportive research department where faculty members are encouraged to contribute to all aspects of the profession. I have no illusions: my work with the AMS and TPSE Math did not get me tenure or promotion to full professor. It was considered a favorable part of my dossier, but my research is the *sine qua non*" (Chap. 25, p. 378). Each faculty member needs to assess his or her institutional culture and the expectations for professional advancement, whether for tenure or for reappointment (either working toward tenure or as part of a renewable fixed term lecturer position). A mentor should play a crucial role in this regard, both in helping to understand institutional expectations and in providing guidance.

The dean has a responsibility to ensure that reappointment, merit, tenure and promotion expectations are clearly articulated and communicated. When a new faculty position is proposed, a dean should talk with the department about how the individual will contribute to larger departmental goals; this conversation should go well beyond the usual (and sometimes pressing) concerns of research and teaching needs. This conversation could include a discussion about how open the department is to work in mathematics education or how a new faculty member could contribute to course and curriculum development, or, more broadly, to the scholarship of teaching and learning. If a recent departmental or general education review includes a recommendation for curriculum revision, a dean should use this as an opportunity to work with the department to shape a position description that emphasizes this interest or experience. This discussion must be in the context of the departmentally-formulated norms for tenure and promotion, where they exist, or in the context of institutional norms. The expectations for the new faculty member need to be made explicit in the position description, in the interview, and in subsequent reviews of the faculty member. As departmental leadership changes, the dean will need to ensure that the personnel or review committee knows that this individual was hired with a certain set of expectations; a clear, written, permanent paper trail is essential in this case. A dean could also use a department's interest and strength in mathematics education to advocate for a new hire, perhaps with a joint appointment in a school of education. In this case, the dean should work carefully with the other dean so that both the candidate and the departments have a clear and consistent understanding of tenure and promotion expectations. These expectations should be clearly articulated so that there are no surprises during the tenure and promotion process. (Of course, such an approach benefits all tenure-track faculty.) If the appointment is a joint appointment, this articulation should address which department has responsibility for leading the tenure and promotion process and how input from the other department will be utilized in that process.

3.4.3 *Evaluating Work in Mathematics Education: Recommendations and Examples*

How should mathematicians' efforts in mathematics education be encouraged, supported and rewarded? The chapter by Umland and Black in this volume addresses this issue. This chapter is a good resource for departments looking for a framework for assessing a wide variety of education-related work. Its authors describe methods for evaluating these contributions. They also reflect on the challenges:

> Unfortunately, many faculty in departments of mathematical sciences are unaware of the complexity of the problems, the dire need for mathematical experts to be involved, and the difficulty of such experts to find meaningful ways to make contributions to these problems. ... Typical promotion and tenure guidelines do not address (or inadequately address) the types of scholarly work discussed here. Even when such work is recognized as appropriate, adequate methods for evaluating it are often lacking (Chap. 9, p. 129).

These authors are clear that academic leaders can and should play a significant role in encouraging (or pushing) departments to have these discussions, outline their standards and expectations, and identify how and when to reward the diversity of faculty work.

We also agree with McCallum's suggestion (2003) that departments looking for examples should consider the guidelines developed by the University of Arizona College of Science for promoting mathematics faculty whose work is in mathematics education (1992). The university received a National Science Foundation Recognition Award for the Integration of Research and Education in 1997. The guidelines were one of two successful initiatives highlighted in the university's application, which stated:

> The University recognizes that science and mathematics educators face special obstacles to career enhancement because traditionally more weight has been given to research than to education. ... In 1992, the College of Science formally adopted guidelines and procedures for evaluating faculty members who play a substantial role in mathematics and science education. The guidelines put educational issues on a par with research expectations, establishing standards of national reputation and impact in the educational arena. ... [These guidelines and implementation] have had broad impact in terms of dissolving boundaries between teaching and research (University of Arizona 1996).

The University of Arizona College of Science guidelines highlight the distinction between research, teaching, and service for faculty whose interests lie in mathematics education; more specifically, the guidelines state:

> Traditional categories (research, teaching, service) may be inappropriate for evaluating science and mathematics educators because the lines between the categories are often blurred. If these categories are to be used, however, caution must be exercised to avoid assigning creative scholarly work to the service or teaching category (where it ordinarily receives less weight in the overall process) simply because it is different from traditional research (University of Arizona College of Science 1992, p. 2).

This document provides guidelines for evaluating the research, teaching and service work of faculty who have a substantial role in mathematics education. The document also outlines procedures, including the expectation of a written agreement between the faculty member and the chair regarding the portion of time to be spent on work in mathematics education.

3.5 Conclusion

As academic leaders, we see major changes affecting higher education, including increased attention to the assessment of student learning and a growing scrutiny of faculty workload and "productivity." With change comes uncertainty and risk. As Dan Butin (2016) of Merrimack College noted, "[a]cademics tend to be risk-averse." Deans, on the other hand, must be comfortable with assessing the relative risks of stasis and change, with determining when change is necessary, and with leading necessary change.

We argue that with change, there is also opportunity. In the context of mathematics teaching and learning, academic leaders should have the institutional perspective and resources to bring mathematics faculty and mathematics education faculty together to have productive conversations and to collaborate on strong projects. They can use the call for assessment as an opportunity to strengthen these collaborations. Deans should encourage departments to explicitly discuss how the tripartite work of the faculty should be recognized and rewarded.

The dean's perspective, advocacy, and, in some cases, resources should serve as driving forces for recognizing work in mathematics education as situated firmly in the mission of a department of mathematics and as a vital aspect of a mathematics department's teaching mission. Good work is already happening at many institutions. Deans can and should celebrate this work locally and publicize it widely.

References

AAC&U. (2016). Trends in learning outcomes assessment: Key findings from a survey among administrators at AAC&U member institutions. Resource document. Association of American Colleges and Universities. Retrieved April 15, 2016 from http://aacu.org/sites/default/files/files/LEAP/2015_Survey_Report3.pdf

AAUP. (n.d.). What do faculty do? Resource document. Retrieved March 30, 2016 from https://www.aaup.org/issues/faculty-work-workload/what-do-faculty-do

Ball, D., & Forzani, F. (2007). What makes education research "educational"? *Educational Researcher, 36*, 529–540.

Bass, H. (2005). Mathematics, mathematicians, and mathematics education. *Bulletin of the American Mathematical Society, 42*(4), 417–430.

Boyer, E. (1990). *Scholarship reconsidered: Priorities of the professoriate*. San Francisco, CA: Carnegie Foundation for the Advancement of Teaching. Jossey-Bass.

Butin, D. (2016, January 13). So you want to be a dean? *Chronicle of Higher Education*. Retrieved March 31, 2016 from http://chronicle.com/article/So-You-Want-to-Be-a-Dean-/234900

Gold, B., Keith, S., & Marion, W. (Eds.). (1999). *Assessment practices in undergraduate mathematics*. Washington, DC: The Mathematical Association of America.

Hutchings, P. (2010). Opening doors to faculty involvement in assessment. National Institute for Learning Outcomes Assessment. Occasional Paper #4. Retrieved March 30, 2016 from http://www.learningoutcomeassessment.org/documents/PatHutchings.pdf

MAA. (2010). Guidelines for undertaking a self-study in the mathematical sciences. Resource document. Retrieved June 12, 2016 from http://www.maa.org/sites/default/files/pdf/ProgramReview/MAA-SelfStudyManual.pdf

McCallum, W. (2003). Promoting work on education in mathematics departments. *Notices of the American Mathematical Society, 50*(9), 1093–1098.

Ralston, A. (2004). Research mathematicians and mathematics education: A critique. *Notices of the American Mathematical Society, 51*(4), 403–411.

Steen, L. (Ed.). (2006). *Supporting assessment in undergraduate mathematics*. Washington, DC: The Mathematical Association of America.

Stratford, M. (2015, October 20). Upping the pressure on accreditors. Inside Higher Ed. Retrieved June 10, 2016 from https://www.insidehighered.com/news/2015/10/20/obama-administration-plans-executive-action-higher-education-accreditation

Sultan, A., & Artzt, A. (2005). Mathematicians are from Mars, math educators are from Venus: The story of a successful collaboration. *Notices of the American Mathematical Society, 52*(1), 48–53.

U. S. Department of Education. (2006). A test of leadership: Charting the future of U.S. higher education. Resource document. Retrieved June 10, 2016 from http://www2.ed.gov/about/bdscomm/list/hiedfuture/reports/final-report.pdf

University of Arizona College of Science. (1992). Procedures for evaluation of faculty who play a substantial role in pre-college mathematics and science education. Resource document. University of Arizona Department of Mathematics. Retrieved June 12, 2016 from http://math.arizona.edu/~mathedconf/willoughby.pdf

University of Arizona. (1996). Integrating research and education: Promotion & tenure guidelines and biology education. A proposal to the National Science Foundation Office of Science and Technology Infrastructure. Resource document. Retrieved June 12, 2016 from http://www.biology.arizona.edu/raire/proposal.html

Part II
Benefitting Pre-Service and In-Service Teachers and Graduate Student Instructors

Chapter 4
Effects of a Capstone Course on Future Teachers (and the Instructor): How a SoTL Project Changed a Career

Curtis D. Bennett

Abstract In this chapter, I revisit my first scholarship of teaching and learning project as a 2000–2001 Carnegie Academy for the Scholarship of Teaching and Learning scholar. I describe my experience as a pure mathematician taking on a pedagogical research project and the effects of this project and of doing the scholarship of teaching and learning on my teaching and career. The project studied student development in a novel mathematics capstone course for future teachers. Student teams worked on semester-long mathematics research problems, while simultaneously completing a content-heavy course on how advanced mathematics informs the teaching of high school (and earlier) mathematical subjects. The course changed behaviors of the students by giving voice to students with prior negative classroom experiences. In addition, one student had a surprising change in attitude towards proof and its value to secondary mathematics teachers. Working through the context of the original study, I reflect on the effects of the course on the students and of the project on the next 15 years of my career.

Keywords Scholarship of teaching and learning • SoTL • Mathematics teachers • Mathematics capstone • Doing mathematics

4.1 Introduction

Scholarship of teaching and learning (SoTL) dates back to the 1990s and Ernest Boyer's (1990) book *Scholarship Reconsidered*. In order to expand the nature of scholarship, Boyer argued that scholarly teaching should be valued as an important

---

MSC Codes
97B50
97DXX

C.D. Bennett (✉)
Department of Mathematics, Loyola Marymount University,
1 LMU Drive, Los Angeles, CA 90045, USA
e-mail: cbennett@lmu.edu

J. Dewar et al. (eds.), *Mathematics Education*, Association for Women in Mathematics Series 7, DOI 10.1007/978-3-319-44950-0_4

aspect of what faculty do. He noted that a professor's teaching is scholarly in nature as it involves reading texts, preparing appropriate examples, analyzing arguments, comparing and judging pedagogical methods for teaching topics, etc. Over time scholarly teaching evolved to the scholarship of teaching and learning with faculty as expert teacher-practitioners capable of generating pedagogical scholarship in the discipline. Such practitioner scholarship brings valuable insights and contributions to pedagogical knowledge. In Dewar and Bennett (2015), *Doing the Scholarship of Teaching and Learning in Mathematics,* (*Doing SoTL*), SoTL is defined as:

> the intellectual work that faculty members do when they use their disciplinary knowledge (in our case, mathematics) to investigate a question about their students' learning (and their teaching), submit their findings to peer review, and make them public for others to build upon.

As a pure mathematician whose research focus is in the areas of groups and geometries and combinatorics, I have no formal training in pedagogical research. However, I have always put great effort into my teaching, and thus I am perhaps the perfect example of a mathematician who should be interested in SoTL.

4.2 The Carnegie Scholars Program

In 1999, the Carnegie Foundation for the Advancement of Teaching invited its first mathematicians as Carnegie Scholars in the Carnegie Academy for the Scholarship of Teaching and Learning (CASTL) program (Carnegie Foundation for the Advancement of Teaching n.d.). The purpose of the program was to "create a community of scholars, diverse in all the ways that matter in teaching and learning, whose work will advance the profession of teaching and deepen student learning" (Carnegie Foundation for the Advancement of Teaching 1999). Over the last 16 years, SoTL in mathematics has gained significant traction as evidenced by Mathematical Association of America (MAA) contributed paper sessions and minicourse offerings on SoTL in mathematics from 2006 to 2016, and the 2015 publication of *Doing SoTL* in the MAA Notes series.

With funding from the Pew foundation, the CASTL scholars program welcomed its first cohort of 15 scholars in the summer of 1998. During its first 4 years, the CASTL scholars program brought together over 100 scholars from more than 20 disciplines. Each cohort first received a 2-week introduction to SoTL. After the initial training the scholars conducted a SoTL project during the academic year. The following summer they returned for a 2-week program during which they presented their results, discussed how to move forward with their work, and met with the incoming cohort of scholars. Initially, the call for applications to the program targeted particular disciplines. Mathematicians were first invited in 1999, and that year the mathematics scholars were Peter Alexander from Heritage College, Thomas Banchoff from Brown University and then president of the MAA, Bruce Cooperstein from the University of California, Santa Cruz, and Anita Salem from Rockhurst

University. The mathematics scholars in the 2000 cohort were Jack Bookman from Duke University, John Holcomb then from Youngstown State University, Marilyn Repsher from Jacksonville University, and the author, then at Bowling Green State University.

A major goal of the program was to help foster an expanding notion of scholarship along the lines of Ernest Boyer's call (1990). In addition to the scholars program, the Carnegie Foundation worked with disciplinary societies, including the MAA, with disciplinary scholars chosen to help this endeavor. After the first 4 years of the program, the funding mechanism changed. Consequently, there was no 2002 cohort, but two more cohorts of CASTL scholars were recruited in 2003 and 2005 under the theme of liberal learning. Three mathematicians, Mike Axtell from Wabash College, Jacqueline Dewar from Loyola Marymount University (LMU), and the author (now at LMU) were part of the 2003 cohort, and one mathematician, Michael Burke from the College of San Mateo, was in the 2005 cohort. This chapter will focus on the SoTL work I completed as part of my 2000 CASTL Scholar fellowship, its findings, and the impact of that work on me, on my students, and on my career.

4.3 My CASTL Proposal

I first became aware of the CASTL program when Cooperstein recommended I apply to the 2000 cohort. In reading the application materials, I discovered that I needed a teaching project for the submission, which led me to looking at my capstone class for future teachers.

The capstone course arose as a result of a 1997 change in teacher certification requirements in Ohio. Both mathematics and mathematics education faculty at Bowling Green State University (BGSU) had concerns about the existing program. The education faculty perceived the content of the real analysis class (primarily taught as a graduate preparation course) as inappropriate for future secondary teachers. As a required course in the secondary mathematics education program, education professors saw real analysis as a roadblock for future teachers. Meanwhile, BGSU mathematics faculty felt students should be exposed to the mathematical rigor of real analysis and wanted to add three units of mathematics to the program. The two groups compromised by adding three more units of mathematics content to the program and replacing the real analysis requirement with a mathematically rigorous capstone course for secondary mathematics teachers that would make explicit connections between college and high school mathematics. As a member of the departmental curriculum committee, I volunteered to develop the new course.

I chose to design the course around mathematical topics appropriate for future teachers, such as solving a cubic equation (and the invention of complex numbers), proofs of the irrationality of π and e, and the impossibility of trisecting an angle. The class focused on the mathematical topics needed to prove these results and how to apply the mathematical understandings gained to issues in Grades 7–12. I piloted the course in 1998 with ten individually selected students. The pilot went well (for a new course), but I was disappointed, as I felt that the students didn't really experience *doing* mathematics. In addition, I didn't tie the mathematics to the high school curriculum as well as I had wanted to.

The students in the first iteration produced notes for the class, and I used these notes for the class in both the fall and spring terms of 1999. After discussing my concerns regarding students having the experience of doing mathematics with my colleague David Meel, I decided to add semester-long mathematics research projects that students would complete in teams. These projects were to be independent of the content of the class. However, they would (ideally) be on some problem in mathematics that would prove interesting to future secondary teachers. I piloted the projects during the fall term of 1999, and I was happy with the outcome, if not my student evaluations (some students objected to the heavy workload and were unhappy about their grades). This class and these projects became the centerpiece of my successful application to the 2000 CASTL Scholars program, leading to my undertaking a study of the projects and the underlying course. While on sabbatical leave at Michigan State University (MSU), I arranged to teach the course as a capstone for their secondary education program in the fall 2000 semester.

Once I was named a CASTL fellow, panic set in. Being a research mathematician focusing on group theory and combinatorics with little formal education in understanding teaching and learning, I was about to embark on a project with little idea how to research what I wanted to know. I was extremely fortunate to have friends, colleagues, and a brother, who knew a lot more about doing this type of work than I did, and I often turned to them for help with my project. In addition, while I was at MSU, I had the great fortune to be able to ask for help from, and collaborate with, a number of mathematics education specialists, and of course, my Carnegie colleagues also helped me out tremendously. While there are too many to mention all of them individually, I owe special thanks to David Meel and Daniel Chazan, both researchers in mathematics education, who provided assistance on the course and the project at different stages.

4.4 The CASTL Research Project

As frequently happens with SoTL projects taken on by disciplinary scholars with no experience in studying teaching and learning, my project idea was ill-defined and naively constructed. Although my intent was to "prove" that the capstone research projects helped students think like mathematicians, the Carnegie project ended up being a more comprehensive look at the capstone course (Bennett 2003).

After my 2-week summer initiation to SoTL, I felt a little more prepared for the project itself. I applied for human subjects research approval for the project from both BGSU, my home institution at the time, and MSU, where the research was to be done. I then had (subject to student consent) approval to record all discussions with students during office hours on course material, record class meetings as appropriate, collect student surveys, conduct student interviews, and collect and report on student work.

My initial goal was to show that students thought more like mathematicians at the end of the course than they did at the beginning and to somehow attribute this change to the research projects students completed in the class. In hindsight, this was hopelessly naïve. However, this starting point led me to collect a great deal of data on what was happening during the course, leading to my eventual CASTL project, the production of an electronic course portfolio (Bennett 2003).

Fourteen students took the course, 13 of whom were prospective teachers (the other changed his major the summer before the class). Two of the students had done very well in previous mathematics classes and two of them reported low confidence in mathematics. Twelve students were traditional 4th year seniors, one was a 5th year senior, and the last was a senior who had returned to college after working for several years.

From the first assignment, a mathematical biography together with four other prompts, I learned that most of the 14 students believed: "we study mathematics to learn how to solve problems, and because it is useful." However, on a separate pre-survey, few students could give examples of applications of higher-level mathematics. Some students reported bad experiences in other mathematics courses, and at least two claimed they needed a grade better than C in the class to be allowed to student teach. The pre-survey also revealed that most students considered mathematics problems to be interesting only if they corresponded to applications (either real-world or expanding on a previously learned concept).

Evidence collected hinted at an underlying issue: many of the students in the capstone course appeared to feel that they could not take part in discussions about mathematics in the college classroom. In discussing his willingness to ask questions in a post-course interview (the entire class participated in one interview), Ron (all names are pseudonyms) mentioned that students know what the "pecking order" of a class is. The subtext of his comment was that in many other courses, "weaker" students are at the bottom of the pecking order and are afraid to ask questions because it would be considered "silly." What stood out in their discussions was that in the capstone class the students felt they could ask questions they always had about mathematics and mathematics teaching. Alan stated this most clearly, explaining how in previous courses, "there were so many things I never knew (about mathematics)... I always wished I could ask them, but I never asked them." These statements indicated how students' previous classroom cultures promoted avoidance behaviors in the future teachers. Covington (1992) documented such avoidance behaviors in adolescents in mathematics classes.

Few of the students perceived themselves as "doers of mathematics" at the start of this class, and most felt little authority to alter the discussion in any class. These future teachers had had experiences in which they learned to sit passively in the classroom for fear of looking foolish or stupid. When the course portfolio (Bennett, 2003) was produced, I was unaware of Boaler's (1999) work on personal identity and mathematics learning —and Cobb, Gresalfi and Hodge's (2009) paper on normative and personal identities had yet to be written. Thus I lacked the knowledge of the vocabulary from the pedagogical literature to express the results of students' prior experiences with words such as agency and personal identity. But looking at the student interview comments today, those are the terms I would use.

By the end of the capstone class, these students felt empowered to take part in and direct mathematical discussions as well as saw themselves as "doers" of mathematics. At least three times during the semester, the students stopped me in class to ask for more detailed discussion (and outlines of the proofs) of mathematical topics that were only tangentially related to the course content, including Gödel's incompleteness theorem, the transcendence of $2^{\sqrt{2}}$, and infinitesimal arithmetic. Moreover, Alan suggested that he felt that "we as a class decided… what was going to come up the next day." While other students might not have used these exact words, they clearly felt more in control of the class than usual. This change in "agency" was likely also affected by a new perception of themselves as "doers" of mathematics as evidenced by two student groups that referred to themselves as "mathematicians" in their project papers (Bennett 2003).

The post-course interview discussion about how some students felt when the class went off-topic showed me the important roles that pecking order and personal identities played in the classroom. John, the most successful student heading into the class (and one whose personal identity included asking questions in mathematics classes previous to this) started the discussion by stating that when the class got, in his words "sloppy," it was less fun. Other students agreed, but had a different sense of "sloppy." As it turned out, for John, "sloppy" meant when we were discussing issues related to educational policy (e.g., tracking in mathematics in high school mathematics classes) or any time the discussion was not about mathematics. On the other hand, for Jim, a student who had been less successful previously, "sloppy" meant any time the class strayed from what he needed to know for the upcoming homework set, particularly when the mathematics was more complicated. Meanwhile Neal, a student closer to Jim in background, suggested that the different kinds of tangential discussions were important as they "reach out to different kinds of people." For more detailed analysis of the change in conversations, see Bennett (2003). This better understanding of what was happening in the classroom conversations and the importance of student agency and personal identities were major revelations of the SoTL project for me—even though I then lacked the language to describe them.

A second key outcome of the project was in presenting what the SoTL literature refers to as "a vision of the possible," in this case a story of something unusual that happened in a class. The first week of class, Neal, came to my office to ask about how he could succeed in the class. Neal commented to me that he had always been pretty good at mathematics until he hit the "proof stuff." Moreover, he was very worried about his grade in the class, because he needed a B in the class to qualify for student teaching. I was quite concerned. I had often seen pre-service teachers like Neal in my upper division mathematics classes at BGSU. While they would often persevere and make it through the class with lots of hard work on both of our parts, never had such a student left class with an appreciation of "the proof stuff." On the pre-survey question: state the most interesting mathematics problem he had ever worked on, Neal wrote, "To be honest, I can't remember—probably very largely due to the fact that my success in mathematics has not been spectacular in the past 2 years—ones that I thought were interesting I always did wrong."

By the end of the class, Neal had a much richer view of mathematics and the importance of proof. On the final examination problem about deriving the quadratic formula and using it in classrooms, Neal's response included the statement, "we need to help students envision the steps of the proof [as] useful as the end itself." This answer appeared to embrace the notion of proof as something important. Further evidence of a change in Neal came from an unexpected avenue. To help me see the class as it unfolded, not just how I perceived or planned it, I obtained copies of student notes from Neal and one other student at the end of the semester. Neal's notes included marginalia, that is, comments he wrote down in class that were not on the board. The other student's notes were essentially a copy of what was written on the board each day. Neal's added comments went through an astonishing transition. Coding of his marginalia showed that in the first 4 weeks of the class, it consisted almost entirely of trivia, brief statements about mathematicians, numerical coincidences, etc. On the other hand, at the end of the term, the marginalia contained an amazing number of statements about teaching mathematics. Moreover, the point during the semester at which the change appears to occur coincided with a discussion in my office hours about how I, the professor, saw proofs inform teaching.

Another change I saw in Neal was his perspective on mathematics as problem-solving versus calculating. At the post-course interview (I conducted an individual interview with Neal), when asked how the course (and project) would influence his teaching, he stated:

> So one thing I definitely want to apply as a teacher is to try to instill the idea with kids, with students, encourage them to bounce ideas off of each other. So in creating lessons, in creating discussions, group work when bringing up challenging ideas, see what kind of questions they have. Trying to spark an interest with the students themselves, but letting them do the exploration so they can create a sense of ownership – perhaps.

The pre- and post-survey showed him to be more mathematical in his perspectives on what aspects make up good mathematical problems and how mathematics is done (Bennett 2003). I do not know whether this change in attitude led to a dramatic difference in Neal's teaching, but the statement suggests that Neal would be more likely to create a classroom where the normative behavior is for students to explore and discover mathematics.

What I took away from this SoTL investigation was a firmer understanding of many of the aspects of this capstone course in mathematics that I found valuable. At the start of the investigation, I knew that the student projects "worked." Unfortunately, I had no idea what that meant. Consequently, I moved from wanting to show what worked in my class to giving a detailed description of what was happening in my class, and why I valued it. In the taxonomy of SoTL questions (Huber and Hutchings 2005), I moved from a "what works" question to a "what is" question. Disentangling various aspects of the course to decide what created the value, however, was (and remains) far more difficult. The students seemed to agree that there was great benefit to having a project that they spent most of the semester on, as it made it easier for them to wrestle with the mathematical question rather than looking immediately for a solution. However, to me the greatest successes in this class were helping the students discover that they could be doers of mathematics and giving them the experience of a classroom where the normative behavior for all students was to ask questions.

These results had a dramatic effect on me as a teacher. Today I am far more aware of the importance of focusing on student attitudes, behaviors, and concerns beyond the course material being taught. While my student evaluations saw little change, I believe that I am a far better teacher than I was before doing this work. That said, I don't believe that I do a significantly better job in getting students to learn the mathematical content of the course, but rather I am much improved in helping students understand mathematics as a discipline and seeing themselves as doers of mathematics.

This work also gave me a deeper understanding of and respect for those engaging in pedagogical research in mathematics. In part, this came from reading the deeper pedagogical literature as opposed to what was popularized, but more importantly it came from engaging in some of the research methods myself. Trying to code data taught me both the care with which such analysis is done to ensure transparency and the difficulty in doing so. Doing this work also gave me a better understanding of the *language* of mathematics education and taught me to see it as more than jargon. For example, prior to my SoTL work, I saw *triangulation* as a word used to give a veneer of scientific methodology for pedagogical research. However, through working on my project, I came to a new understanding: triangulation of data means that multiple types of evidence or research methods lead to the same conclusions. Consequently, in pedagogical research triangulation makes results more likely to be reproducible (and valid). I also became aware of the language of mathematicians that can lead to misunderstandings in conversations with non-mathematicians. I once used the word "fundamental," meaning foundational, but was surprised that my education colleague took the word to mean the underlying mathematics was simple. This understanding of the communication differences between mathematics education and mathematics practitioners has proven vital to my collaboration with mathematics education specialists.

4.5 Effect the Project Had on my Career

Doing SoTL work, particularly with a focus on the preparation of future secondary teachers, has affected me profoundly in multiple ways. It has led to a larger network of colleagues, my changing jobs, expanding my scholarly work in multiple directions, and national recognition. It has also created complications for my efforts to continue my mathematical research.

As a faculty member at BGSU, which has a PhD granting program in mathematics, one of my primary responsibilities was to participate in and support our PhD program in my research field of group theory and combinatorics. My colleagues there were supportive of my doing SoTL as an addition to my traditional mathematics.

I too was concerned that it not detract too much from my mathematics research. Thus, while I was conducting the SoTL project on my sabbatical, I worked hard to publish traditional mathematics papers too. Even after I left BGSU, doing traditional mathematics research has been important to me. Splitting time between two (or more) endeavors makes it harder to keep up with any one of them, and it is fair to say that my focus on SoTL has decreased my productivity in traditional mathematics.

Despite this decrease in mathematics research productivity, I am grateful for the many opportunities SoTL work has brought me. Initially I had some difficulty in getting BGSU to support my application to the program. Once the CASTL Fellowship was announced, however, I received a great deal of administrative support. Consequently, I was able to organize a SoTL conference at BGSU.

As part of the CASTL program, I was introduced not only to other mathematicians in the SoTL movement, but also to interesting scholars in a variety of fields. Consequently, I had the opportunity to meet with historians, biologists, physicists, sociologists, and others who were interested in SoTL, some of whom were also prominent in their research fields. As a result of one of these connections, I met Deborah Ball and was invited to participate in her education seminar at the University of Michigan, which improved my thinking about teaching and learning. Later, she suggested I apply to her workshop on teaching mathematics for elementary education majors to be held after the Joint Mathematics Meetings in 2002. My new professional connections also led to me being invited to many teaching and learning conferences and building a network of colleagues in SoTL. And perhaps most significantly, as a result of my participation in the Deborah Ball workshop, I met Jacqueline Dewar, leading to a job offer from Loyola Marymount University (LMU) and a significant future collaboration.

The reception of SoTL by the mathematics department as LMU was very different from that at the BGSU department. At BGSU, my work was appreciated by my colleagues and the administration, and had I stayed, I am convinced it would have been accepted as part of my portfolio for promotion to full professor. However, I perceived that many of my colleagues viewed it as of lesser importance than my traditional mathematics. I was very fortunate that my BGSU mathematics education colleagues were very welcoming of my interest and helped me discover and understand the existing literature.

When I arrived at LMU in 2002, the entire department embraced my interest in SoTL. In my first year, the CASTL program put out a new call for applications, and I was able to apply jointly with Dewar, who was new to SoTL at the time. Thereafter, we started a SoTL brown bag group at the university, and faculty members from across the university took an interest, including faculty members from most of the science disciplines as well as psychology and economics. In addition, the Center for Teaching Excellence at LMU was quick to embrace our forming such a group. In contrast, the Center for Teaching at BGSU did not embrace my efforts to expand SoTL (likely because prior to my joining the CASTL program I had little contact with BGSU's Center for Teaching). I saw the difference in the responses to my SoTL work as reflective of the administrations' attitudes at the two universities.

However, today I believe this had more to do with the size of the institution and administrative bureaucracy (BGSU is roughly 5 times the size of LMU) than specific goals of the administration. Moreover, by the time I arrived at LMU, I was less naïve and more knowledgeable about how to position myself and my work with the administration.

The CASTL project and my focus on future teachers also created professional and career opportunities for me. Professionally, embarking on this work opened up new avenues of scholarship and national recognition for me. For example, I am sure my SoTL work played a significant role in my winning the MAA Haimo award, and it has led to my presenting minicourses at JMM and at various universities. In addition, I have been fortunate enough to find a number of collaborators in this work, and at this time, SoTL works are an important part of my publication record, including the MAA book *Doing SoTL* in Mathematics (Dewar and Bennett 2015). But it isn't the recognition or opportunities for publications that drive me. As with my mathematics research, my publications are a consequence of my interest. The difference is when I work on SoTL, the focus of my interest is on student learning, and when I work on my mathematics research, the focus is on more traditional areas of mathematics.

Pursuing SoTL also means I have less time to do mathematics research. Since so much in my career has changed over the last 16 years, capturing the full effect of splitting my time between endeavors is impossible. However, striking a balance between doing mathematics and doing SoTL is difficult and, at times, frustrating. As all teachers know, doing a good job in the classroom will absorb as much time as we allow it to. Consequently, even before engaging in SoTL work, I would have to divide my time between class preparation and scholarship. I find that engaging in SoTL makes that partitioning of time much more difficult, because the line between SoTL and class preparation is often blurry. This contrasts with my experience in a doctoral granting department where mathematics research and directing dissertations and theses support each other. The goal of directing doctoral students was for them to successfully solve a mathematics problem, and thus such work supplemented my research. At a primarily undergraduate institution like LMU however, such teaching and (mathematics) research synergies are less common. This makes it more challenging to manage two distinct areas of work.

Perhaps the most dramatic effect of my undertaking SoTL work has been expanded career opportunities for me. As previously mentioned, my SoTL work led very directly to my current position at LMU. However, it also led to administrative opportunities. As a result of my work, I was invited to apply for the Center for Teaching Excellence director position at LMU, which I did not follow up on. Later, I was asked to serve as Associate Dean for Faculty Development in the Seaver College of Science and Engineering at LMU, a position I held for 4 years. While the SoTL work was not the primary reason for my selection, it contributed to the dean's view that I would be a good fit. My SoTL experiences also contributed greatly to my success in that position. Having already worked with scholars in other disciplines in SoTL helped me to work with LMU colleagues in other disciplines. In addition, I was much better prepared to pursue grant opportunities that involved revising courses and to plan faculty development activities.

For me the benefits of working on issues of teaching and learning have far outweighed the costs to my research program. I see teaching and research as two equally valuable aspects of being a mathematician. In pursuit of my vocation as a teacher, contributing more widely to the teaching and learning enterprise is a mission for me. Moreover, I have found in my collaborations with pedagogical specialists and researchers that we all improve the enterprise of teaching, and that we do this best when collaborating together with respect for the expertise that each of us brings.

References

Bennett, C. (2003). Advanced mathematics for secondary teachers: Course portfolio, The Gallery of Teaching and Learning, Carnegie Foundation. http://gallery.carnegiefoundation.org/gallery_of_tl/castl_he.html, (mirrored at http://myweb.lmu.edu/cbennett/portfolio/Portfolio.htm).

Boaler, J. (1999). Participation, knowledge, and beliefs: A community perspective on mathematics learning. *Educational Studies in Mathematics, 40*, 259–281.

Boyer, E. (1990). *Scholarship reconsidered*. New York: Carnegie Foundation for the Advancement of Teaching, Jossey-Bass.

Carnegie Foundation for the Advancement of Teaching. (1999). *CASTL higher education brochure*. Menlo Park, CA: Carnegie Foundation for the Advancement of Teaching.

Carnegie Foundation for the Advancement of Teaching. (n.d.). CASTL. Retrieved June 24, 2016, from http://web.archive.org/web/20010410103822/http://www.carnegiefoundation.org/CASTL/index.htm

Cobb, P., Gresalfi, M., & Hodge, L. (2009). An interpretive scheme for analyzing the identities that students develop in mathematics classrooms. *Journal for Research in Mathematics Education, 40*(1), 40–68.

Covington, M. V. (1992). *Making the grade*. Cambridge, England: Cambridge University Press.

Dewar, J., & Bennett, C. (2015). *Doing the scholarship of teaching and learning in mathematics*. Washington, DC: Mathematical Association of America.

Huber, M. T., & Hutchings, P. (2005). *The advancement of learning: Building the teaching commons*. San Francisco, CA: Jossey-Bass.

Chapter 5
By Definition: An Examination of the Process of Defining in Mathematics

Elizabeth A. Burroughs and Maurice J. Burke

> *"What is a good definition? For the philosopher or the scientist, it is a definition which applies to all the objects to be defined, and applies only to them; it is that which satisfies the rules of logic. But in education it is not that; it is one that can be understood by the pupils."*
>
> —Poincaré (1914/1952, p. 116.)

Abstract Our work as mathematics education researchers in a department of mathematical sciences rests upon a foundation of reflections on our own teaching practices and on the mathematical practices that our preservice mathematics teachers are expected to learn. In the nexus of these reflections sit issues relevant to the learning and teaching of mathematics, and the understanding of these issues requires sustained and systematic research that includes and goes beyond reflecting on practices. We present an example of a teaching episode focused on preparing secondary mathematics teachers to better understand mathematical definitions and the process of creating mathematical definitions. We then examine some historical developments in mathematical practice related to definitions and the defining process and relate these developments to challenges entailed in teaching practice. We conclude with examples of researchable areas embedded in these issues and comment on the impact of our work.

Keywords Mathematical definitions • History of mathematics • Secondary mathematics teacher preparation

---

MSC Codes
97B50
97Gxx

E.A. Burroughs (✉) • M.J. Burke
Department of Mathematical Sciences, Montana State University,
PO Box 172400, Wilson Hall 2-214, Bozeman, MT 59717-2400, USA
e-mail: burroughs@montana.edu; mburke@montana.edu

J. Dewar et al. (eds.), *Mathematics Education*, Association for Women in Mathematics Series 7, DOI 10.1007/978-3-319-44950-0_5

5.1 Introduction

In this chapter, we offer an example of one type of mathematical work we engage in as members of a mathematics education research group within a department of mathematical sciences that includes mathematics, statistics, and mathematics education as separate but related disciplines. Our research focuses on the preparation of secondary mathematics teachers and the expertise required for teaching school mathematics. The research and teaching aspects of our academic careers are intertwined: our work is informed but not determined by our own teaching. We take mathematics teaching as a scholarly unit of study that goes beyond our own reflective practices.

Mathematics education research is rooted in the scholarship of two disciplines: mathematics and educational theory. It is grounded in mathematics content through the study of curriculum and mathematical practice. It is generally carried out through social science research methods, including both qualitative and quantitative analysis, but it also includes the study of mathematical ideas. In this chapter we address a research area—the use of definitions by undergraduate secondary mathematics teaching majors—by describing some researchable issues that arise from reflections on both teaching practice and mathematical practice.

Our interest in the topic of mathematical definitions stems from classroom observations made during a study of student conceptions of limits after a standard engineering calculus treatment. These informal observations led us to feel uneasy about a traditional mathematical pedagogy in which instructors provide students crucial mathematical definitions with no attention paid to the broader context of defining. In such a pedagogy, novices do not grasp the nuances of definitions, nor are they given the opportunity to do so, and they find themselves floundering in their own attempts to draw valid conclusions and follow inferences made by their teachers. This observation led us to reflect on our own practices in the mathematical preparation of secondary mathematics teachers.

Reflective teaching can take many forms: *action research* (Mertler, 2009) and *lesson study* (Hurd and Lewis, 2011) are two somewhat formal methods for deep analysis of one's own teaching, and we teach both methods to students in our master's program for mathematics teachers. But reflective practice also occurs on a more regular and less formal basis within our own teaching.

An example of this kind of informal reflection is presented in Sect. 5.2. We describe a teaching episode regarding mathematical definitions and frame its outcomes within a broader reflection on mathematical practices that ultimately determine much of what we hope to achieve in teaching practice. Mathematical practice is complex and consists of at least five historically evolving components: language, sets of accepted statements, sets of accepted reasonings, sets of questions selected as important, and meta-mathematical views, which includes standards of proof and definitions and beliefs or claims about the scope and structure of mathematics (Kitcher, 1984, p. 163). For the purposes of this chapter, we limit ourselves to reflecting on some historical developments within the last component, that is, meta-mathematical views, since these developments are especially useful in identifying critical issues for teaching practice regarding definitions. These historical reflections are presented in Sect. 5.3.

Mathematics education research is often situated in issues at the nexus of teaching practice and mathematical practice. In Sect. 5.4 we conclude the chapter with examples of such research and a description of some of the impacts of our work.

5.2 Teaching Activity Focused on Definitions and Defining

The activity focused on defining a "square" in the context of hyperbolic geometry, where there are no quadrilaterals satisfying the definition "… four right angles and four congruent sides." It engaged our undergraduate students in constructing definitions, examining their equivalence or non-equivalence, and carefully evaluating their usefulness in different settings.

5.2.1 Student Conceptions

In designing the activity we started with some preliminary notions about our students, their mathematical conceptualizations regarding definitions, and goals we thought were important in the defining process. We had in mind the following hypothetical proof, similar to ones we had received from former students:

Statement to be proved: The diagonals of a square have equal lengths.
Proof: Let ▭ ABCD be a square.
AC and BD are diagonals of the square by definition of diagonal.
AC = BD by definition of square.

From the perspective of researchers in the field (Przenioslo, 2004; Tall and Vinner, 1981), such a student might lack a sense of the role of definitions in mathematics, placing personally held concept images for *square* on par with the mathematical definition in determining what is implied "by definition of square." Tall and Vinner describe a concept image as "all the cognitive structure in [an] individual's mind that is associated with a given concept" (1981, p. 151). Suppose the student's textbook defined a square as a rectangle with equal sides. The student might be assuming that this implies any property of rectangles that is part of a personally held concept image, including everything previously proved about rectangles, all become part of a definition "bundle" for the term *square*. Hence, the student concludes that properties like *diagonals are equal in length* or *diagonals bisect each other* are true of squares "by definition," because they are true of rectangles. Properties like *diagonals are perpendicular* are not true by this definition of square but need to be proved. Notice that if the textbook defined a square as *a quadrilateral which is both a rectangle and a rhombus*, these latter properties would not have to be proved either, according to such reasoning.

The hypothetical student's way of reasoning indicates a misunderstanding of the specificity and minimality intentions of mathematical definitions (see Sect. 5.2.2). Specificity is lacking because two individuals may have quite different lists of mathematical properties of rectangles in their own concept images, rendering the necessary conditions stipulated in any mathematical definition dependent on an individual and hence indeterminate. Minimality is lacking because the only properties necessary in the first definition of square ("a rectangle with equal sides") are the assumption of congruent sides and the necessary conditions for being a rectangle stipulated by the definition of rectangle.

5.2.2 Mathematical Considerations

In designing the activity, we also included preliminary ideas we had about mathematical practice regarding definitions. In mathematics, proving is a process of *warranting* statements as logically valid conclusions and a process of *explaining* or making sense of why a statement is true based upon prior notions that are understood (Weber, 2008). These prior notions are typically definitions, axioms, and previously proved statements. Hence, definitions play a crucial role in proof and are often strategically selected from a set of competing definitions in order to simplify proofs.

> Mathematical definitions are "equivalent" when they determine the same set of elements or processes within a particular mathematical system. In this activity, we wanted students to understand when two definitions are equivalent and to understand that the choice between equivalent definitions is not arbitrary, but strategic. We also intended them to understand that mathematicians have choices in deciding between competing non-equivalent definitions, such as whether to define *trapezoid* as a quadrilateral having at least one pair of parallel sides or as a quadrilateral having exactly one pair of parallel sides (Usiskin and Griffin, 2008, p. 27).

We began with the notion that a definition must be specific:

> Specificity. A mathematical definition provides unambiguous conditions for identifying and classifying elements of interest that exist in the system.

We continued with three additional commonly used criteria impacting decisions about mathematical definitions (Usiskin and Griffin, 2008).

> Minimality. A definition tries to minimize redundancies in the stipulated properties (p. 37).
>
> Generality. A definition generalizes to or is easily extended to other closely related systems. For example, when defining the *interior of an angle* in Euclidean geometry, the definition based on the intersection of two half-planes nicely extends to hyperbolic geometry (p. 40).
>
> Hierarchical efficiency. A definition fits into a nested set of definitions to avoid redundancy of proofs within the system. For example, when defining *rectangle*, the definition "parallelogram with a right angle" is preferred over "quadrilateral with four right angles" since a property proved for all parallelograms can be inferred for all rectangles (p. 37).

To these, we added a fifth criterion, particularly important in teaching.

Referential clarity. A definition has easily identifiable examples and is closely related to its natural-language usage. For example, Euclid's definition of a square as a quadrilateral with four right angles and four equal sides is easier to unambiguously visualize and connect to natural-language usage than defining it as a quadrilateral with four congruent angles and two adjacent sides with the same length.

We did not make this list explicit to students before the activity; rather, we relied on their experiences within the activity to create an understanding of the need for these five criteria.

5.2.3 *The Teaching Episode*

The undergraduate geometry course uses selected chapters from Reynolds and Fenton (2008) *College Geometry Using the Geometer's Sketchpad* and relies on students' use of dynamic geometry software to provide interactive, manipulable models of Euclidean geometry. These software programs provide a setting for exploration, generalization, and refutation of properties of geometric objects. We introduce students to the Poincaré disk model of hyperbolic geometry through a dynamic web applet called "Non-Euclid" (https://www.cs.unm.edu/~joel/NonEuclid/NonEuclid.html). Our students use their knowledge of Euclidean geometry to discover things that are true in both Euclidean and hyperbolic geometries and things that are different. They thereby gain a critical appreciation of the Parallel Postulate in the geometry that is studied in secondary mathematics, where they will be teaching.

Approximately two-thirds of the way through the course, when students had completed a study of Euclidean geometry and had just been introduced to hyperbolic geometry, we asked them to find as many alternate definitions of a square in Euclidean geometry as they could. They generated the following list.

1. A square is a quadrilateral with four 90° angles and four congruent sides in which the intersection of the diagonals is 90°.
2. A square is a rectangle with four congruent sides.
3. A square is the result of reflecting a right isosceles triangle across its hypotenuse.
4. A square is a rhombus with at least one right angle.
5. A square is a quadrilateral with diagonals that are angle bisectors and at least one right angle.
6. A square is a quadrilateral whose diagonals form four congruent isosceles triangles.
7. A square is a quadrilateral with four congruent angles and a pair of adjacent sides congruent.
8. A square is a four-gon with all equal angles and all sides of equal length.
9. A square is a quadrilateral with four congruent sides and four congruent angles.

This list was not meant to be exhaustive; rather, it was the result of whole-class brainstorming. The class agreed that these were "equivalent definitions." The lack of specificity in definition (3), which seems to suggest the points on the hypotenuse are also on the square, initially did not bother the students.

In the course of the discussions, students occasionally proposed definitions as equivalent and then rejected them based on "counterexamples," which in the context of this activity meant examples of quadrilaterals that satisfy one of the proposed definitions but not the other. For instance, one student proposed the definition that a square was a quadrilateral with diagonals that were "equal and perpendicular to each other." Others responded with counterexamples, i.e., quadrilaterals that they agreed were not squares by Euclid's definition, yet fit the student's definition. For all students the ultimate test of equivalence was whether the definitions determined the same set of quadrilaterals, and most were satisfied with informal justifications based on the lack of counterexamples.

In the next class period, students investigated their nine definitions when applied in hyperbolic geometry; they were already familiar with using the Poincaré disk model of the hyperbolic plane represented in the web applet Non-Euclid. Already knowing that no quadrilaterals in hyperbolic geometry satisfied Definitions 1 and 2, they discovered that the other definitions, equivalent in Euclidean geometry, led to different classes of quadrilaterals in hyperbolic geometry. They categorized them as follows (See Figs. 5.1 and 5.2):

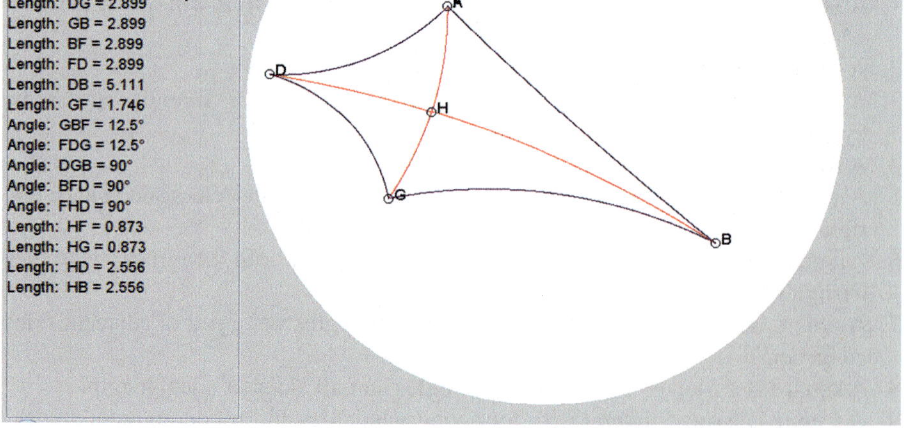

Fig. 5.1 A quadrilateral in the Poincaré disk that satisfies the definition of SQT1

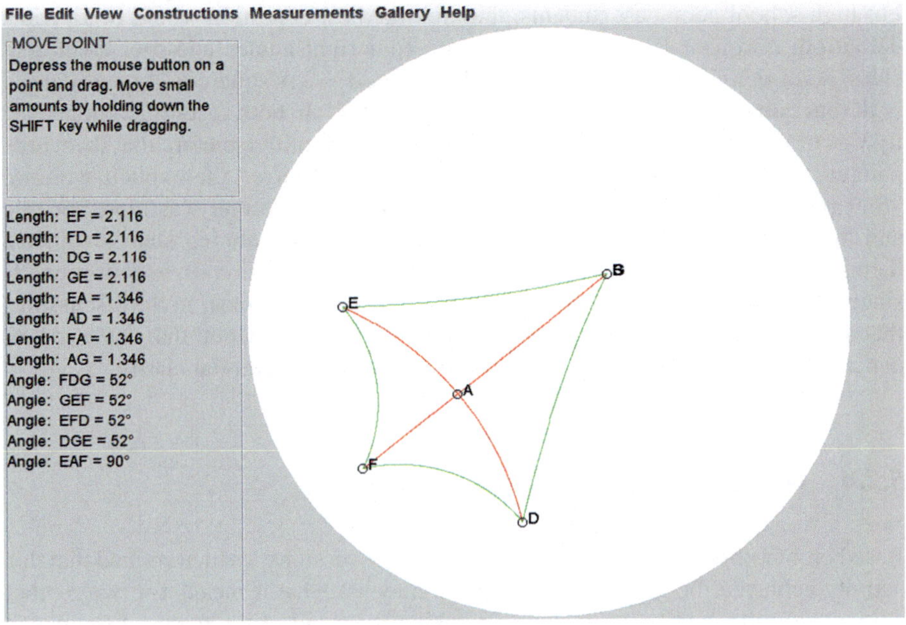

Fig. 5.2 A quadrilateral in the Poincaré disk that satisfies the definition of SQT2

- Definitions 1 and 2 determine objects that do not exist in hyperbolic geometry.
- Definitions 3, 4 and 5 determine objects the class named "Special Quadrilateral Type 1" (SQT1). These are rhombi with at least one right angle. These quadrilaterals are equilateral and have two pairs of opposite angles congruent. Their diagonals are not equal in length but are perpendicular bisectors of each other. Their opposite sides are parallel.
- Definitions 6, 7, 8 and 9 determine objects the class named "Special Quadrilateral Type 2" (SQT2). These are quadrilaterals with four congruent sides and four congruent angles. These quadrilaterals are equilateral and equiangular. Their diagonals are equal in length and perpendicular bisectors of each other. Their opposite sides are parallel.

We asked the class: what is the proper definition of a square in hyperbolic geometry? Students conducted a technology-enabled investigation of properties of SQT1 and SQT2. The students decided that SQT2 possessed more of the symmetry they associate with the word square. Thus, from the point of view of deciding which definition generalized better, the "equal sides and equal angles" definition won out. However, the students decided that neither of these figures should be called a *square* since the word was already enshrined in natural-language usage as a figure with four right angles.

We asked students to decide upon the best definition of a square in Euclidean geometry. After some discussion, they decided the answer depended on context.

For high school geometry students, they felt the best definition would be Euclid's definition: A square is a quadrilateral with four right angles and four congruent sides. As a definition for themselves, they preferred "A square is a quadrilateral with four congruent angles and four congruent sides." In both cases referential clarity was of utmost importance—they appreciated the intuitive picture that the words painted. For high school geometry students, they preferred the explicit wording "four right angles" as opposed to "rectangle" since the former was more specific and direct; moreover, the word "rectangle" conjured up oblong shapes. But for themselves, clarity and generality won out, preferring definition (9) to definition (2) since rectangles do not exist in hyperbolic geometry. It was clear in the discussions that students recognized the value of minimality but understood that some redundancies were acceptable as long as they contributed to referential clarity.

5.2.4 Discussion of the Teaching Episode

Based on our observations of student work and discussions, we determined that this activity achieved the goal of clarifying for students what it means for two mathematical definitions to be equivalent. By taking them out of the realm of Euclidean geometry and asking them to apply definitions in hyperbolic geometry, the activity revealed how a set of equivalent definitions in Euclidean geometry generalized to non-equivalent definitions in hyperbolic geometry. The activity also seemed to have helped the students to develop an understanding of how considerations of specificity, minimality, generality and referential clarity work with the system of axioms and other contextual factors in the process of mathematical definition making. Indeed it is fair to say that the activity led students to a greater appreciation of Poincaré's attitude given at the beginning of this chapter. Finally, the activity clarified the notion that depending on the axioms, a definition might not define anything; it also drew students' attention to interdependence of definitions and axioms in establishing mathematical terrain.

The activity, however, left things unsettled with respect to many of our other goals. It did not lead students to an appreciation for the hierarchical efficiency criterion. It did not elicit a formal strategy for proving when two definitions are equivalent—we had assumed students learned such a strategy in the foundations of mathematics course they had taken as a prerequisite of the geometry course. It was not clear whether and to what extent our students could articulate a distinction between a definition and an axiom, even though they had increased sensitivity to the role of axioms in formulating a definition. Finally, even though students had developed a greater appreciation for the specificity and minimality traits of definition, it was not clear how our students would assess such proofs as the one given in Sect. 5.2.1 regarding diagonals of a square.

This episode illustrates how easy it can be for mathematics instructors to take for granted student understanding of the critical nuances of a mathematical idea where such understanding does not exist. As we reflect on the guided-exploration pedagogy of the activity in which an instructor poses strategic questions but otherwise lets students take ownership of the activity and the critique of emerging ideas, we can identify many didactical issues, including what instructors can and do conclude from the multifaceted information resulting from the activity. Tasks involving student discovery and exploration, though engaging, might not generate anticipated outcomes because of cognitive barriers, insufficient preparation, or simply failure to notice. In the latter case, we wondered why the class did not generate definitions of *square* embodying traits of Saccheri and Lambert quadrilaterals, which they had recently encountered in the course.

Following a guided exploration, students often welcome their instructor's input, and a reflective teaching practice might take advantage of this to address deficiencies like those mentioned above. By direct questioning in an instructor-led discussion, the instructor can assess the student understanding of critical aspects of the concepts involved. In the case of our students, they had grappled with issues at a sufficient depth for us to follow the activity with an instructor-led discussion on mathematical practices. Such a follow-up was used to clarify traits attributed to good definitions, including the five criteria we identified, and retroactively apply them to the students' own work in the guided exploration. Other appropriate follow-up lessons could require students to evaluate the proof given by the hypothetical student. Or, while honoring students' use of counterexample reasoning in assessing the equivalence of definitions, the instructor could take this natural opportunity to review methods for formally proving two definitions are equivalent.

To conclude the chapter here might give the reader an idea of how we approach our mathematics teaching. But our work as scholars in mathematics education goes beyond our teaching. In spite of our best efforts, reflective teaching practice tends to leave us with more questions than answers. When put in perspective with analysis of mathematical practices from which they gain their relevance, the questions accentuate a need for systematic study—that is, they invite mathematics education research.

5.3 Mathematical Practices Surrounding Definitions

The mathematical act of defining has evolved over the course of the history of mathematics. Our investigation of history highlights three challenges: the challenge of rigor; the challenge of clarity; and the challenge of axiomatic systematization. We find parallel challenges in teaching *defining* in a modern sense, where the mathematical practice of defining has emerged from the practice of mathematics by mathematicians. We identify researchable areas within each challenge.

We delve into these historical developments in mathematical practice regarding definitions intending to provide the background necessary to stake out a vantage point from which to view the mathematics education research questions we identify. As we turn to history, we note that when talking about the deep meaning of terms used by people in ancient times and cultures, we are standing on thin ice. We adapt

Wittgenstein's rope metaphor (Wittgenstein, 1953/1986, Inv. 67): think of the meanings of a term for our predecessors as many threads of meaning twisted and braided together throughout history from them to us, with possibly no thread running the full length of the rope. This is particularly important as we talk about what the ancient Greeks thought about "definition." It would be nice to fit their definitions into modern categories found in philosophical discourse (Gupta, 2015; Robinson, 1954/1962) such as *descriptions*, *stipulations*, *explications*, and *lexical*. However, it is clear from the scant primary sources that exist that these categories don't quite fit. Furthermore, ancient mathematicians, scholiasts, and philosophers were inconsistent in how they used the terms ascribed to the Euclidean axiomatic system: *definition*, *postulate*, and *axiom* (Szabó, 1978, pp. 222–223). Since historians and philologists argue over what the Greeks meant by these terms, the account we are giving necessarily includes room for debate.

5.3.1 The Challenge of Rigor

Definitions have played a primary role in the rigorization of mathematics. Thales (ca. 640–542 BCE), the acknowledged founder of Greek mathematics, is credited with having proved five basic theorems about angles, triangles, and circles. Historians debate whether he could have actually deduced these theorems or if he used more empirical, inductive, and intuitive arguments. But either way, the propositions about angles had to distinguish between angles formed by two straight lines and angles formed by other "lines" (i.e., curves), such as horned angles formed by intersecting a circle with a straight line (and discussed in *The Elements* of Euclid). The word "angle" may have been used indiscriminately to refer to both kinds of angles, just as there is evidence that Greeks at the time of Thales considered the deltoid-shaped figure formed by three intersecting circles to be a triangle. For his "theorems" to be credible, empirically or deductively, when sensory evidence and common usage in language dictated that they were patently false, Thales would have had to employ a defining process to place restrictions on the meaning of the terms he used and eliminate the obvious counterexamples. Specifically, he would have had to distinguish the meanings of angle and triangle from their common usages. Poincaré said it best: "Exactness cannot be established in the arguments unless it is first introduced into the definitions" (1914/1952, p. 124).

The earliest mathematical definition on record is attributed to Pythagoras (ca. 580–500 BCE, alleged to have been tutored by Thales): A point is "unity having position." The classicist James Gow concluded, "To them [the Pythagoreans] is probably due the introduction of definitions of some kind and the use of ordinary deductive proofs in geometry" (1884/2010, p. 153). More forcefully conjectured, Szabó claims: "I am inclined to think that the 'earliest mathematical principles' were the definitions. The foundation of mathematics as a deductive science seems to have started, in the historical development, by formulating definitions" (1964, p. 127).

To say the least, the practice of defining terms and using definitions in logical arguments played a pivotal role in the transformation of mathematics in classical times (600–200 BCE) from an intuitive practical activity of scribes, merchants, and surveyors to a deductive science. We note that K–12 teachers are challenged to help students make a similar transition in the rigorization of their own mathematical reasoning over the course of their precollege educations. Thus as we consider our undergraduate secondary mathematics teaching majors who are to become teachers of mathematics, we identify some researchable areas that arise at this juncture:

- The influence of the natural language foundation on most precollege mathematical language development, and the impacts of various curricular and pedagogical interventions on that development.
- The role of examples and counterexamples in the heuristics for developing precise definitions, and the relationship of precision in definition to the level of rigor in the reasoning of students who are beginning to learn mathematical language.

5.3.2 The Challenge of Clarity

By the time Euclid systematized geometry (ca. 300 BCE), definitions and other hypotheses were the hallmark of mathematics and philosophy. The Greek word for "hypothesis" is often simultaneously used for definitions and foundational principles (Szabó, 1978). It is significant that Plato, using "hypothesis" in this sense, seems annoyed when he writes:

> "I fancy that you know that those who study geometry and calculation and similar subjects, take as hypotheses the *odd* and the *even*, and *figures*, and three kinds of *angles*, and other similar things in each different inquiry. They make them into hypotheses as though they knew them, and will give no further account of them either to themselves or to others on the ground that they are plain to everyone. Starting from these, they go on till they arrive by agreement at the original object of their inquiry (Republic VI/510 c-d, in Szabó, 1978).

Plato seems to think of the referents of these words "the even" or the "three kinds of angles" as ideal objects and is annoyed that these objects are assumed to be understood when given only the briefest description, or none at all. They are given without an explanation of their essential properties.

Consider an example. Euclid defined *point* as "that which has no part." Euclid states his definition at the beginning of his text, his first premise before any postulates or axioms are given, and intends it to be taken as a "primary premise," probably in Aristotle's sense of primary premise—a basic truth to be assumed. It is stated as a hypothesis without the justification Plato would have liked. To Aristotle, and likely to Euclid in many cases, such a definition is stated as a declarative sentence asserting the essential identity of a point as an idealized object within the category of spatial objects: every point shares that essence and everything that shares that essence is a point.

Definitions that are unclear as to their status as mathematical claims or ambiguous in their meanings were common in mathematics until the late 19th century.

For example, Euclid defined a number as a "multitude of unities," which excludes one from being a number. Newton refers to Euclid in his own definition:

> By number we understand, not so much a Multitude of Unities, as the abstracted Ratio of any Quantity, to another Quantity of the same Kind, which we take for Unity. And this is threefolds integer, fracted, and surd: An Integer, is what is measured by Unity; a Fraction, that which a submultiple Part of Unity measures; and a Surd, to which Unity is incommensurable" (Newton 1769, p. 2).

Here the definition is not a stipulation, but rather an explication of Newton's meaning of *number*. It is an explication that, today, would be considered too vague. One can say it is an intuitive definition of *real numbers*, but it is too imprecise to be of any use in settling issues regarding limits and continuity and other questions: for example, "Are all surds algebraic numbers?" as asked by Leibnitz, or "Do the least upper bound property and the nested set principle hold for numbers given by this definition?" as asked by Bolzano, Cauchy, Weierstrass, Dedekind and others (see Bressoud, 2007, for a historical development of these issues).

By the end of the nineteenth century, definitions had become stipulations, to use Poincaré's term, "baptizing" with a name mathematical constructs of the stipulator's design, thereby giving those constructs an identity. In many cases the stipulations are nominal or notational in the sense that they merely decree certain notations, words or expressions to be synonymous with a notation, word, or phrase, thereby providing convenient abbreviations. Either way, unlike ancient Greek views, modern mathematical definitions do not assert identity as a truth founded upon the apprehension of some referent's essential nature. Definitions identify a mathematical construct that satisfies conditions of logical consistency and eliminability. Like axioms and theorems, they are formulated from previously introduced terms, defined or undefined, so that, in the end, all theorems can be logically reduced to statements given strictly in terms of the undefined terms of the axiomatic system. Unlike axioms, definitions do not stipulate a truth value for statements given strictly in the undefined or primitive terms of the system. Instead, they stipulate an identity for constructs described in those primitive terms.

We note that our students—future teachers—come to us with a naïve realism coupled with many natural language usages of most terms related to the mathematical concepts they are to learn. Our students often accept the conventional definitions given to them by teachers and texts as facts describing some ideal but specific world, or even simply as arbitrary laws that hold within the context of school mathematics. Researchable areas salient to the preparation of mathematics teachers are:

- The optimal use of intuitive definitions and meta-mathematical justifications of definitions, for engaging students in the defining process that concludes in a stipulation, without giving the impression that the justifications are proofs and textbook definitions are theorems; and the extent to which students prefer the intuitive definitions or descriptions as the "real meaning" and try to assimilate mathematical definitions to these meanings instead of the other way around.
- The cognitive developments in students over the course of K–12 education, and appropriate teaching methodologies and curriculum needed for moving them

from the notion of a mathematical definition as a factual assertion about real objects, to the notion of a definition as an assertion about an ideal object, and finally to the notion of a definition as the formation and naming of a mathematical construct that is arrived at through agreements within the mathematical community.

5.3.3 The Challenge of Axiomatic Systematization

In the Greek axiomatic system, particularly for Euclid, definitions came before postulates presumably because the fundamental entities in the postulates had to be identified. In other words, these fundamental entities did not depend in any way upon the axioms for their essential definitions. This creates interesting contrasts with the modern axiomatic system. Euclid seems to take for granted the existence of the things he defines, except in a few cases such as the infinitude of primes, where he proves, by contradiction, the existence of a prime number that does not belong to an arbitrarily given finite set of primes. In modern axiomatic systems, definitions follow the postulates, and it is considered good practice, at least since Leibnitz, to demonstrate that definitions are consistent with the postulates by proofs of "existence within the system." This is most often done by producing an example, before or right after the definition is given.

The mathematical practice of Euclid's time, as depicted above, leads to interesting variations in the standards for definitions. For example, Euclid classifies quadrilaterals in one of his early definitions given before the postulates:

> Of quadrilateral figures, a *square* is that which is both equilateral and right-angled; an *oblong* that which is right-angled but not equilateral; a *rhombus* that which is equilateral but not right-angled; and a *rhomboid* that which has its opposite sides and angles equal to one another but is neither equilateral nor right-angled. And let quadrilaterals other than these be called *trapezia* (*Elements* I.19).

The first thing to note is that the genus of quadrilaterals is divided into disjoint species or categories, somewhat fitting Aristotle's ideal for the defining process. The first four categories are given as essential definitions while the remaining quadrilaterals are left undifferentiated and stipulated to be referenced by the name *trapezia*. The second thing to note in the above definition is that Euclid was not trying to give an efficient definition, per his axioms, such as "a square is equilateral with a right angle," knowing that he would be able to prove all of the angles were right angles. This suggests that for Euclid, it is not an *accidental quality* of squares, in Aristotle's sense, to have four right angles (Aristotle, *Posterior Analytics*, trans. 2002, Part 22). It is an *essential quality*. The third thing we note is that the above definition is inefficient within a deductive system. Modern textbooks present students with an intentionally hierarchical system that classifies quadrilaterals in such a way that makes the set of proofs of all their standard properties very lean.

Therefore, the ideal of giving minimal conditions needed in a definition, though a common value for ancient and modern practices, meant something different in ancient practice from what it means in modern practice. In modern axiomatic systems, the minimal conditions criterion decouples the definition from any dominant concept image or "essence" and is allowed to rely on logical efficiency within an axiomatic system. Consequently, the hierarchical organization of definitions in the ancient axiomatic system is different from that found in the modern.

A common feature of ancient and modern practices is the institutionalizing effect of successful axiomatizations. Similar to the influence set theory has on modern axiomatizations, Euclid's *Elements* had a huge institutionalizing effect on mathematics, even to the present. Some of his definitions were so institutionalized that they became laws not to be questioned. Simon Stevin in his 1585 *L'arithmetique* felt the need to give a philosophical and mathematical argument for changing Euclid's definition of number so that *one* could be considered a number. He concluded it should be a number, but conceded on somewhat Platonic grounds that *zero* should not be a number, even though he admitted that *zero* was useful in calculation and notation (Klein 1968, pp. 191–197).

Another practice common to ancient and modern axiomatic systems is the value placed on definitions that extend other definitions to new terrain and are fruitful in quickly enabling meaningful deductions in those domains. In this respect, the definition of proportion usually attributed to Eudoxus (*Elements*, V.5) is the most important definition given in the *Elements*. In this definition, Eudoxus extended the notion of being *in proportion* to include the ratios of incommensurable magnitudes by simply saying that two ratios were in proportion provided they were not out-of-proportion, i.e., one ratio was not greater than the other or less than the other. Eudoxus' definition is so fruitful, deductively speaking, that it has had a huge influence and inundates mathematical practice with double *reductio ad absurdum* proofs of proportionality showing one ratio is (a) not larger than the other and (b) not smaller than the other, and hence the two ratios are in proportion.

The axiomatic systematization of mathematics has implications for teaching. As mathematics education researchers we know that in geometry, for instance, students are very affected by visual concept images and often assume squares are not rectangles or rhombuses or parallelograms because of dominant images they associate with the terms (van Hiele, 1957/2004), much like the images suggested in Euclid's classification of quadrilaterals. Even a figure's alignment on the page where it appears affects students' judgment of its identity (Battista, 2009). Students often do not take advantage of hierarchical properties of figures to simplify their arguments or draw immediate conclusions from textbook definitions that are hierarchical. Students don't naturally check for consistency or existence when considering definitions. And they don't naturally inquire about alternative definitions since their belief that the definitions are facts is bolstered by the institutional status bestowed on them as statements from their textbooks to be memorized.

These observations, coupled with our analysis of relevant mathematical practices, open up further researchable areas:

- The curricular and pedagogical supports needed for developing in preservice teachers the ability to create and critique definitions for purposes of systematizing mathematics using criteria such as specificity, minimality, hierarchical

organization, clarity, and generalizability, given Poincaré's admonition quoted at the start of this chapter.
- The development of an understanding that textbook definitions are conventions, some of which become institutionalized, while balancing this with an understanding that defining is a vital creative process in mathematics.

5.4 Mathematics Education Research and Impacts of Our Work

The teaching episode illustrates the potentials and limitations of reflective *teaching practices* for shedding light on a range of issues whose relevance emerges from *mathematical practice*. Simply put, constructing a knowledge base that adequately addresses those issues we identified, and many others we did not mention for lack of space, requires much more than reflective teaching practice. Systematic research is required. Indeed, research related to each of the researchable areas we identified can be found in mathematics education journals. For instance, Edwards and Ward (2004) and Vinner (1977) provide two straightforward examples related to the challenge of clarity. These two studies reveal that many undergraduate mathematics majors think mathematical definitions are theorems, laws, or facts, and many frequently resort in their proofs to personal definitions extracted from mathematical experiences and intuitions instead of formal mathematical definitions.

Our work in a department of mathematical sciences is built on a foundation of trust with our colleagues from statistics, applied mathematics, and theoretical mathematics. We share the trust that our expertise as scholars of mathematics education is distinct from theirs as mathematicians or statisticians, and that our expertise is necessary for the robust research, teaching, and outreach programs in the department.

When considering our scholarly work within a department of mathematical sciences, we note that we reflect on the impact of our work much as any researchers might reflect on the impact of theirs: there are scholars within our subdiscipline who read and cite our work; when we publish in practitioner journals, there is an immediate application of our work; and, like all active scholars, we use our expertise in making decisions about the teaching emphasis and curriculum for courses we teach. Perhaps unlike mathematicians who join a mathematics department with the expectation of research in mathematics and then later contribute to mathematics education, our positions were appointed with the expectation that our research contributions would be in mathematics education. A natural outlet for our work is in practitioner journals—that is, journals whose intended audience is teachers or others who expect to use the results of mathematics education research in their own teaching. Our department considers these contributions equally valuable as those that appear in research journals.

This particular work that we have done surrounding definitions has impacted the work of others in mathematics education. For example, a colleague was pursuing research on student conceptions of infinite series in calculus. Our focus on definitions helped him to pinpoint some cognitive troubles students were having in understanding that infinite series required new definitions, and that students were not attuned to this underlying need.

The nature of our mathematics education research is solidly grounded in mathematics. For instance, this work in definitions stemmed from observations of students in mathematics courses, but quickly turned to an examination of the features of mathematical definitions themselves. Some work in mathematics education is grounded in sociological aspects of schooling or in the psychological aspects of learning; our research is concerned with teaching practice and the *doing* of mathematics. We view the content of school mathematics as a worthy focus of academic study, grounded in mathematics itself, and find satisfaction in the relevance of our research to mathematics classroom practices.

References

Battista, M. (2009). Highlights of research on learning school geometry. In T. Craine & R. Rubenstein (Eds.), *Understanding geometry for a changing world: 71st yearbook* (pp. 99–108). Reston, VA: The National Council of Teachers of Mathematics.

Bressoud, D. (2007). *A radical approach to real analysis*. Washington, DC: Mathematical Association of America.

Edwards, B., & Ward, M. (2004). Surprises from mathematics education research: Student (mis) use of mathematical definitions. *American Mathematical Monthly, 111*(5), 411–424.

Gow, J. (2010). *A short history of Greek mathematics*. New York: Cambridge University Press. Original work published 1884.

Gupta, A. (2015). Definitions. Online encyclopedia. The Stanford Encyclopedia of Philosophy. Retrieved Apr 22, 2016, from http://plato.stanford.edu/index.html.

Hurd, J., & Lewis, C. (2011). *Lesson Study step-by-step: How teacher learning communities improve instruction*. Portsmouth, NH: Heinemann Publishing.

Kitcher, P. (1984). *The nature of mathematical knowledge*. Oxford, UK: Oxford University Press.

Klein, J. (1968). *Greek mathematical thought and the origin of algebra (trans: Eva Braun)*. Cambridge, MA: MIT Press.

Mertler, C. (2009). *Action research: Teachers as researchers in the classroom*. Thousand Oaks, CA: Sage Publications, Inc.

Newton, Isaac (1769). *Universal arithmetic, or, a treatise of arithmetical composition and resolution* (trans: Raphson, J. and S. Cunn, Ed.). Retrieved Apr 22, 2016, from https://archive.org/details/universalarithm00wildgoog.

Poincaré, H. (1952). *Science and method*. New York: Dover. Original work published 1914.

Przenioslo, M. (2004). Images of the limit of function formed in the course of mathematical studies at the university. *Educational Studies in Mathematics, 55*(1-3), 103–132.

Reynolds, B. E., & Fenton, W. E. (2008). *College geometry using the Geometer's Sketchpad* (Preliminary ed.). Emeryville, CA: Key College Press.

Robinson, R. (1962). *Definition*. Frome, U.K.: D.R. Hillman & Sons. Original work published 1954.

Szabó, Á. (1964). The transformation of mathematics into deductive science and the beginning of its foundation on definitions and axioms. *Scripta Mathematica, 27*(1), 113–139.

Szabó, Á. (1978). *The beginnings of Greek mathematics*. Dordrecht, Holland: D. Reidel Publishing Company.

Tall, D., & Vinner, S. (1981). Concept image and concept definition in mathematics, with special reference to limits and continuity. *Educational Studies in Mathematics, 12*(2), 151–169.

Usiskin, Z., & Griffin, J. (2008). *The classification of quadrilaterals: A study in definition.* Charlotte, NC: Information Age Publishing, Inc.

van Hiele, P. M. (2004). The child's thought and geometry. In T. P. Carpenter, J. A. Dossey, & J. L. Koehler (Eds.), *Classics in mathematics education research* (pp. 60–66). Reston, VA: National Council of Teachers of Mathematics. Original work published 1957.

Vinner, S. (1977). The concept of exponentiation at the undergraduate level and the definitional approach. *Educational Studies in Mathematics, 8*(1), 17–26.

Weber, K. (2008). How mathematicians determine if an argument is a valid proof. *Journal for Research in Mathematics Education, 39*(4), 431–459.

Wittgenstein, L. (1986). *Philosophical investigations (trans: G.E.M. Anscomb).* Oxford: Basil Blackwell Ltd.. Original work published 1953.

Chapter 6
Characterizing Mathematics Graduate Student Teaching Assistants' Opportunities to Learn from Teaching

Yvonne Lai, Wendy M. Smith, Nathan P. Wakefield, Erica R. Miller, Julia St. Goar, Corbin M. Groothuis, and Kelsey M. Wells

Abstract Exemplary models to inform novice instruction and the development of graduate teaching assistants (TAs) exist. What is missing from the literature is the process of how graduate students in model professional development programs make sense of and enact the experiences offered. A first step to understanding TAs' learning to teach is to characterize how and whether they link observations of student work to hypotheses about student thinking and then connect those hypotheses to future teaching actions. A reason to be interested in these connections is that their strength and coherence determine how well TAs can learn from experiences. We found TAs can connect observations and future teaching, but that the connections vary in quality. Our analysis suggests future revisions to TA development programs, which we discuss in the conclusion.

Keywords Post-secondary professional development • Mathematics teacher growth • Reflective practitioner • Graduate teaching assistants • TA development

---

MSC Codes
97B99
97B50
97-xx
97Axx
97B40

Y. Lai (✉) • N.P. Wakefield • E.R. Miller • J.S. Goar • C.M. Groothuis • K.M. Wells
Department of Mathematics, University of Nebraska-Lincoln,
203 Avery Hall, PO BOX 880130, Lincoln, NE 68588-0130, USA
e-mail: yvonnexlai@unl.edu; nathan.wakefield@unl.edu; erica.miller@huskers.unl.edu; s-jstgoar1@math.unl.edu; corbin.groothuis@huskers.unl.edu; kelsey.wells@huskers.unl.edu

W.M. Smith
Center for Science, Mathematics & Computer Education, University of Nebraska-Lincoln,
251 Avery Hall, PO BOX 880131, Lincoln, NE 68588-0131, USA
e-mail: wsmith@unl.edu

J. Dewar et al. (eds.), *Mathematics Education*, Association for Women in Mathematics Series 7, DOI 10.1007/978-3-319-44950-0_6

6.1 Introduction

One theory for how instructors learn from their own and others' teaching experience is that learning occurs through deliberately connecting future teaching plans and prior experience. Specifically, instructors create opportunities to learn when they articulate future actions in terms of observations based on previous experience (e.g., Hall & Horn, 2012; Horn, Kane, & Wilson, 2015). Under this theory, enhancing the ability to learn from experience requires both improving how instructors conceive of teaching and tightening connections between future plans and current thinking.

Our goal is to improve TAs' ability to learn by reflecting on their experiences. We report on a study of novice mathematics graduate student teaching assistants (TAs), who were teaching college algebra and intermediate algebra and were all enrolled in a seminar as part of a TA development program. The program aims to help TAs to teach from the principles that: (a) student learning occurs through the student's lens, and observation of student learning occurs through the observer's lens; (b) understanding the experiences that shape students' thinking is important to teaching; and (c) learning occurs through building on prior knowledge. Our study explored the question:

> How do these TAs connect observations and beliefs about their students, hypotheses about student thinking, and proposed next teaching actions?

We open this chapter with one TA's reflections about her students' learning, based on a paper written as part of the TA development program. We use this TA's work to illustrate how we model TA thinking so as to study the opportunities they created to learn from experience. After describing our model for TA thinking, we discuss the literature informing our work and the context in which this study took place. We then describe how we collected and analyzed data to study TA thinking. Finally, we describe paths of TA thinking that we found useful in considering how to improve our TA development program. We reflect on our future actions in terms of observations and beliefs about TAs and hypotheses about TA thinking.

6.2 Modeling TA Thinking

6.2.1 *One TA's Reflection*

At the time of the study, TA12 was a first time instructor who taught College Algebra. She had recently assigned this quiz problem:

> *Determine whether the following function is a rational function:*

$$f(x) = 1 - \frac{6}{x^3} + \frac{-2x}{x+4}.$$

If it is a rational function, write it in the form $f(x)=\dfrac{p(x)}{q(x)}$, *where p(x) and q(x) are polynomials.*

On the same quiz, she had also asked her students to express the following as a single fraction:

$$1-\frac{6}{8}+\frac{-4}{6}.$$

In a report of her students' performance, TA12 first explained that she had designed the quiz purposefully: the fraction expression is equivalent to evaluating the function *f* at $x=2$. She then observed that most students simplified the expression in mathematically valid ways; a typical solution was:

$$1-\frac{6}{8}+\frac{-4}{6}=1-\frac{3}{4}+\frac{-2}{3}=\frac{12}{12}-\frac{9}{12}+\frac{-8}{12}=\frac{12-9-8}{12}=\frac{-5}{12}.$$

However, many students—including some who had performed a valid calculation for fractions—simplified the rational expression as follows:

$$f(x)=1-\frac{6}{x^3}+\frac{-2x}{x+4}=\frac{1-6+-2x}{x^3(x+4)}.$$

She hypothesized that students' thinking about rational functions did not draw on their experiences with fractions:

> While there are some students who struggle with combining these fractions, most of my students are able to do so successfully. That shows me that they are familiar with and able to use fraction operations, so the root of the misconception in this case is not that they have misconceptions concerning the fraction operations ... For some reason, the introduction of variables into the fraction numerator and/or denominator causes a breakdown in their reasoning, which I believe is the root of the misconception. (TA12 Final paper, p. 4)

In TA12's interpretation, students have productive knowledge to build upon, because they can work with closely related numerical expressions in mathematically valid ways. At the same time, students may separate their knowledge of fractions from their knowledge of rational expressions. TA12 then speculated how her own teaching or others' instruction may enforce this separation, calling out the role of emphatically distinguishing operations with numbers from operations with variables (e.g., stressing that $3+2=5$, but $3a+2b\neq 5ab$). TA12 thus hypothesized that students may benefit from experiences in which they explicitly connect operations on fractions with operations on rational functions.

As a next step, TA12 proposed to hold a structured conversation with her students. TA12 scripted a hypothetical conversation, a portion of which follows:

T: Now, even though we used mathematical operations specifically to combine fractions, the truth is that we can use these mathematical operations with any ratio. What exactly are the mathematical operations we used to combine the fractions?

S: First we found a common denominator. Then we changed each individual fraction so that it had the common denominator. Last we added together the changed numerators to get one faction.

T: Exactly! Let's see if we can use those same operations to solve the problem we started with. How could we find a common denominator?

S: I don't really know.

T: How did you find a common denominator of (1, 3, and 4)?

S: I multiplied them together.

T: Exactly! So, you actually did this previously, but we could find the common denominator in our problem by multiplying the denominators together. Just because they have variables in them doesn't change our process. What would our new common denominator be?

S: $x^3(x+4)$

T: Correct! So, think back to the second operation you said you used for combining fractions: "change each individual fraction so that it has the common denominator." How do you think we could do this with our problem with variables?

S: We could figure out what we need to multiply each fraction by to get the common denominator!

TA12 envisioned guiding the student in identifying fraction operations used while working with different denominators, and then building concrete connections between fractions and rational expressions. TA12 emphasized that plan was not to stick to the script but rather to ask "purposeful, guiding questions" that allowed as much as possible for "the students to… generate as much knowledge on their own" (TA12 Final paper, p. 8).

6.2.2 Modeling TA Thinking

To explain how we model TAs' thinking about instruction, we use TA12's reflection as an example. Our model has four components: data, student thinking, hypothesis, and future teaching actions. Figure 6.1 displays this model. We define *data* to consist of written and oral expressions made by students that are observed by an instructor. In the case of TA12, the data are her students' performance on a quiz. *Student thinking* is an interpretation of the data. For example, TA12 interprets the combination of mathematically valid work with fractions and mathematically incorrect work with rational expressions as an indication that her students did not draw on their knowledge of

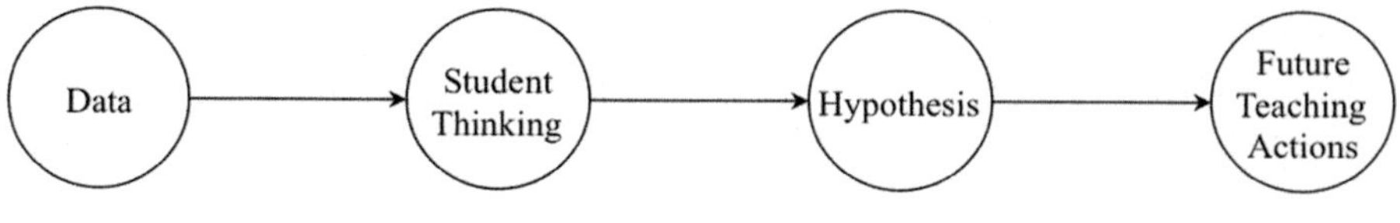

Fig. 6.1 Model for TA thinking

fractions when working with rational expressions. A *hypothesis* is a conjecture about likely experiences that have shaped or could shape the student thinking. TA12's hypothesis is that students may benefit from explicit connections between operations on fractions and on rational functions. *Future teaching actions* describe how the instructor might work with students in the future, given their interpretation of student thinking. TA12 proposed to hold a structured conversation in which she would guide students toward describing properties of rational functions based on properties of fractions, and then give students an opportunity to use these properties.

We use TA12's reflection as an example because it shows how the components of the model fit together, even if there are places where the reflection can be improved or may be unrealistic. The interpretation of the data is reasonable: students are not applying their knowledge of numeric fraction operations to fractions that have variables. The hypothesis addresses the interpretation directly: TA12 interprets that student do not use their knowledge of fractions when working with rational functions, even though this knowledge is useful, and so TA12 proposes that students construct and then use parallels between fractions and rational expressions. The future teaching actions are envisioned to elicit the relevant similarities between fractions and rational expressions. There are places where the dialogue may seem contrived or where the instructor may be appearing to do too much of the students' work. TA12's interpretation that the students separate their knowledge of fractions from rational expressions may be overly simplistic. However, in making these judgments, we should keep in mind that TA12 is a first time instructor, and that on the whole, the components do lead from one to the other. As we discuss later in this chapter, there are examples of TA reflections whose components are not as well connected.

6.3 Literature Informing the Study

In this section we summarize the literature informing our model for TA thinking and the rationale for drawing on results from K-12 teacher education and professional development. The goal of our model is to describe TAs' claims about future teaching actions as potential opportunities to learn. Thus, our model connects two literature bases: one on argumentation and the other on teachers' opportunities to learn from teaching.

6.3.1 Literature on Argumentation

Toulmin (1958) is a foundational reference about modeling argumentation. Toulmin originally created his model to analyze legal arguments, and it has since been used for other fields, including mathematics education (e.g., Inglis et al., 2007). The three key components of Toulmin's model are: the grounds, the claim, and the warrant. The grounds are the evidence on which the claim is made, and the warrant is the reason that the grounds support the claim. Toulmin uses the following claim as an example: "I am a British citizen." Possible grounds for this argument include, "I was born in Bermuda;" a warrant could be, "British law states that persons born in Bermuda are British citizens." In our model, we consider both the data and the interpretation to be the grounds of the TA's argument. The future teaching actions are a claim about what instruction may be beneficial. The warrant is the hypothesis about experiences that might shape or have shaped the students' understanding. The way we map our model to Toulmin's components is consistent with the cognitive theory that learning involves the interpretations a person ascribes to their experiences and the inferences made from these interpretations (see Thompson, 2016 for an overview of this theory, which is based on work of the psychologist Piaget).

6.3.2 Literature on Opportunities to Learn from Experience

Our model is also shaped by studies of K-12 teachers, especially the research of Horn and colleagues (e.g., Hall & Horn, 2012; Horn et al., 2015). Their work focuses on describing the "opportunities to learn" that teachers create in conversation about student data and previous experiences, including how some opportunities may be stronger than others.

6.3.2.1 Opportunities to Learn

In the theory developed by Horn and her colleagues described in the papers cited above, learning opportunities for improving one's teaching are strongest when: teachers marshal observations and stances about teaching experiences to mobilize themselves for future plans, and these plans represent skillful teaching. "Stances" refer to what the teachers believe is important for them to know about learning and teaching, and how to come by this knowledge. To put this in terms of our model, when the data and interpretation are strongly linked to the hypothesis and future teaching actions, in a way that is consistent with what is known about teaching quality, there is greater opportunity to learn.

6.3.2.2 Features of Skillful Teaching

In our view, which is consistent with the writing of Horn and colleagues, skillful teaching includes: responsiveness to and respect for student thinking; providing opportunities for students to articulate their thinking and respond to others' thinking; maintaining cognitive demand (e.g., if an assigned question is challenging, the teacher helps the student work on the question without stripping away the difficulty); focusing students on core mathematical ideas, especially the meaning behind expressions and procedures; and inclusiveness (all students are attended to).

Additionally, when students work on problems based on real-world scenarios, the instructor helps the students understand the real-world context, how mathematics could model this context, and develop common terms to refer to key ideas in the context. Our views are informed by studies of teaching complex tasks (e.g., Jackson et al., 2013; Stein et al., 1996); studies linking qualities of teaching to student outcomes (Learning Mathematics for Teaching Project, 2011); and studies that identify and describe components of tasks of teaching (Boerst et al., 2011; Sleep, 2012).

6.3.3 Parallels Between K-12 and Post-Secondary Education

Commonalities between pre-calculus courses at the undergraduate level and the high school level make it reasonable to hypothesize that results from K-12 teacher education and professional development are also promising for development of instructors of undergraduates. After all, K-12 teachers and undergraduate course TAs share some challenges, especially when TAs teach a course such as college algebra. Students typically enroll in college algebra through requirement rather than by choice; often they are placed in the course through a combination of assessments and previous coursework. Students are likely to need many opportunities to break unproductive habits. College algebra is also a gateway to many courses needed for scientifically-oriented careers.

We now discuss the particular TA development context in which we collected data on TAs' thinking.

6.4 Context, Data, and Method

6.4.1 Context

The TAs in this study were enrolled in a seminar on teaching and learning mathematics at the post-secondary level. The TAs all taught college algebra or intermediate algebra in sections of approximately 40 students, consisting primarily of first-year college students. With few exceptions, every college algebra or intermediate algebra TA participates in the seminar. Each TA is the sole instructor for his or her section. The lessons in all sections feature small group discussions and small group work led by the TAs. Each beginning TA teaches only one section of the course; in subsequent years TAs would teach two courses in the fall and one in the spring. There are a few adjunct instructors (mostly former or current high school mathematics teachers), who teach college algebra, but TAs teach the majority of the sections. The adjunct instructors do not participate in the TA development program.

The seminar met two hours per week in the fall semester, when the study was conducted. To help develop language for reflecting on teaching, TAs read educational literature describing examples and theories for understanding student learning. These include Erlwanger's (1973) classic account of a child's arithmetic understanding, and Tsay and Hauk's (2013) exposition of constructivism. In the seminar, TAs were asked to discuss teaching experiences in terms of the readings.

6.4.2 Data

We collected final papers written by all 16 TAs enrolled in the seminar. In these papers they were asked to (a) report on student performance on a quiz they assigned, (b) interpret student thinking in the quiz performance, (c) hypothesize about experiences that contributed to the students' thinking, and (d) propose future teaching actions to refine student thinking. In the assignment, the TAs were asked to focus on interpreting student work that was not mathematically valid.

6.4.3 Rationale

Recall that our aim is to study TAs' thinking, and that we model TA thinking with four components (as shown in Fig. 6.1): data, interpretation, hypothesis, and future teaching actions. The intention of the assignment was for TAs to hypothesize how or why students may have found their way to mathematically invalid reasoning, and for TAs to describe future teaching actions that are built on productive ways of thinking and provide settings where new ways of thinking might be useful. The assignment is designed to elicit TA thinking for each component and how it related to the previous component: (a) data (b) interpretation (c) hypothesis (d) future teaching actions.

6.4.4 Analysis

We analyzed the TAs' papers in two parts. First, we examined the components of the model represented, including whether TAs articulated the components and their connections clearly or if components were missing or conflated with other components, and to what degree they represented skillful teaching (as described in Sect. 6.3.2.2). Second, we examined the internal consistency or inconsistency (i.e., at least two components contradict each other) of the components. As discussed previously (in Sect. 6.2), TA12's paper is an example of an internally consistent paper. The paper written by TA13 (discussed in more detail in Sect. 6.5) provides an example of inconsistency as well as conflated components. TA13 interpreted that students see equations as "a string of symbols to memorize." Later TA13 stated several hypotheses, including "When my students see an equation, their first thought is that they have to memorize it" and that students were "uncomfortable" with function notation (TA13 Final paper, pp. 1–2). Thus the hypothesis and interpretation are conflated. Furthermore, she then described future teaching actions that deliberately avoided addressing or using function notation. These future teaching actions are inconsistent with the hypothesis and interpretation, because they seek to address students' use of function notation without opportunities for students to use function notation.

We compared the consistency and connectedness of papers relative to each other, rather than to an external standard. The reasons for this approach were twofold. First, to our knowledge, there is no widely-accepted rubric for judging the coherence of pedagogical argument, though there are theories about the components of such an argument (which we used as a foundation for this study, as discussed in Sect. 6.3.2). Second, it is a well-established cognitive science result that people are more reliable comparing impressions against one another than judging an impression of quality in isolation (Laming, 1984; Thurstone,1927). The relative comparisons can then be used to sort objects into categories of relative quality and identify attributes contributing to the impression of quality (e.g., McMahon & Jones, 2015). We classified papers into "high", "medium", and "low" connectedness by consensus, in which at least four of the chapter authors weighed in on each paper, with more authors discussing controversial papers. Highly connected papers articulated all four components with internal consistency. Papers with medium connectedness conflated components (for instance, TA13) and were not entirely internally consistent. Low connectedness papers did not specify the reasoning between each component, for instance leaping from data to future teaching actions (as is the case with TA02, to be discussed further in Sect. 6.5).

During this analysis, we discovered a highly connected paper that did not feature a coherent argument. Although we agreed that the TA attempted to connect all components, and we also agreed on the specific weaknesses of the argument, we disagreed on the plausibility of the connections, and we never resolved our disagreement. We called this type "highly connected with low coherence." We classified two other papers in this way.

6.5 Results

We asked: How do TAs connect observations and beliefs about their students, hypotheses about student thinking, and proposed next teaching actions? Table 6.1 summarizes the TAs' papers by category. Figure 6.2 shows the connectedness paths exhibited in our data. We now illustrate each path with an example paper. To retain anonymity of the TAs, we use the same pronoun, "she", to refer to all TAs. (Of the 16 TAs in the cohort studied, nine were female.)

Table 6.1 TA papers by category

Connectedness	Which TAs' final papers exhibited this connectedness
High	TA04, TA05, TA08, TA11, TA12, TA15
High, with low coherence	TA02, TA10, TA14
Medium	TA01, TA06, TA13, TA16
Low	TA03, TA07, TA09

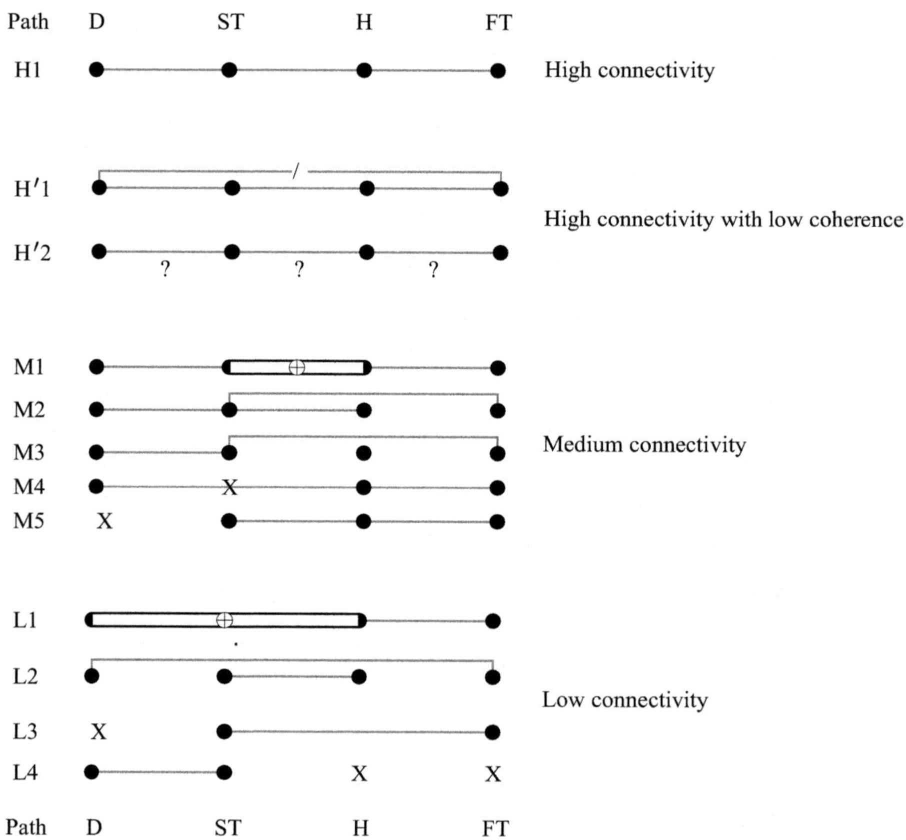

Fig. 6.2 Connectedness of TA papers. Key*:* D=data, ST=student thinking, H=hypothesis, FT=future teaching actions, X=component absent, ⊕ =conflated components, •—•=link was attempted and satisfied criteria, $\underset{?}{\bullet\!-\!\!-\!\bullet}$ =no consensus from research team on whether link is plausible, • – / – • =future teaching actions do not plausibly address student thinking identified in data

6.5.1 *Illustrations of Paths of TAs' Thinking*

6.5.1.1 High Connectedness (H1 in Fig. 6.2; TA12)

TA12's paper (described in Sect. 6.2) is highly connected. She described all four components, and the components and her reasoning from one to the next were internally consistent.

6.5.1.2 High Connectedness with Low Coherence (H'1; TA02)

TA02 posed the quiz problem, "How would you find the y-intercept of a function f? Explain why your method gives the y-intercept." Several students responded similarly to: "You would plug in a zero into the x. By putting zero into the x, the y-intercept would be the only thing left." TA02 concluded, "It seems that they view 'plug in 0 for x' as a way to get rid of the x, rather than a consequence of the fact that if a point is on the y-axis, its x-value must be 0" (TA02 Final paper, p. 2). As a result, TA02 hypothesized, "They see the graph and the equation as two distinct objects—related, because you can sketch the graph given the equation, but not exactly representing the same mathematical relationship. The confusion about how to find the y-intercept is probably a special case of this disconnect" (p. 2). TA02's interpretation and hypothesis are plausibly linked to the data.

TA02 proposed that in the future, she would design a worksheet that asked students to sketch a graph of a given function, complete an input/output table, and sketch vertical lines on the graph. The intention would be for students to experience finding outputs of a function both using its defining equation and using intersecting the graph of the function with vertical lines. While the worksheet does link to the hypothesis, it did not plausibly address the TA's interpretation of student thinking. The first worksheet question asked students to graph a function defined by an equation. However, assuming TA02's interpretation of student thinking, the student most likely would struggle with graphing. It is possible that the remainder of the worksheet would have helped the student recognize the connection between graphs and equations, but this assumes that the student would be able to begin the work. We classified this paper as an example of an H'1 pathway where despite plausible links between components, the argument still does not hold together.

6.5.1.3 Medium Connectedness (M1; TA13)

Using data from a unit exam question involving revenue, profit, and cost (denoted $R(n)$, $P(n)$, $C(n)$ respectively), TA13 observed:

> ... many of them seemed to think that an equation was a string of symbols to memorize, as opposed to something that they were capable of understanding or even constructing on their own. ... During the exam I had students raise their hands and tell me, 'I forgot the formula from class!' They wrote down things like $P(n)=R(n)+C(n)$, $P(n)=C(n)-R(n)$, ... This was very surprising because I am confident that every one of them has an intuitive

> understanding of the concept of 'net gain'. This reminded me very much of the situation described in the paper *Mathematics in the Streets and in Schools* [Carraher et al., 1985], in which kids were perfectly capable of doing arithmetic in the marketplace, but when handed pencil and paper and asked to work the same problems out symbolically, were frequently flummoxed. (TA13 Final paper, p. 1)

To address the students' conception of equation, TA13 proposed to use a story about a renter saving up money for an upcoming vacation to derive a formula involving rent R, living expenses L, monthly income I, savings S, and the cost T of round trip plane tickets. As the students arrived at expressions to solve the problem, TA13 would ask students to justify their findings. TA13 explained that this task

> requires students to create something, as opposed to manipulating a formula that is given … It would get students used to the idea that equations are not divinely inspired, can be written by ordinary people, and used as shorthand to describe events that are entirely understandable (p. 4).

Prior to describing these future teaching actions, TA13 put forth several hypotheses, including students' discomfort with the terms "revenue" and "profit," with function notation, and with the notion of inputs and outputs of a function. However, these hypotheses do not link plausibly to the future teaching actions; TA13 notes that she purposefully designed the worksheet to avoid function notation, even if there are variables used. There is also no mention of revenue and profit.

TA13 proposed one more hypothesis that she emphasized as the most probable cause: "When my students see an equation, their first thought is that they have to memorize it" (p. 1). TA13 described other situations where schooling mandated memorization because the information was in some sense arbitrary, such as naming the 50 states and their capitals. Although this last hypothesis does connect to the future teaching actions proposed, it only restates the interpretation of student thinking.

TA13 interpreted student thinking in a way that was consistent with the data. The future teaching actions are connected to the interpretation of student thinking and the data. Although she attempted to describe experiences that shaped student thinking, the only applicable hypothesis restated the description of student thinking. In other words, TA13's paper conflated interpretation of student thinking and hypothesis. For these reasons, we categorized TA13's paper as an example of medium connectedness.

6.5.1.4 Low Connectedness (L2; TA03)

On a quiz given by TA03, students were unable to articulate the difference between the word "constant" and the phrase "constant rate of change." Concerned that the students may not understand that these terms denote fundamentally different ideas, TA03 allowed students time to discuss the difference between the terms in groups. However, confusion persisted. TA03 attributed this misunderstanding to lack of precision in language and grammar, leading students to gloss over verbal differences between the two terms. TA03 went on to suggest that lack of precision causes students to group together similar looking functions, even to the extreme of "glossing over the distinction between lines with slope zero and lines with nonzero slope" (TA03 Final paper, p. 1). Generally, TA03 was concerned that imprecise language leads to confused mathematical thinking.

TA03 then proposed that in the future, she would ask students to graph the monthly profits of two businesses in a story problem, one whose monthly profits are a constant function of time ($1000 each month), and the other whose monthly profits have a constant (nonzero) rate of change over time ($1000, $1200, $1400, etc.). TA03 reasoned, "Get students to admit that the second business is much different than the first; in fact, it's much better! Then, and only then, broaden out the discussion to include the actual words 'constant' and 'constant rate of change'. ... By building a common starting point through discussing which business is doing better than the other, the teacher can buy themselves enough goodwill to introduce the more abstract terminology" (p. 3). This activity targets the confusion encountered in class with regard to the two terms. However, the activity neither addresses precision of language or precision in a students' view of functions in any significant way. That is, this activity does not build on the hypothesis. Hence, the future teaching actions are connected to the data, but not to any component between data and future teaching actions, despite an effort on the part of TA03 to do so. We speculate that one source of this issue for TA03 is that she genuinely believes that addressing precision of language in general would solve many problems. Perhaps this view pervaded her thinking so strongly that TA03 struggled to identify a hypothesis that provided more guidance for future teaching actions.

6.5.2 *Summary of Findings*

We modeled TAs' final papers as a practical argument with four components. We found that components could be present, absent, or conflated, and we found that connections could be present, absent, internally consistent, or internally inconsistent. In some cases, we found TAs connected components that were non-adjacent in our model without connecting adjacent components. We also found one final paper in which the research team arrived at consensus on weaknesses of the TA's argument but could not arrive at consensus on whether links were plausible. In total, our data of the 16 TAs' final papers exhibited 12 paths in four categories. We illustrated one example path for each category.

6.6 Reflections on TA Education

As we specifically analyzed TAs' written reflections to an assignment from the TA pedagogy course, we detected different levels of TA reflection, as well as different patterns of components and connections. We were able to categorize papers as *high connectedness, high connectedness with low coherence, medium connectedness,* and *low connectedness.*

We believe these results are valuable for TA development. First, in terms of research, they extended theory from the K-12 teacher education literature on opportunity to learn to post-secondary instruction. As far as we know, theorizing on opportunity to learn in the context of TA development is novel. Second, more practically, our results support research into TA learning by describing ways in which TAs may, or may not, connect their experiences to future teaching.

Learning from experience is the goal of many TA development programs, but as the research in K-12 teacher education and our own results show, the potential for TAs to learn from experience can vary. What we have added to this conversation is particular examples of how TA learning opportunities can vary, even when the TAs are asked to do similar tasks. TAs whose papers were categorized as *high connectedness* were able to clearly articulate the four assignment components, as well as explicit links and connections among the components. TAs in this category have illustrated their capacity to act as reflective practitioners, and use their understanding of student thinking to support student learning. TAs whose papers were categorized as *high connectedness with low coherence* were able to articulate components, but the links were weaker or implicit. These TAs were on their way to becoming more reflective as teachers: they have the components, but need to learn to better articulate connections or links among those components. Given the limitation of analyzing written reflections, we can only conclude the TAs did not write about the connections among the components; it may be the TAs did see those connections, but need further practice in expressing teaching reflections in writing.

Other TAs were still at a stage in which written reflections did not capture the type of components and connections intended, but instead revealed the struggles of novice instructors trying to make sense of student thinking and determining how to respond. These TAs at the *mid* and *low connectedness* prompt us to think about how we might better support TAs in being explicit in their writing, and help TAs to both see and express connections among the components. When we do not see explicit links or components, we do not always have enough information to judge whether the omission was truly a reflection of the TAs' maturity as a practitioner, or instead, a reflection of the TAs' skills in reporting their thoughts and actions in a written reflection. We sometimes find TAs who are pursuing doctorates in mathematics profess to "not be good at writing" and who struggle to express their thoughts in coherent written paragraphs.

In considering what we have learned and others who might learn from our experience, we turn to our model of TA thinking. In this meta analysis, the data are the TAs' final papers. We interpret that while TAs are invested in helping their students learn and are committed to helping students construct knowledge, their proposed teaching actions do not always align with their interpretations of student thinking. We hypothesize that seminar discussions, in which TAs practiced describing their experiences in terms of literature on constructivism, fostered the TAs' dedication to giving students experiences to develop their own mathematical knowledge. Across the cohort of TAs, we saw this evidenced in their proposed teaching. We also hypothesize that these seminar discussions did not support TAs in selecting hypotheses or connecting components because they emphasized the components rather than the connections. Several TAs suggested hypotheses that they did not design instruction to address. In each of these cases, the hypotheses were general statements about students' ways of doing mathematics that would be difficult to mediate in the span of a lesson, rather than hypotheses that specifically applied to student performance on quiz problems.

We propose that in the future, TAs continue to read literature that encourages them to see the value of students discovering mathematics. We also propose that TAs hold seminar discussions in front of their peers, where they make explicit the connections between future teaching actions to hypotheses. In these discussions, the facilitator and peers would help revise one TA's hypotheses and teaching actions to be better defined and more strongly connected. This public revision is reminiscent of discussions between mentor and mentees to design action research, where the goal is to define addressable research questions and design data collection and analysis that address the research questions. In this analogy, research question is to hypothesis as data collection and analysis are to future teaching actions. Holding a public discussion aligns with research on K-12 teachers' learning suggesting that when groups of teachers reflect on experience, they create a "collective zone of proximal development" (Engeström, 1987) where they learn more than would be possible individually.

Finally, we comment that it is reasonable to wonder whether having TAs focus on student error is productive: will that reinforce deficit-views of student thinking? In our experience, many TAs enter graduate school believing that students' mathematical thinking is either right or wrong, and student learning can be accomplished by exposure to "right" ways. For instance, when discussing why students might struggle with composition of functions, TAs at the beginning of the semester have often proposed, "The students just need to learn the rule." We have found that discussing student errors has helped TAs move away from black and white judgments of student thinking. The TAs' final papers, even those of low connectedness, displayed more potential for sensitivity to student thinking than at the start of the semester. We also have the impression that the cohorts have become more sensitive to student thinking over time, which we attribute in part to new graduate students entering a culture where a critical mass of TAs hold a more constructivist orientation. We are optimistic that this trend will grow in the future. To continue encourage this trend to continue, we propose two changes. The first is to focus seminar discussions and assignments on unexpected correct solutions to complex tasks, and the second is focusing TAs' hypotheses more explicitly on what about the student thinking can be built on in future teaching actions.

References

Boerst, T., Sleep, L., Ball, D., & Bass, H. (2011). Preparing teachers to lead mathematics discussions. *Teachers College Record, 113*(12), 2844–2877.

Carraher, T. N., Carraher, D. W., & Schliemann, A. D. (1985). Mathematics in the streets and in schools. *British Journal of Developmental Psychology, 3*, 21–29. http://dx.doi.org/10.1111/j.2044-835X.1985.tb00951.x.

Engeström, Y. (1987). *Learning by expanding: An activity-theoretical approach to developmental research*. Helsinki, Finland: Orienta-Konsultit.

Erlwanger, S. H. (1973). Benny's conception of rules and answers in IPI mathematics. *Journal of Children's Mathematical Behavior, 1*(2), 7–26.

Hall, R., & Horn, I. S. (2012). Talk and conceptual change at work: Adequate representation and epistemic stance in a comparative analysis of statistical consulting and teacher workgroups. *Mind, Culture, and Activity, 19*, 240–258. http://dx.doi.org/10.1080/10749039.2012.688233.

Horn, I. S., Kane, B. D., & Wilson, J. (2015). Making sense of student performance data: Data use logics and mathematics teachers learning opportunities. *American Educational Research Journal, 52*(2), 208–242.

Inglis, M., Mejia-Ramos, J. P., & Simpson, A. (2007). Modelling mathematical argumentation: The importance of qualification. *Educational Studies in Mathematics, 66*(1), 3–21.

Jackson, K., Garrison, A., Wilson, J., Gibbons, L., & Shahan, E. (2013). Exploring relationships between setting up complex tasks and opportunities to learn in concluding whole-class discussions in middle-grades mathematics instruction. *Journal for Research in Mathematics Education, 44*(4), 646–682.

Laming, D. (1984). The relativity of "absolute" judgments. *British Journal of Mathematical and Statistical Psychology, 37*, 152–183.

Learning Mathematics for Teaching Project. (2011). Measuring the mathematical quality of instruction. *Journal of Mathematics Teacher Education, 14*, 25–47. http://dx.doi.org/10.1007/s10857-010-9140-1.

McMahon, S., & Jones, I. (2015). A comparative judgement approach to teacher assessment. *Assessment in education: Principles, policy and practice, 22*(3), 368–389. http://dx.doi.org/10.1080/0969594X.2014.978839.

Sleep, L. (2012). The work of steering instruction toward the mathematical point: A decomposition of teaching practice. *American Educational Research Journal, 49*(5), 935–970.

Stein, M. K., Grover, B. W., & Henningsen, M. (1996). Building student capacity for mathematical thinking and reasoning: An analysis of mathematical tasks used in reform classrooms. *American Educational Research Journal, 33*(2), 455–488.

Thompson, P. W. (2016). Researching mathematical meanings for teaching. In L. D. English & D. Kirshner (Eds.), *Handbook of international research in mathematics education* (pp. 435–461). New York: Taylor & Francis.

Thurstone, L. L. (1927). A law of comparative judgment. *Psychological Review, 34*, 273–286.

Toulmin, S. E. (1958). *The uses of argumentation*. New York, NY: Cambridge University Press.

Tsay, J.-J., & Hauk, S. (2013) Constructivism. In S. Hauk, N. M. Speer, D. Kung, J.-J. Tsay, & E. Hsu (Eds.), *Video cases for college mathematics instructor professional development*. Retrieved June 19, 2016, from http://collegemathvideocases.org.

Chapter 7
Lessons Learned from a Math Teachers' Circle

Gulden Karakok, Katherine Morrison, and Cathleen Craviotto

Abstract In this chapter, we describe our experience running the Northern Colorado Math Teachers' Circle (NoCOMTC), founded in 2011. The goal of the NoCOMTC is to improve middle school mathematics teachers' mathematical and pedagogical content knowledge through interactive mathematical problem-solving professional development sessions. Our leadership team is an effective collaboration between university mathematics and mathematics education professors and middle and high school mathematics teachers. In this chapter, we describe our leadership team's journey from founding the NoCOMTC through four academic years of monthly evening mathematics teachers' circle sessions and three residential summer immersion workshops. We also discuss our recently initiated student circle program. We focus on aspects that were essential to forming and sustaining our program. In addition, we highlight lessons we have learned while planning and facilitating both mathematical problem-solving sessions and activities designed to help teachers' implementation of problem solving.

Keywords • In-service mathematics teachers • Math teachers' circle • Middle school mathematics • Problem solving • Professional development

---

MSC Codes
97B50
97D50
97A99

G. Karakok • K. Morrison (✉)
School of Mathematical Sciences, University of Northern Colorado,
501 20th Street, Campus Box 122, Greeley, CO 80631, USA
e-mail: gulden.karakok@unco.edu; katherine.morrison@unco.edu

C. Craviotto
Boulder Valley School District, 1020 Arapahoe Avenue, Berthoud, CO 80513, USA
e-mail: cathleen.craviotto@bvsd.org

J. Dewar et al. (eds.), *Mathematics Education*, Association for Women in Mathematics Series 7, DOI 10.1007/978-3-319-44950-0_7

7.1 Introduction

Inspired by a shared interest in giving back to the local community through work with teachers, Cathleen Craviotto and Gulden Karakok founded the Northern Colorado Math Teachers' Circle (NoCOMTC) in 2011 together with three local district teachers. Katherine Morrison joined this leadership team in 2013 shortly after she began working at University of Northern Colorado. While all three authors had been involved in different forms of professional development in the past, we were particularly inspired by the Math Teachers' Circle (MTC) model. We thought this model would provide us the opportunity to share our personal excitement about mathematical problem solving while stimulating teachers' interest in solving problems. In addition, we believed that this professional development model could support our local middle school mathematics teachers' implementation of problem solving in their classrooms by way of their personal experience in our sessions. This support for mathematics teachers was especially important for our local school district where all of the middle schools rank low both in students' mathematics achievement and growth as measured by state mathematical assessments. The goal of the NoCOMTC is to respond to these issues by improving middle school mathematics teachers' mathematical and pedagogical content knowledge (i.e., specific knowledge of how to teach mathematics) through interactive mathematical problem-solving sessions.

To achieve this goal, we have been offering monthly evening problem-solving sessions and annual weeklong residential summer workshops for middle school teachers. In these settings, the participants collaboratively solve challenging mathematics problems and discuss solutions and problem-solving strategies. In addition, we facilitate discussions on ways to implement these problem-solving tasks in participants' classrooms. The following quote from one of our summer workshop participants captures the learning process in our MTC's problem-solving environment. Similar to the reactions of many of our participants, this participant made connections between this experience and students' experience:

> I struggled, I felt uncomfortable, I was anxious and I was nervous. BUT … I LEARNED a lot about math topics and concepts that I was not necessarily familiar with. I was able to understand concepts that were presented to me in college that I never understood before and this is because I was encouraged to discover. I made sense of the things we were doing because I was able to think, communicate, and "digest" the new information in my own way… I think this is all the same for my students! I need to teach them to problem solve, discover, and explore information in their own way. I need to allow them plenty of time to process and discover, and I need to make sure that I guide and facilitate (vs. tell).

7.2 Our Experience Starting

Math Teachers' Circles (MTCs) are an American adaptation of an idea that originated in Bulgaria and Russia. Math Circles began over 100 years ago as a way for professional mathematicians to work on problem solving with secondary students.

Emigrants who had been inspired by these sessions as teenagers then initiated circles in the US. Many teachers began transporting students to Math Circle activities in the Bay Area, but the teachers were not allowed to participate in these exciting problem-solving sessions. Soon they requested that Math Circles be created for teachers too, so they could have a similar experience. In 2006, the American Institute of Mathematics (AIM) began its first Math Circle for teachers and, in 2007, AIM began to organize workshops to teach others how to run MTCs (Math Teachers' Circle n.d.).

At the 2011 Joint Mathematics Meetings, Craviotto attended sessions featuring MTCs and was inspired to explore this form of professional development for teachers. Her love of problem solving and her past experiences teaching middle school students, where she was frequently impressed by their flexible thought processes, drew her to this work and got her excited about sharing these experiences. Also, she had substantial experience working with pre-service teachers, as this is the primary target population of University of Northern Colorado (UNC), where she was a mathematics professor. Thus, she set out to recruit a leadership team and create a local MTC. At the time, Karakok was a newly hired mathematics education professor in the mathematics department at UNC. She had significant experience facilitating professional development for in-service teachers and teaching mathematics content courses for pre-service teachers. Karakok was also excited by the problem-solving focus of the MTC model, and together they invited two local middle school mathematics teachers and a district mathematics coach, forming the initial leadership team. As Craviotto was leaving the university setting in May 2013, they invited newly hired mathematics professor Morrison to replace her in the faculty role on the team. (Despite no longer being at the university, Craviotto continues to be a key member of the team.) Morrison had previously been involved in teaching summer mathematics courses for in-service teachers and was excited to work with the middle school teacher population.

The initial leadership team attended a weeklong workshop at AIM on *How to Run a Math Teachers' Circle* in 2011. This workshop was an exciting time for the team to coalesce, to get to know each other better and to enjoy doing mathematics together. We gained firsthand experience with the MTC model as participating teams worked together on challenging mathematics problems. This experience helped our team learn about each other's mathematical thinking and communication. The workshop also helped us learn how to best capitalize on the strengths of our team members to structure our circle activities. For example, as we worked on mathematics problems we experienced the frustration of not having enough think time. We discussed the importance of explicitly supporting think time, then having each person share their approach, and discussing all of our approaches. We learned how to create a collaborative space for all of us to solve problems, respecting each of our different mathematical backgrounds and experiences. These experiences helped us to create a similar environment for our participants in which they can collaborate and learn from each other's distinct mathematical backgrounds.

The following academic year (2011–2012) we focused on fundraising to support running a residential summer workshop and building infrastructure to run academic-year sessions and recruit local teachers. We managed to secure sufficient funding to have a summer workshop in the summer of 2012; however, quite surprisingly, we did not have enough applicants (only three) to run a workshop. To address this recruitment problem, we cultivated stronger connections with our local district personnel and teachers and implemented monthly evening sessions. Faculty members on our team reached out directly to local district administrators, asking them to advertise these sessions to their teachers. The two middle school teachers on our team personally recruited teachers and advertised the program at all district events. The district mathematics coach on our team made arrangements to provide district professional development hours for participants, which has been a great incentive for teachers.

Since 2012–2013, we have been able to run six academic-year evening sessions annually and a weeklong summer residential workshop, which we have opened up to teachers statewide. For each of these sessions, we met regularly as a team to discuss the content of mathematics problems, which were then facilitated by the faculty members. In addition, the faculty members continued to work with the university foundation office to write multiple grant proposals to support our activities. We have secured enough funds annually to provide dinners for participants at each evening session, cover the weeklong summer workshop expenses such as room and board for each participant, books, supplies and materials, and stipends for the leadership team.

In the 2015–2016 academic year, with the continued success of our circle, we expanded our program to host a Math Circle for students. This expansion was proposed by one of our local teacher leaders who wanted us to do more for the local school district. We piloted six sessions and experienced tremendous enthusiasm from attendees. As a result of this strong interest from students and financial support from the community, we ran our first three-day Student Math Circle summer camp in June 2016, for students in grades four through eight.

7.3 Structure of Our MTC

We design and lead all of our problem-solving sessions to focus on enhancing the content and problem-solving process knowledge of our teacher participants. We strategically choose problems that will foster a general enthusiasm for mathematics. In particular, the problems that we select are frequently initially challenging and exciting for us as well, so we greatly enjoy planning together and engaging in problem solving with the teacher participants each session. This same sentiment has been expressed by many of our teachers, "I really enjoyed doing real math again!! It gets boring just doing 7th grade math and never being challenged. I gained an appreciation for the importance of developing problem solving skills."

Since the choice of problems is instrumental to our success, we use problems that we find inherently interesting and that fall into three different categories, namely those focused on *content, advanced content* and *problem-solving strategies. Content problems* refer to problems that are directly related to the content of the Common Core State Standards of Mathematics (National Governors Association 2010) at the middle school level. *Advanced content problems* include content that is not necessarily at the middle school level, but may build on topics from that level and extend them beyond the typical middle school curricula. The third type focuses on problem-solving strategies and approaches rather than on any particular content. These strategies are inspired by and aligned with research studies in the area of problem solving (e.g., Carlson and Bloom 2005), the Common Core State Standards for Mathematics (CCSS-M), and professional mathematicians' perspectives (e.g., Tanton n.d.; Zeitz 1999). Our frequent guest facilitator, Paul Zeitz, has been instrumental in bringing such problems to our summer workshops. In the Appendix, we give examples of these problem types, and in the following subsections we describe how we implement them in our monthly evening and summer workshop sessions.

7.3.1 Monthly Evening Sessions

During each academic year since 2012, we have had six monthly evening sessions on Mondays, each lasting two and a half hours including dinner. Each session is designed around a content or problem-solving strategy theme. The session begins with a warm-up problem that serves as an ice-breaking activity for teachers to get to know each other and re-energize after a long workday. Also the warm-up problem eases participants into exploration of the main topic or theme of subsequent problem(s).

One of the content themes we explored was divisibility. The session began with warm-up problems such as "If the five-digit number 5DDDD is divisible by 6, then what must be the value of D?" and "If A is odd, which ordered pairs (A, B) will make the three-digit number 4AB a multiple of 4? What about making it a multiple of 8?" These problems led participants to explore divisibility rules, facilitated discussions on why these rules work, and prepared them to approach the more challenging *Social Security Number* problem listed in the Appendix. These problems directly connect to the two domains of the Number System and Expressions and Equations in the standards for grades four through eight.

At the end of each session, the last 15 min are reserved for a wrap-up discussion on connections to teachers' curricular practices. This is important as many teachers feel that, due to increased content expectations of the district's implementation of the Common Core State Standards, they do not have sufficient time in their classes to facilitate problem solving. We typically engage teachers in brainstorming about possible ways to modify the session's problems to integrate them into their curriculum and enable *differentiation* (i.e., identification of appropriate entry and exit points for students at different ability levels) in their classrooms. We also encourage them to think about possible connections that they can make to the Standards for Mathematical Practice as outlined in CCSS-M while implementing such problem-solving lessons. For example, during the wrap-up session of the aforementioned *Social Security Number* problem, we highlighted the specific domains and their standards as well as the relevant mathematical practice standards such as "attend to precision," "look for and make use of structure," and "look for and express regularity in repeated reasoning" (National Governors Association 2010, pp. 7-8).

7.3.2 Summer Workshop

Our summer workshops bring together teachers from various geographical regions in Colorado with differing mathematics and teaching backgrounds. Similar to our evening sessions, we strive to create an encouraging environment that allows teachers to think deeply about mathematics, problem-solving strategies, and their teaching practices. There is palpable excitement throughout the week as participants get to know each other professionally and personally and as they develop a deeper understanding of the various aspects of mathematical problem solving. One participant summarized this idea by stating, "This workshop was an amazing experience for me! Not only was it a great way to form new relationships with other educators in Colorado, but a great way to understand how to better help foster problem-solving skills."

Each of our three residential summer workshops has followed the same overall structure suggested in the AIM *How to Run a Math Teachers' Circle* workshop. We provide some detail of our morning and afternoon sessions first. These sessions mainly focus on mathematical problem solving and have remained consistent in style throughout all our three workshops. However, the evening sessions have changed significantly over the three years. We detail the evolution of these evening sessions next.

The morning and afternoon sessions engage teacher participants in solving challenging mathematical problems in a group setting, with the last 30 min of each afternoon dedicated to reflection and discussion of connections to their curricula. We have followed a number of guiding principles in our choice of problems for these sessions. The first day problems were designed to elicit a variety of problem-solving strategies (see sample problems in the Appendix). These problems enabled full group discussions of these strategies and of the types of problems that evoke the use of different strategies. Having teachers notice or use different problem-solving strategies on the first day served two purposes: it prepared them to implement these strategies on different problems throughout the week and helped us facilitate discussion on collaborative mathematical problem-solving norms.

Morning and afternoon sessions on other days typically consisted of *content* and *advanced content* problems, similar to those shared in the Appendix. We sought a balance between these types of problems throughout the week so that teachers were engaged in problems directly related to their curricula but also experienced solving problems with advanced content to see extensions of the content that they teach. Even problems that had direct connections to middle school curricula were targeted at a higher level than typical middle school problems in order to challenge and provide teachers with authentic problem-solving experiences.

We have also tried to provide balance among the mathematical domains covered by these problems. Each year we made sure that problems were distributed across CCSS-M domains such as the Number System (e.g., fractions, place value, divisibility), Geometry (e.g., transformational geometry, lattice polygons, taxicab geometry), Ratios and Proportional Relationships, Expressions and Equations (e.g., patterns and expressions), and Statistics and Probability (e.g., mean-absolute deviation). In addition to this domain-specific balance, we also incorporated problems that require more hands-on exploration with concrete models such as playing cards with the *Card Shuffling* problem (see Table 7.1) and geoboards for the *Pick's Theorem* problem (see the Appendix). These hands-on active sessions keep the teachers more engaged and energized and are especially valuable in the afternoons.

Table 7.1 Problems of the weeklong *Card Shuffling* theme

Session time and problem type	Session focus
Monday afternoon (Problem-solving strategies)	*Card shuffling:* A perfect shuffle is a card shuffle that is accomplished by dividing a deck of cards into two equal piles, and perfectly interleaving the two piles, while keeping the top card on top. Explore what happens when you perform perfect shuffles with a deck of cards.
Monday evening around campfire (Content)	*Pass the ball:* Some number of participants are spread evenly around a circle. One person starts with a ball. A "step" is then chosen that will tell which person the ball is passed to next; for example, a step of 2 means the ball is passed to the person 2 people away clockwise from the person with the ball. Repeat this until the ball returns to the starting person. For which numbers of participants and which steps will the ball be touched by everyone before it is passed back to the starting person?
Tuesday afternoon (Content)	*Threading Pins* (Driscoll 1999): A number of pins are spread evenly around a circle. A thread is tied to some starting pin. A "step" is then chosen that will tell what the next pin is that the thread is looped around. The thread is then looped tightly around a next pin so that the step between the first two pins equals the step between the next two pins and so on. The process is continued until we return to the starting pin. If some pin has not yet been used, the process starts again with a new thread. Notice that for 5 pins with a step of 2, we only need one thread, while 6 pins with a step of 3 requires three threads. How many pieces of thread will be needed in general?
Wednesday afternoon (Advanced content)	*Introduction to modular arithmetic*: Topics included: addition and multiplication tables mod 7 and well-definedness of these operations; simple equation solving; powers of 10 mod n and powers of 2 mod n (calculated with attention paid to well-definedness).
Thursday morning (Advanced content)	*Fractions and decimals in different bases*: Topics included: correspondence between fractions and decimal expansions, e.g., for which fractions will the decimals terminate, for which will they repeat, how many digits may be required before the decimal expansion of $\frac{m}{n}$ terminates (all questions were posed initially in base 10, then in other bases); connection between length of the repeated portion of the fraction $\frac{1}{n}$ and powers of 10 mod n; connection between length of the repeated portion of the fraction $\frac{1}{n}$ when expanded in base 2 and powers of 2 mod n
Friday morning (Advanced content)	*Card shuffling and connections to powers of 2 mod n*: Overview: returned to card shuffling and the question "How many perfect shuffles are required to return a deck of n cards to its original order (for n even)?"; found that the formula for position of card k after one shuffle was $2k \bmod n-1$; found that the first power of 2 that was congruent to 1 mod n gave the necessary number of shuffles; solved modular equations to find cards that returned to their original position before the rest of the deck, explored zero divisors and inverses.

For our third summer workshop (2015), we collaborated with the Rocky Mountain Math Teachers' Circle (RMMTC), and were inspired to implement a weeklong theme for some of our sessions based on RMMTC's previous positive experience with this method. In Table 7.1, we provide the sequencing of problems that we used in our 2015 summer workshop in order to develop the mathematical content and problem-solving strategies to solve the *Card Shuffling* problem. On the first day, this problem engaged teachers in posing questions and investigating different problem-solving strategies such as modeling with mathematics. The *content* type problems, *Pass the ball* and *Threading Pins*, prompted discussions on concepts such as the greatest common factor (GCF), the least common multiple (LCM), and relatively prime, which are directly related to middle school content. In addition, these problems prepared teachers to make connections to advanced content and problems related to modular arithmetic, which were needed to solve the main problem of *Card Shuffling*.

This sequence of themed problems enabled our participants to see connections across different mathematical representations (e.g., embodied (physical), concrete, algebraic) they encountered. One participant, for example, highlighted this aspect of the week by stating the importance of "seeing how, even as an adult, manipulatives add to the math experience and …supported my learning." These thematic problems provided participants with the experience of perseverance in problem solving. They were well received because of the balance between *content* and *advanced* content problems. Participants were excited as so many connections emerged across the week, and some noted:

> I liked the purposeful connection between multiple problems. How different tasks can be selected and sequenced is something that I will be thinking about more carefully in the future. This type of intentional plan can only help to deepen conceptual understanding.
>
> Today [the last day] tied mod arithmetic into decimals, fractions, and long division. So many base concepts that can absolutely tie it together at a much deeper level for me and my students. Awesome! It was a combination of many loose ends until today where it all tied together and brought closure.

Similar to the morning and afternoon problem-solving sessions, we implemented a thematic approach to our two-hour evening sessions during our 2015 summer workshop. This approach evolved from the feedback we received in previous summers regarding the evening sessions. In our first summer workshop (summer 2013) during these evening sessions, participants played mathematical games, further investigated problems from that day, or had informal discussions of their curricula. These sessions were relatively unstructured in 2013 because we hoped this would enable stronger community building among the teachers. However, post-workshop surveys and our follow-up meetings with teachers indicated that participants desired more structured evening sessions centered on pedagogical discussions. For example, one participant stated, "Spend some time discussing how the problems we did can connect with our common core content." Another suggested, "Have teachers bring their math curriculum and work to make problems more about process."

As a result of these suggestions, we redesigned our second and third summer workshop evening sessions to focus on connections to curriculum. We structured the sessions to help participants develop or modify tasks and lesson plans around problem solving using research-based activities. This was a different approach from what we had experienced in the AIM *How to Run a Math Teachers' Circle* workshop, and seemed to be a valuable addition.

For the evening sessions we had participants read and discuss an article on the Task Analysis Guide developed by Smith and Stein (1998). Participants found this article valuable as it outlines various levels of cognitive demands of mathematical tasks and the roles of such tasks in the learning and teaching of mathematics. Following this discussion, we asked teachers to modify or develop a task to promote problem solving in their curriculum. After identifying and modifying (or developing) a task, participants generated a lesson plan to implement this task.

At our third summer workshop, we further structured the lesson-planning component of the evenings by introducing teachers to the eight Mathematics Teaching Practices outlined in *Principles to Actions* (NCTM 2014). At the end of each afternoon problem-solving session, we facilitated discussions in which participants shared examples of moments when they experienced an implementation of any of these teaching practices while they were engaged in solving problems during the day. This gave teachers the opportunity to reflect on their daily experience as learners, but also had them "put their teacher hats back on" to discuss implementation of problem-solving tasks in their classrooms. As one participant noted, "To see the practices in action and then identify them helped me become more aware of what they are and how they may look in my own classroom."

During evening sessions when participants modified or designed a task and a lesson plan for their classrooms, they discussed how to implement the eight teaching practices while teaching their lesson. These revised evening sessions were well received by participants as evidenced by their related written comments on the post-workshop survey.

> As a teacher, the evening sessions were wonderful because we could collaborate and create a real, usable task. I also love that we had time to draft, revise, get feedback, revise again and create a high quality task. Most of the time teachers do not have the chance to create, edit, [and] revise their work for high quality tasks.

7.4 Impacts

7.4.1 Impacts on Participants

Throughout our sessions we have witnessed teachers' excitement to engage in problem teachers' excitement to engage in problem solving and willingness to implement problem solving in their teaching. The following quotes from summer workshops' post-surveys provide examples of participants' learning gains in mathematical problem-solving skills and strategies:

> I was not familiar with parity before but that proved to be quite helpful in solving problems. I also experienced an increase in perseverance (Participant, summer 2013).

It was reinforced to use visuals to solve problems as well as to work backwards. These are both strategies that I'd heard of and used before but this week reminded me to use them more often (Participant, summer 2015).

All participants indicated a desire to implement problem-solving activities with their students or plans to shift their teaching practices to provide their students more opportunities to engage deeply with mathematics. Below are selected quotes from participants from summer workshops' post-surveys:

It was so great to be able to just do math, talk about doing math and the best part, learn ways to use it to get kids excited about solving math problems (Participant, summer 2014).

I feel I haven't been pushing or allowing my students to work ahead or challenge themselves, and ask the right questions on justifications (Participant, summer 2015).

In addition, each summer some participants commented on how our workshop helped them to understand their students' issues regarding the learning of mathematics. For example, one participant shared this struggle during the workshop:

[The workshop] was very challenging. It has been decades since I've been put to task to solve harder math problems, much less put so much thought into it, which makes me realize how much I need to be aware of this for my students (Participant, summer 2014).

Participants made further connections between their experience as learners during the day and their teaching in general:

This workshop really focused on putting the participant in the learner role. This experience was then a point for reflection and planning from a teacher role [in the evening sessions]. Most workshops only have someone in a teacher role thinking about, but not experiencing the learner role. This workshop gave me work as a learner that was challenging at my understanding level. … I was able to take my deeper conceptual understanding and make more meaningful and appropriate connections to what I teach because of this increase in my own understanding (Participant, summer 2015).

Participants also compared their summer workshop experience to their other professional development experiences, highlighting the opportunities we provided for them to actively engage in mathematics:

Every workshop that I have ever attended, at some point, became boring, but this workshop was mentally intensive and fun. I was never bored (Participant, summer 2013).

Differences in the other PD workshop are that we actually got to be the student and do the expected tasks instead of being told how to implement the tasks (Participant, summer 2014).

Participants have shared how they implemented our MTC problems and problem-solving activities in their classrooms. For example, at the beginning of the second summer workshop, three teachers described three different implementations of ideas from the first summer workshop. One of these teachers had given problems from our summer sessions to her students, and these students then worked on these problems throughout the school year. She asked her students to keep problem-solving journals to record their progress on each problem. Another teacher explicitly tried to improve her question-posing practice after noticing how the facilitators of the summer workshop were posing questions throughout each problem-solving session. The third teacher posed a problem each week for students to work on, and as a class they discussed solutions at the end of the week.

7.4.2 *Impacts on Leadership Team Members*

Working with each other and with the teachers has had a significant impact on us. As we have planned sessions and workshops and written grants, we have developed a camaraderie and mutual trust among the three of us that has sustained our work. We have different research backgrounds and teaching experiences. These varied backgrounds have enabled us to provide meaningful and challenging mathematics problems and implement research-based teaching practices at our sessions. For example, Craviotto's knowledge gained from writing grants for similar programs helped us to secure funding for the first two years of the program, and also helped Karakok and Morrison learn how to write such grant proposals through her mentorship. Karakok's background in mathematics education research provided support to structure sessions related to curriculum. She also shared with Craviotto and Morrison her approach (based on Stein et al. 2008) to implementing productive mathematical discussion in her mathematics courses. Morrison's experience working in programs on problem solving with in-service teachers in Nebraska helped to inform our decisions on problem selection. This reciprocal mentorship has enabled us each to grow significantly and has kept the work continually fresh and enjoyable over our past five years together.

The process of facilitating problem solving with teachers has also directly affected our teaching. One of the most challenging parts of running the MTC has been finding appropriate, interesting high-ceiling low-floor problems (that is, problems with easy entry points for all, but no significant upper bound on how far students can progress) for teachers to investigate. Since Craviotto and Morrison have taught university courses such as problem solving, discrete mathematics, and number theory, we could utilize problems from our MTC in our courses and vice versa. Additionally, the process of mentoring struggling teachers and challenging strong teachers gave all three of us more insight into differentiating instruction, and that in turn improved our university teaching. Furthermore, watching the excitement and perseverance of teachers as they solved problems reinvigorated us as teachers and motivated us to continue this work.

Our MTC program has also had a positive effect on some of the authors' careers. For example, Karakok has begun to engage in some research on the impacts of different activities in Math Teachers' Circles, and her work with the MTC led directly to new district professional development work. Morrison was asked to take a national leadership role in the Math Circles community, and as a result has dramatically increased her connections to other Math Circle organizers.

7.5 Challenges, Sustainability and Support

So far, we have been able to sustain up to six evening sessions each academic year and an annual residential summer workshop for our circle. One challenge we continually face is attracting more local teachers and sustaining the participants' attendance at our monthly evening sessions. Our local district has a high teacher turnover rate especially among middle school mathematics teachers, with many teachers leaving the district each year and new teachers and district personnel hired each year. Consequently, each year we need to re-establish our connection with the district personnel and advertise our program to the new teachers. Another challenge is securing funding. We submit grant proposals to multiple sources each year to sustain our program activities for the following year and usually secure only enough funds to cover our program's expenses for 1 year, requiring us to spend time on grant-writing every year.

To overcome these challenges and also reach more teachers, we recently began collaborating with another Colorado MTC. We ran the 2015 summer workshop with the RMMTC, and we again held a joint summer workshop in 2016. These collaborative efforts have helped us to recruit more teachers and increase participants' numbers for monthly sessions of both programs. In addition, this collaboration enables us to conserve funds for future summers. More importantly, leadership team members of both circles are sharing problems, discussing effective implementation methods and occasionally visiting each other's monthly circle to facilitate sessions.

Our department has helped sustain our program for the past four years by providing classrooms, administrative support, and some materials and supplies. The faculty has also shown significant support for our program, with a number of faculty members facilitating our monthly evening sessions. We have established this faculty buy-in by frequently discussing the success of the program at department meetings. Additional institutional support has been provided by our university's foundation office.

Since fall semester 2015, we have expanded our program to provide a monthly student circle, which has proven promising for our sustainability efforts. We decided to pilot six sessions for fifth through eight grade students and run them concurrently with our evening sessions for teachers. In these concurrent sessions students do mathematical problem solving in groups in one room while teachers explore the same problem(s) in another room. After an hour, the teachers join the students for dinner and a discussion. During this time, students informally present their solutions and strategies on the problem to the teachers, and then teachers join the discussion. After 30 min, the students depart and the teachers continue to work on related problems. We believe the conversation between the teachers and the students at these sessions is helpful for both. Teachers are able to see that students can engage in deep mathematical problem solving of a challenging nature and hear the students' rich mathematical conversations. The middle school students have enthusiastically appreciated these sessions, and we have noticed an increase in the number of teachers, especially new teachers, at our evening sessions, because many teachers are bringing their students to the student circle.

In June 2016 we facilitated our first 3-day summer camp (non-residential). Quite unexpectedly, it attracted 52 local fourth through eighth grade students. We organized two classrooms of problem solvers based on grade level and prior experience in attending our evening sessions. We also had two undergraduate pre-service elementary or secondary mathematics teachers and one mathematics education doctoral student in each classroom to be mentors for students throughout the workshop. The pre-service teachers were excited to work directly with students on problem solving. They learned more about the student population they will teach and how to implement problem-solving activities for their future students. Similarly, our graduate students, who will be future faculty members, gained experience working with pre-service teachers and students. While the camp was a great success, we will be reflecting on our experiences and improving the camp based on journal entries from the middle-school students and the mentors.

7.6 Future Goals and Concluding Remarks

Over the years, our circle has grown and changed in ways we did not envision at the start. In particular, when we first began our circle at AIM, we fully expected to primarily serve teachers from our local school district. We made plans based on the assumption that we would establish a cohort of 30–40 local teachers who would come to our evening sessions and participate in our summer workshop. We never established a local base this large. As a consequence, we invited teachers from across our state, which enriched our summer workshop in unexpected ways. We had not anticipated that our evening summer workshop sessions would be tailored to helping teachers reflect on utilizing problem solving in their classrooms. These changes occurred as we reflected on the strengths and weaknesses of each workshop and carefully considered feedback from participants. Our efforts to be responsive are evidenced by a comment from a participant: "I felt like we were heard and not just talked at! We led the learning as much as the leaders."

Similarly, when we began to plan for our MTC, we did not envision creating a student circle. It appears that we are filling an important niche as our student enrollment continued to grow during the pilot year. In addition, the student circle has reinvigorated our teacher circle by bringing new teacher participants to the circle and adding a new level of motivation for teachers to incorporate problem solving in their classrooms. We fully expect our student circle and summer problem-solving camp to continue for the next few years. We expect to grow and adapt both circles in a continuing effort to be receptive to the needs of local students, local teachers and teachers across the state.

Appendix

The following table gives examples of problems used in the MTC sessions falling into three different categories, namely those focused on *content, advanced content* and *problem-solving strategies.*

Table A.1 Sample problems used in Northern Colorado MTC sessions

Problem type	Sample problems
Content	*Social Security Number*: My friend has a very unusual social security number. Each of the digits 1 through 9 is used exactly once, and if you write out the digits of her number as $a_1a_2a_3\ a_4a_5\ a_6a_7a_8a_9$ then the number formed by the first i digits is always divisible by i. For example, the number $a_1a_2a_3a_4a_5$, i.e. $10{,}000a_1 + 1{,}000a_2 + 100a_3 + 10a_4 + a_5$, is divisible by 5. What is my friend's social security number?
	Fabulous Fractions (MTC website): Find four different decimal digits a, b, c, and d so that $\frac{a}{b} + \frac{c}{d} < 1$ and is as close to 1 as possible. Prove that your answer is the largest such number less than 1.
	Folding Triangles (Posamentier and Schulz 1996): A triangle is cut at random from a piece of paper and a vertex is folded to the midpoint of the opposite side. What figures can result, and what determines which one appears?
	Pentominos (Liljedahl et al. 2007): A pentomino is a shape created by the joining of five squares such that every square touches at least one other square along a full face. If a pentomino is placed somewhere on a 100's chart will the sum of the numbers it covers be divisible by 5? If not, what will the remainder be? Explain how you can know "quickly"!
Advanced content	*Taxicab Geometry* (Krause 1987): What does a circle look like in taxicab geometry? In this context, what should the value of π be? Do taxicab equilateral triangles always have 3 angles that measure 60°? Do taxicab isosceles triangles always have congruent base angles?
	Billiards (MTC website): Imagine a pool table with pockets only in the corners. Start with a ball at the bottom left corner moving up at an angle of 45°. The ball bounces off each side of the pool table until it finally reaches one of the corners and rolls into the pocket. For example in a 5 by 10 table (width 5 and height 10), the ball will hit and bounce off the middle of the right side and then fall into the pocket on the top left corner. We count this as 1 bounce. We do not include going into the pocket as a bounce. Suppose a pool table is a by b and a ball is at the bottom left corner moving up at an angle of 45°. How many total bounces will the ball have before going into the pocket? Which pocket will the ball finally land in?
	Pick's Theorem: What is the area of a lattice polygon (a polygon whose vertices lie on lattice points)? It is pretty close to the number of lattice points in the interior of the polygon, but this is not exact. The lattice points on the boundary also matter. If you know how many lattice points are inside, and how many are on the boundary, can you determine the area?

Problem type	Sample problems
Problem-solving strategies	*Concentration*: The Participant deals out cards in a 4x4 grid, putting some face up and some face down randomly. Then the Magician deals a few more cards, adding one more row and one more column. The Magician is then blindfolded and the Participant picks one card in the grid and turns it over (i.e., if it was face up, now it is face down, and if it was face down, now it is face up). The Magician takes off the blindfold and is miraculously able to spot the altered card. How?
	Shortest Path (MTC website): Find the number of shortest paths from the lower left corner to the upper right corner of a 4×3 grid.
	Basic Takeaway: A set of 16 pennies is placed on a table. Two players take turns removing pennies. At each turn, a player must remove between 1 and 4 pennies (inclusive). The winner is the last player who makes a legal move. See if you can find a winning strategy for one of the players. Try to prove that your strategy works. And, always, try to generalize!

References

Carlson, M. P., & Bloom, I. (2005). The cyclic nature of problem solving: An emergent multidimensional problem-solving framework. *Educational Studies in Mathematics, 58*(1), 45–75.

Driscoll, M. (1999). *Fostering algebraic thinking: A guide for teachers grades 6–10*. Portsmouth, NH: Heinemann.

Krause, E. F. (1987). *Taxicab geometry: An adventure in non-Euclidean geometry*. New York: Dover Publications.

Liljedahl, P., Chernoff, E., & Zazkis, R. (2007). Interweaving mathematics and pedagogy in task design: A tale of one task. *Journal of Mathematics Teacher Education, 10*(4-6), 239–249.

Math Teachers' Circles. (n.d.). *Math Teachers' Circle Network*. Retrieved Mar 1, 2016, from http://www.mathteacherscircle.org/about.

National Governors Association. (2010). *Common core state standards for mathematics*. Retrieved Jun 26, 2016, from http://www.corestandards.org/wp-content/uploads/Math_Standards1.pdf.

NCTM. (2014). *Principles to actions: Ensuring mathematical success for all*. Reston, VA: National Council of Teachers of Mathematics.

Posamentier, A. P., & Schulz, W. (Eds.). (1996). *The art of problem solving: A resource for the mathematics teacher* (pp. 188–189). Thousand Oaks, CA: Corwin.

Smith, M. S., & Stein, M. K. (1998). Selecting and creating mathematical tasks: From research to practice. *Mathematics Teaching in the Middle School, 3*(5), 344–350.

Stein, M. K., Engle, R. A., Smith, M. S., & Hughes, E. K. (2008). Orchestrating productive mathematical discussions: Five practices for helping teachers move beyond show and tell. *Mathematical Thinking and Learning, 10*(4), 313–340.

Tanton, J. (n.d.). Curriculum inspirations. Retrieved March 1, 2016, from http://www.maa.org/node/129058/.

Zeitz, P. (1999). *The art and craft of problem solving*. New York: John Wiley.

Chapter 8
Transforming Practices in Mathematics Teaching and Learning through Effective Partnerships

Padmanabhan Seshaiyer and Kristin Kappmeyer

Abstract This is the story of a partnership between a university professor of mathematics (the first author) and a high school mathematics teacher (the second author). It started in 2007 through the Teacher Partnership Program of the Association for Women in Mathematics. They describe a succession of joint projects designed to benefit students from middle school to college. The partnership began with informal visits to one another's classrooms and engagement with a high school mathematics team. The association extended to include a research experiences for undergraduates program in 2009. In 2011 the collaboration continued in the STEM Accelerator Program at the College of Science at George Mason University (GMU). The GMU collaborative work included a summer camp for middle school girls called FOCUS and a transitional program for prospective and incoming college students called STEM Boot Camp. They describe these programs and say something about their impact on the student participants. The authors also reflect on what they have learned from the collaboration over the years and how their partnership has affected their professional lives.

MSC Codes
97A99
97B40
97D40
97I50
97U50
00A09
00A66

P. Seshaiyer (✉)
Department of Mathematical Sciences, George Mason University,
4400 University Drive, MS 3F2, Fairfax, VA 22030, USA
e-mail: pseshaiy@gmu.edu

K. Kappmeyer
Arlington Public Schools - H-B Woodlawn Secondary Program,
4100 Vacation Lane, Arlington, VA 22207, USA
e-mail: kristin.kappmeyer@apsva.us

J. Dewar et al. (eds.), *Mathematics Education*, Association for Women in Mathematics Series 7, DOI 10.1007/978-3-319-44950-0_8

Keywords Teacher partnerships • Mathematics camp for middle school girls • Transition to college mathematics • Research experiences for undergraduates

8.1 Introduction

This is the story of a partnership between Padmanabhan Seshaiyer (Professor of Mathematical Sciences and Director of the STEM Accelerator Program at GMU) and Kristin Kappmeyer (a high school mathematics teacher with Arlington Public Schools), the authors of this chapter. It started in 2007 through the Teacher Partnership Program (TPP) of the Association for Women in Mathematics (AWM) and still continues. The authors tell of a succession of joint projects designed to benefit students from middle school to college level. Emphases of their work together include focusing on student inquiry, connecting mathematics to other disciplines and increasing the diversity of successful STEM students.

Two of the joint projects described in this chapter are part of the STEM Accelerator program at GMU, which Seshaiyer directs. The STEM Accelerator was created by the College of Science in 2011. The program has four major goals: increasing the number of STEM majors, improving retention rates of STEM students, reducing their time to graduation, and helping them join the STEM workforce or continue their education upon completion of their Bachelor's degree in STEM disciplines. It has become a hallmark program for GMU, impacting over 5000 students per semester through various initiatives. These include the undergraduate Learning Assistants tutoring program, which has helped improve retention of STEM majors and transfer students from Northern Virginia Community College. This peer-to-peer tutoring initiative has improved on-time graduation rates for the college.

The STEM Accelerator program runs many initiatives at all levels, ranging from STEM Mania, a summer camp for students in grades 3–5, to STEM teacher professional development programs engaging teachers in effective pedagogical practices through problem solving and lesson study. The STEM Accelerator program received the "Programs That Work" award of the Virginia Mathematics and Science Coalition in 2015 for being one of the exemplary STEM programs in Virginia. It also received the 2016 Programs that Work award for two of their novel initiatives: the STEM Boot Camp and FOCUS camp, which are described in this work.

In Sect. 8.2 the authors describe their initial work in the TPP of the AWM. They started with informal visits to each other's classrooms and engagement with a high school mathematics team. The next two sections describe the STEM Accelerator projects. Section 8.3 concerns a summer camp for middle school girls of color called FOCUS. Section 8.4 is about a transitional program for prospective and incoming college students called STEM Boot Camp. Section 8.5 describes a research experiences for undergraduates program. In each of these three sections, Seshaiyer and Kappmeyer describe the programs and what they know about their impact on the student participants and how each program helped to sustain the collaboration. Finally, in Sect. 8.6 they reflect on what they have learned from the collaboration over the years and how it has affected their professional lives.

8.2 The Teacher Partnership Program of the AWM

The AWM TPP (Hsu et al. 2009) was launched in August 2006 to link teachers of mathematics in schools, museums, technical institutes, 2-year colleges, and universities with other teachers working in an environment different from their own and with mathematicians working in business, government, and industry. Participant activities included electronic communications, teaching projects, classroom visits when feasible, and outside-of-classroom mathematics activities. Mathematicians and K-12 teachers with common interests were matched by the Partnership organizers. The authors were matched in August 2007 (Seshaiyer 2008).

Kappmeyer invited Seshaiyer to make a presentation to the high school mathematics team that she and her colleague, Mark Dickson, had started the previous year. The team met once a week during lunch to work on American Mathematics Competitions (AMC) and Virginia Mathematics League (VML) practice problems. The students competed monthly in the VML contests and annually in the AMC 10 and AMC 12 contests, for students in grade levels 10 and below and grade levels 12 and below, respectively. Seshaiyer offered students insights into some common types of problems, and he gave them some entertaining ways to remember number facts. Two of those students (see Fig. 8.1) earned scores that qualified them to par-

Fig. 8.1 Seshaiyer and Kappmeyer working with the students (Seshaiyer 2008)

ticipate in the American Invitational Mathematics Examination. Both of those students are currently working on their doctorates at the University of Texas, Austin, one in mathematics and the other in government. The mathematics student's research is in geometric topology, on distinguishing "mutant pretzel knots." The government student's research involves how price fluctuations prompt protests and political shifts in resource-dependent nations.

In addition to helping students prepare for mathematics contests, Seshaiyer lectured on the usefulness of mathematics in solving real-world problems. He mentored one inspired student from Kappmeyer's school in her science and engineering fair project, and that will be described in Sect. 8.5.

8.3 FOCUS: STEM Summer Camp for Middle School Girls of Color

The Females Of Color Underrepresented in STEM (FOCUS) Camp was designed to pique the interest in STEM of middle school girls of color by offering them educational and social enrichment with positive role models to build their confidence and enhance their knowledge in STEM. The FOCUS Camp has been offered each summer since 2014. Campers participated in a coordinated set of inquiry-based STEM activities each day over a 5-day period.

An inquiry-based mathematics classroom incorporates well-researched student-centered approaches. They are often driven by questions and the use of thoughtful investigations designed to help students make sense of the information as they develop new understanding (Diggs 2009). Investigations can also help teachers answer common classroom questions from students such as "why do I need to know this?" or "when will I ever use this?" Students need opportunities to explain their thinking processes to the teacher and class, and it is this exchange of ideas that provides the foundation for true understanding of mathematical concepts (Chapko and Buchko 2004). Not only does this process help students think like mathematicians, developing new knowledge by creating or discovering mathematics, but it also helps the educators learn to elicit and analyze their students' thinking.

The camp was run by the STEM Accelerator Program at GMU in collaboration with Girls Inspired and Ready to Lead (GIRL Inc., http://www.girlsinc.org/). GIRL Inc. is a non-profit organization whose mission is to mentor and empower teen girls for future success through promoting academic excellence, leadership skills, community service, a healthy lifestyle, and self-esteem.

8.3.1 FOCUS Camp Activities

Each of the first 4 days of camp was dedicated to science, technology, engineering and mathematics respectively. The fifth day was set aside for the girls to create posters showing what they had learned and to present those posters to their peers, undergraduate and graduate students, faculty and parents. In 2015 this program attracted 74 middle school girls and 20 undergraduate women as STEM counselors. The program grew over 300 % from 2014 when there were only 18 middle school girls and 5 undergraduate women counselors. One of the main reasons for this growth was that participants from 2014 not only wanted to come back, but they also wanted to bring their friends so that they could also be impacted by it.

Nationwide, around 19 % of engineers are women. Negative perceptions of women engineers may play a role in explaining these low numbers (Graham and Smith 2005). Research has also revealed that high school- and college-age women commonly see the STEM environment as "chilly, male-dominated," highly impersonal and unsympathetic to women's unique needs (Morganson and Jones 2010). Therefore an important component of the FOCUS program was career counseling. To excite the students about engineering, the Vice President of Aerospace Corporation, Catherine Steele, came to share her experiences in engineering, including her education and her path to a position of leadership. On the last day of camp, there was a career panel consisting of five prominent women who represented a range of careers in STEM. The panelists gave advice to the participants, judged the participants' poster presentations and provided feedback on the girls' work. See Sect. 8.3.2 for the girls' reactions to the panel.

Each day of camp included a dedicated block of time for building an underwater remotely operated vehicle called a SeaPerch (For the program that originated it, see http://www.seaperch.org/). The students were divided into 18 teams, and each team built their own SeaPerch (see Fig. 8.2). This hands-on building exercise not only helped students learn to work in groups to solve a problem, but it also increased their understanding of the engineering concepts behind the SeaPerch. On the last day of the camp, the 18 teams competed against each other in an underwater mission coordinated at the aquatic center at the GMU recreational center.

Fig. 8.2 FOCUS participants testing their SeaPerch in the lab

Throughout the camp the students were engaged in a variety of activities, each with an emphasis on science and engineering, including an introduction to 3D printing, an investigation of a polymer, modeling the forces of flight, and a drone activity, which is described in more detail next.

The drone activity involved building models of drones using simple items such as paper plates, thumb tacks, popsicle sticks and pencils. Camp participants learned about real drones and their various electronic parts: flight controllers, motors, propellers, electronic speed controllers and all their respective functions. After demonstrating how drones work, the high school presenter (shown in Fig. 8.3) explained how drones can be very useful in identifying poachers who are killing elephants and rhinos in Africa for their tusks and horns respectively.

Many of the students' science and engineering activities were connected to mathematics. In building SeaPerch, the students investigated the relationships among distance, rate and time and different kinds of variation: inverse, direct and joint. In the slime activity the students worked with proportions and percentages of chemicals being mixed. In the drone activity students discovered how to use the Pythagorean Theorem and coordinate geometry to explore uniform motion and average speed. Students solved problems such as:

> A drone identifies a poacher hiding behind a tree 3 miles west of the Alpha Station which receives all communications from the drones and relays it to rangers close by. A ranger is

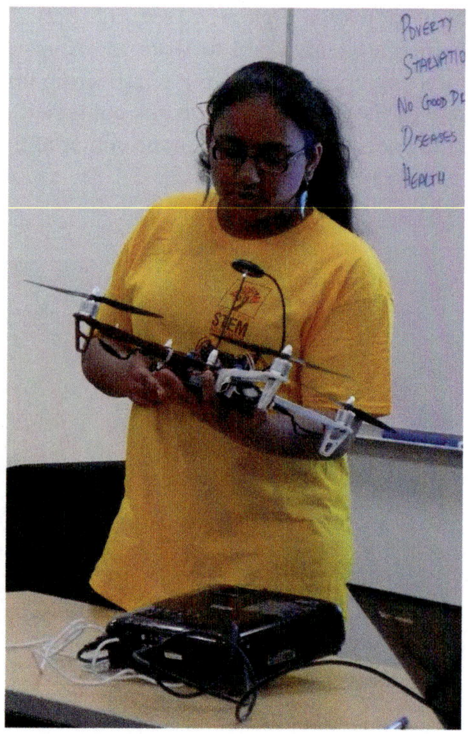

Fig. 8.3 FOCUS drone activity presentation

located 4 miles south of the Alpha Station. If the drone uses a global positioning system (GPS) to communicate to the Alpha Station about the location of the tree, what is the shortest distance the ranger has to travel to catch the poacher that is hiding behind the tree?

Students also learned how GPS works and how it can be used to identify the locations of animals, poachers, and rangers. A related example the students solved was,

Rangers A and B are 3 miles apart. A drone has received information that a poacher is located 2 miles from Ranger A and 2 miles from Ranger B. Is this information enough for the drone to communicate to the rangers the exact location of the poacher?

Such problems provided an opportunity for the students to learn about the concept of a perpendicular bisector that identifies two possible locations for the poachers—something that most students did not expect. This type of hands-on, inquiry-based and conceptual learning not only helped them to appreciate mathematics more, but it also helped them understand how it can be used to solve major real-world problems that impact society.

After hearing about poachers killing elephants and rhinos in Africa, the students brainstormed potential strategies to prevent this harmful practice. Engaging the students in such real-world problems, and showing them how to use the mathematics they have learned in school in pursuit of solutions, made the students' learning more meaningful. The brainstorming led to some powerful ideas such as using cameras on drones to take pictures of poachers and using multiple drones that can communicate and work in teams. These are creative ideas that can potentially impact the development of new technology that can help prevent poaching.

8.3.2 Evaluating the FOCUS Camp Experience

The advice offered by the members of the career panel on the last day of camp had a great impact on the girls. The panelists' comments and feedback on the girls' poster presentations were very inspirational, and the question and answer session that followed the panel motivated the girls to think about their own goals and aspirations. The panelist who spoke last asked the girls to stand up and state their own dreams. There was such overwhelming response to the question that the girls were asked to form a line to the microphone. The line extended to the back of the auditorium. Their dreams included becoming cancer researchers with the hope of ending cancer, environmental scientists working to stem global warming, medical doctors serving local and overseas communities, and engineers of all types working to design and maintain the built environment. It was a special moment that inspired great confidence in the future.

Seshaiyer and Kappmeyer evaluated some aspects of the effectiveness of the week-long summer camp using focus group interviews and pre- and post-surveys. The post-survey questions focused on whether the camp helped them (1) to have a better idea what they will do after graduating from high school and what their career goals are, and (2) to decide to take different classes in school from what they had previously planned. Also, the authors wanted to know if the camp increased students' interest in studying STEM in college or made them more confident in their ability in projects and activities. Many of these questions were explored with statements to which the students responded: "a great deal," "moderately," "slightly," or "not at all." For every question of this type on the survey about 50 % of the students responded "a great deal," about 30 % of the students responded "moderately," and about 15 % responded "slightly." Students were also asked to rate the FOCUS camp overall and more than 89 % of the participants indicated that the camp was much better than they had expected. At least in the short term, the camp had an impact.

8.3.3 The Authors' Roles in the FOCUS Camp

Kappmeyer received an Arlington Community Foundation STEM Workforce Development Fellowship in 2015 to work with the FOCUS camp. Her primary goal was to increase a bank of STEM-related problems for her and her Arlington Public School colleagues to use in the years ahead. To this end she developed lessons on sequences of nucleotides that form a unit of genetic code in DNA and RNA and on global positioning systems. She is continuing this work by developing a unit on ocean acidification with the chemistry and biology teachers at her school. The teachers plan to embed challenging mathematics in regular and Advanced Placement science lessons. Another goal of her Workforce Development Fellowship was to increase the district's outreach to students who are under-represented in STEM fields. These goals are consistent with those of the FOCUS camp organizers.

Kappmeyer contributed to many aspects of the FOCUS program. She greeted presenters, helped with activities, improved communication among camp counselors, and encouraged the camp participants. She even prepared a crossword puzzle that incorporated the week's activities to engage the students on the last day. She also was an informal evaluator, creating a written record of each activity over the course of the week. Her thorough notes helped the organizers of the camp who were conducting sessions, and therefore could not attend every activity. Her report served as anecdotal evidence of the positive outcomes of the program.

Seshaiyer was the coordinator of the FOCUS camp along with two other STEM Accelerator faculty members. He worked with them to recruit students, communicate with parents, book rooms on campus for each session, recruit undergraduate student counselors, and coordinate FOCUS sessions for each day of the program. He worked with presenters, ordered necessary materials for each session and provided other important logistical support necessary for the event to be a success. Seshaiyer worked with Kappmeyer to coordinate the informal evaluation that she was conducting for each activity during the week. He was also responsible for securing a grant from Northern Virginia Community Foundation's women's giving circle that supported the undergraduate counselors. He visited all sessions to make sure that they were going as planned. He was a co-presenter in the Drones and Math session, where he worked with the participants. Finally, he was responsible for completing a final report of the FOCUS event that was successfully approved by the women's giving circle.

8.4 STEM Boot Camp for Students Beginning College

In July 2015, they collaborated on the second annual STEM Boot Camp offered by the STEM Accelerator program at GMU. The STEM Boot Camp is a 1-week pre-college program for incoming freshmen and high school students who are dual-enrolled in GMU. The Boot Camp program gives students a preview of the first semester of college by introducing them to content from introductory gatekeeper classes such as calculus, general chemistry, cell biology, and physics. The students engage in hands-on laboratory projects and other skills needed to be ready for college, such as how to study, take an exam, and manage their time. The Boot Camp program also makes students aware of undergraduate research opportunities.

The goals of the Boot Camp are to improve retention of students in STEM majors and to increase the likelihood of successful completion of STEM degrees in 4 years. Results from the first and second cohorts (2014 and 2015) of incoming first year GMU students show that over 83 % of those who declared a STEM major are currently in their major. In the Boot Camp the authors participated in, most of the students in the calculus group were planning to major in engineering. In keeping with national trends, the biology group had the highest proportion of women and the calculus group had the lowest.

There were 36 students enrolled in calculus. Preparation varied widely among these students, and therefore the group was subdivided. Another STEM Accelerator faculty member worked with a small group of students, reviewing pre-calculus topics so that they could qualify for placement in calculus in the fall. Seshaiyer and Kappmeyer worked with a larger group of students, introducing some of the essential ideas in calculus. In their group of 25 students there were three women, two of whom were African-American. During the weeklong Boot Camp, each of the three young women approached Kappmeyer to discuss their career aspirations. They were identifying with the female instructor, who reflected on her own undergraduate civil engineering education in which she had not a single female instructor.

Bernard L. Madison's article "Mathematics at the School-to-College Transition" (Madison 2016) called attention to differences between high school and college instruction in mathematics. He claimed, "The transition from school to college mathematics is one of the most troublesome in all of U.S. education. More students report difficulty succeeding in college mathematics than in any other discipline, and more students are dissatisfied with their college mathematics than with any other subject." He added, "The problem is not the lack of agreement in topics covered; it is more too much agreement and too much overlap." Indeed he noted that "80–90 % of enrollments in college mathematics courses are in courses whose content is taught in high schools," with the geometry, algebra, trigonometry and calculus sequence dominating mathematics in both venues.

In their collaboration, the authors saw first-hand that while high school and college calculus content may be the same, instruction can vary dramatically. GMU does not allow the use of graphing calculators in entry-level calculus classes. However, graphing calculators are used extensively in high school AP Calculus classes. The AP Calculus course description reads, in part, "Technology should be used regularly by students and teachers to reinforce the relationships among the multiple representations of functions, to confirm written work, to implement experimentation, and to assist in interpreting results." (CollegeBoard 2012) In addition, college-level courses place more responsibility on the student, and they move through the material at a faster pace.

The planning of the Boot Camp classes took into account the "too much of the same thing" effect that Madison (2016) described. Some of the students had already taken AB or BC Calculus in high school. Others had not even taken an introduction to calculus. Every effort was made to challenge each student at his or her own level.

8.4.1 STEM Boot Camp Activities

Seshaiyer and Kappmeyer's class of calculus students worked in self-selected groups of 6 to 8. As a result, the groups were heterogeneous, which elevated the discussion at each table. The authors worked to challenge each group at an appropriate level, even writing different final exams for different groups. Throughout the week, they posed open-ended problems such as "How can we find the area of this amorphous shape?" and "How can we find the rate of change of this quantity at this one instant in time?" The walls of the classroom consisted of floor-to-ceiling white boards (see Fig. 8.4). The students worked on the walls as they hashed out their thoughts and then presented their work to their classmates. This inquiry-based format allowed for a variety of approaches to problem solving, with each student thinking creatively and collaboratively. The final exams were exploratory problem sets that required students to apply all that they had learned that week. Time was allotted for students to present their thinking to the entire class. Students gained an important glimpse of what they should expect in college. They worked on their communication skills and developed poise and self-confidence as the week progressed. They learned as much, if not more, from one another as they did from the instructors.

The STEM Boot Camp students had many opportunities to learn about GMU resources such as the Learning Assistant system, which provides help and reinforcement outside of STEM classes. As they do during the academic year, graduate students support the students throughout the week, helping them learn how to study effectively in groups. In a question and answer session with GMU faculty, students were encouraged to talk with their professors if they found themselves struggling. Students also attended presentations about how to find research opportunities and develop relationships with mentors. The STEM Boot Camp offered rich learning opportunities that went far beyond those of a traditional classroom.

8.4.2 Evaluating STEM Boot Camp

For assessing the impact of the program, students were asked to complete a pre-camp survey prior to their arrival on campus. In 2015, the majority of the students were engineering majors. Ninety-six percent of all the participants were eager or

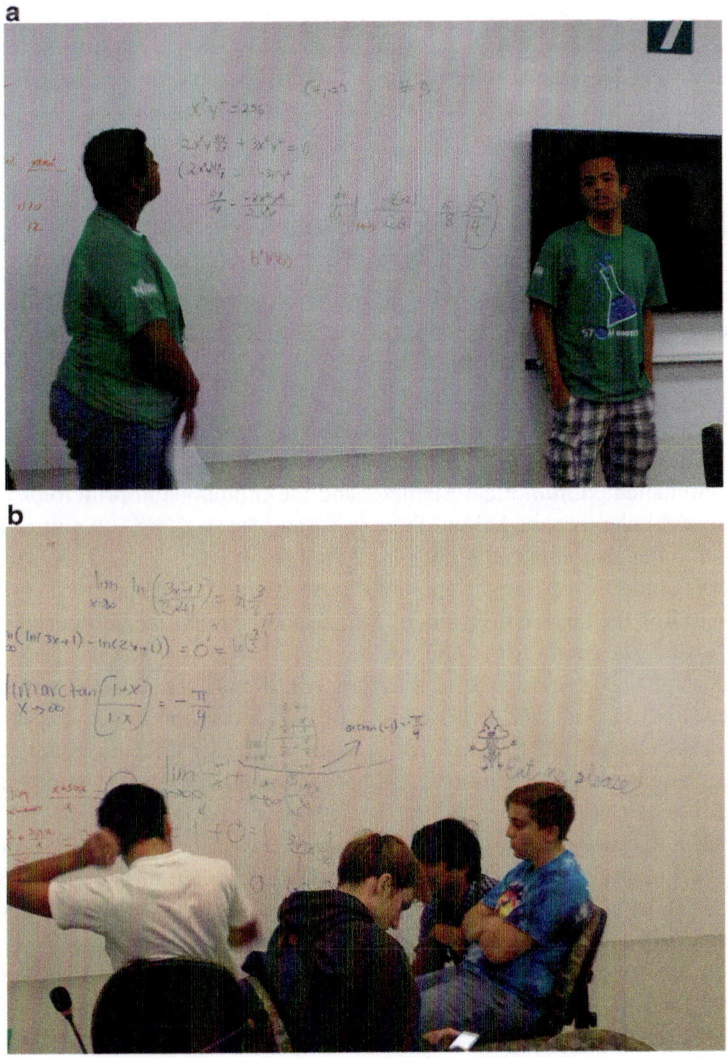

Fig. 8.4 Boot Camp calculus participants (in (**a**) and (**b**)) collaborating and communicating their solutions via white walls

very eager to begin their freshman year. Although they were eager to begin the semester, the confidence level was split between somewhat confident (45 %) and confident (45 %). 93 % believed that their major courses would be difficult or very difficult. Regarding degree completion, 76 % plan to complete their BS degree, 53 % plan to complete a master's degree and 22 % intend to complete a doctorate degree in a STEM discipline. When asked why they registered for the STEM Boot Camp, some of the reasons were to prepare for their hard science classes and to meet people with similar interests. Students also completed post-surveys the last day of camp. Large gains were made in clarifying what is expected of them as students; moderate-large gains were made in skills such as taking notes during lecture, studying for exams, and learning how to manage time. The STEM Boot Camp was held again in July of 2016.

8.5 Undergraduate Research Experiences

In 2009 GMU received grants from the research experiences for undergraduates (REU) program of the National Science Foundation (NSF) and the Department of Defense for a program titled "Multidisciplinary REU Program in Computational Mathematics and Nonlinear Dynamics of Biological, Bio-inspired and Engineering Systems." In this program Seshaiyer and Kappmeyer worked together in new ways.

Eight college undergraduates from across the country spent the summer of 2009 learning sophisticated numerical methods and the fundamental principles that govern the physical phenomena behind their research projects. GMU faculty members, graduate, and undergraduate students were their mentors.

The REU students were four men and four women. They came from varied institutions including the large University of Maryland, the mid-size Southern Methodist University, and the small Alma College. Every effort was made to create a group with a wide range of experience so that they could learn from one another. To add to the breadth of background experiences of the group, Kappmeyer was selected as a high school teacher participant who would work alongside the undergraduates. Her research mentors were Seshaiyer and his undergraduate student Minerva Venuti.

In addition to researching their own topics, learning to use MATLAB® (a computer algebra system widely used for computations in science and engineering), meeting with mentors, and exploring all that the Washington, DC environs had to offer, the REU students attended weekly presentations by faculty members in the GMU Department of Mathematical Sciences covering a wide range of research topics and methodologies. Students gained insight into the application process for graduate school, and they learned about applying for grants to fund their own research in the future. They spent one full day touring the National Institutes of Health (NIH), where they visited the mathematical modeling laboratory and saw how the NIH was using modeling in its research practices to improve public health. They also visited the National Institute of Standards and Technology nearby.

The eight REU students and Kappmeyer spent the first couple of weeks investigating their topics and formulating their own questions. With support from their faculty advisors, the students made roadmaps for reaching their destinations. Their projects included medical applications like "A Mathematical Approach for the Reconstruction of Neural Networks," industrial applications like "A Computational Model for Batten Behavior in Micro Air Vehicles" and applications that transcended a single categorization like "Synchronization and Coherence of Dynamical Systems." Kappmeyer's project was titled "Viscoelastic Modeling of Biological Tissue in an Idealized Cerebral Aneurysm." She worked alongside one of the REU students whose project was "Applying Numerical Methods to Fluid Structure Interactions in Biological Systems." They worked with Venuti and Seshaiyer as their mentors. Venuti's project "Modeling, Analysis and Computation of Fluid Structure Interaction Models for Biological Systems" received three national awards for outstanding undergraduate research. In further recognition, Seshaiyer, Kappmeyer, and Venuti received letters from the Governor of Virginia's office commending their contributions to the study of STEM fields.

The REU program inspired these eight undergraduates to continue their mathematical studies while discovering their own interests. One of them is completing his PhD at the University of Arizona with plans to teach at the university level when he finishes his studies. He said "My REU experience influenced significantly my decision to pursue a graduate degree. I found the research experience interesting and worthwhile. Also, coming from a small undergraduate institution, before participating in the REU, I was not entirely sure that I could make it at the next level. The REU experience showed me that, yes, I could." One student earned a Ph.D. in France researching gene regulatory networks, and she is now working in industry. Another is an NSF Postdoctoral Fellow at the University of Pennsylvania where he is a researcher and instructor, and two other women are currently finishing their Ph.D. studies at the University of Illinois Urbana Champaign and Volgneau School of Engineering at GMU respectively. Clearly, the passion for research was sparked by their REU experiences. Such research experiences also help to develop quantitatively literate citizens and change agents (Seshaiyer 2012) while providing opportunities for students to enhance their real-world problem solving abilities.

8.6 Reflections

In this section the authors reflect on what they have learned over the course of their 9-year partnership, focusing on the activities described in the earlier sections.

The issue of gender and minority participation in STEM fields came home to them when considering the composition of their calculus class in the STEM Boot Camp. Out of 25 students, about 30% were either women or African-American. The authors believe that the encouragement of women and minorities to enter STEM fields happens at the middle and high school levels. Kappmeyer advises her interested young female and minority students to consider engineering fields. She notifies her students of university programs that introduce engineering and computer science in summer camps, and she advises them of universities that offer support for underrepresented groups.

The Boot Camp gave the authors the opportunity to team-teach calculus. Not only did they plan the complete curriculum together, but they also debriefed each day, discussing student progress and how to enhance the following day's lesson. They developed a synchronous teaching style: as one presented review topics, the other spontaneously generated related example problems on the board for students to consider. Over the course of the week, as gaps in student understanding or background became apparent, one of the instructors would generate such examples. In this fashion students were able to integrate their understanding of geometry, algebra, and trigonometry into the new work at hand.

In terms of teaching style, the authors discovered a shared desire to encourage intellectual curiosity. Seshaiyer sees Kappmeyer as a very conscientious teacher, and the Boot Camp students clearly respected her. She has a very effective personal style, and she presents herself as an enthusiastic teacher and researcher. Kappmeyer sees Seshaiyer as a passionate teacher and mentor. His ability to inspire students to ask deep questions and go solve them is commanding. Together they were able to make use of the state-of-the-art Active Learning with Technology room, allowing them to engage the students in twenty-first century skills including communication, collaboration, critical thinking and creativity.

The Boot Camp experience allowed Kappmeyer to see what knowledge, skills, and attitudes university students need to succeed. At the same time it allowed Seshaiyer to see the knowledge, skills, and attitudes of some current Virginia high school students. This exchange encouraged each of them to view mathematics education as a continuum that need not have a sharp boundary between high school and college.

In the REU experience, Kappmeyer's research project built on her undergraduate mentor Venuti's project that involved mathematical and computational modeling of cerebral aneurysms—thin, balloon-like widenings of arterial walls. The rupture of these aneurysms is the most common cause of bleeding into the space between the brain and the skull, resulting in strokes. Predicting the potential of aneurysms to rupture is fundamental to clinical diagnosis and treatment. In her research, Venuti solved coupled partial differential equations for fluids (modeling blood and cerebral spinal fluid) interacting with elastic structures modeling aneurysms. It had been 30 years since Kappmeyer had studied differential equations as a civil engineering student. It was a huge undertaking to relearn the mathematics and computational techniques necessary to understand the biological model of the aneurysm. In becoming the student, she reflected on how exciting it was to have a real problem to motivate her learning. She wanted to share that feeling with her students. When she observed Seshaiyer's interactions with his undergraduate and REU students, she noted that his role was more that of a mentor than teacher. He asked probing questions that led students on a path towards answers to their own questions. He rarely answered a question directly—rather he asked questions that opened new areas of inquiry. When a student really had trouble understanding a concept, he described an analogous situation so that the student could make sense of his or her own problem. Kappmeyer has strived to incorporate such an approach to teaching ever since her REU experience.

In the fall of 2009, she developed a lesson for her Intensified Pre-Calculus classes based on the aneurysm research. Kappmeyer, Seshaiyer and Venuti came to participate in a lesson study, a form of classroom inquiry in which several teachers collaboratively plan, teach, observe, revise and share the results of a single class lesson. The lesson introduced slope fields and ordinary differential equations to the students. It broadly explained partial differential equations to give students a sense of what modeling the idealized cerebral aneurysm entailed. The students had already read excerpts from George Pólya's book *How to Solve It: A New Aspect of Mathematical Method* and applied his four problem solving steps to explorations in class. The steps, simply stated, are: understand the problem, devise a plan, carry out the plan, and look back (Pólya 1945). These are the steps that Kappmeyer applied to her REU project, and in the lesson, she modeled them for her students. While she had previously assigned open-ended projects to these students, her REU research allowed her to see just how far she could push the notion that students should find the mathematics to solve their own problems. In this way, they become mathematicians.

The goal of her REU research project was applying sound methods to investigate three broad areas: mathematics, teaching, and biological modeling. The authors' partnership focused on the educational possibilities of using mathematical modeling to promote active, student-directed learning.

The research into mathematical modeling of cerebral aneurysms continued after the summer REU. Venuti worked on it the following year, as did four of Seshaiyer's undergraduate students. One of Kappmeyer's high school students was inspired to continue working with Seshaiyer on aneurysm modeling in her science fair project that year. Her project won the grand prize in the Northern Virginia Science and Engineering Fair in 2009, and she was invited to the Intel International Science and Engineering Fair that year. The student went on to pursue her undergraduate studies at Stanford University.

Participation in this aneurysm project included one high school student, one high school teacher, four undergraduates, and one graduate student. Five of the seven researchers were women. Two of the seven were African-Americans. Undergraduate research is one way to encourage women and under-represented minorities to pursue multidisciplinary mathematical careers that bridge the scientific and engineering communities.

Seshaiyer's work on these projects with Kappmeyer is based on decades of experience teaching and conducting research in mathematical modeling. His primary goal in working with students has always been to help them appreciate the importance of advanced mathematics and computational techniques to solve complex real-world problems. He strives to give students the necessary foundation in mathematical modeling, to help them analyze the models they create, and to engage them in using appropriate technology to simulate the problem situation. He also employs active learning in class, where the students learn by doing, using cooperative learning and learning by discovery. He motivates students to participate in his classes by asking them questions, giving extra-credit work, and using group-assignments. His classroom assessment techniques help the students to learn about multiple problem-solving strategies and representations (including algebraic, graphical, verbal, pictorial, tabular). Seshaiyer teaches STEM subjects by developing and making explicit for students the ideas they might have already picked up and used in informal settings and other practical scenarios. All of these teaching experiences informed the work he and Kappmeyer did together and were, in turn, reinforced by the successes they shared.

In its 2012 National Curriculum Survey report ACT wrote, "A large gap still exists between how high school teachers perceive the college readiness of high school graduates and how college instructors perceive the readiness of their incoming first-year students" (ACT 2012, p. 4). Policy recommendations from that survey include the recommendation that "K-12 and postsecondary educators must collaborate to ensure that course curricula and classroom materials reflect the skills needed for college and career readiness and that these materials are seamlessly aligned across grade levels and the two systems" (ACT 2012, p. 13). The authors feel very fortunate that the AWM promoted such a collaboration through which they have been able to exchange ideas to enhance the learning experiences for their students.

Acknowledgements The authors would like to thank the National Science Foundation (DMS 1062633), 4-VA Collaborative, Virginia Department of Education, Dominion Foundation, Arlington Community Foundation, Northern Virginia Community Foundation's women's giving circle, GIRL Inc. and the GMU College of Science STEM Accelerator Program for supporting in part the activities described in this chapter. We also want to thank the AWM Teacher Partnership Program for creating a platform for a novel sustainable and effective partnership.

References

ACT. (2012). *ACT National Curriculum Survey 2012: Policy implications on preparing for higher standards*. Retrieved June 17, 2016, from http://files.eric.ed.gov/fulltext/ED542018.pdf.

Chapko, M. A., & Buchko, M. (2004). Math instruction for inquiring minds. *Principal, 84*, 30–33. Retrieved June 17, 2016, from http://www.naesp.org/resources/2/Principal/2004/N-Dp30.pdf.

CollegeBoard. (2012). AP Calculus course description. *Fall, 2012*. Retrieved June 17, 2016 from https://secure-media.collegeboard.org/ap-student/course/ap-calculus-ab-bc-2012-course-exam-description.pdf.

Diggs, V. (2009). Ask—think—create: The process of inquiry. *Knowledge Quest, 37*(5), 30–33.

Graham, J., & Smith, S. (2005). Gender differences in employment and earnings in science and engineering in the US. *Economics of Education Review, 24*(3), 341–354.

Hsu, P.-s., Lenhart, S., & Voolich, E. (2009). Linking teachers and mathematicians: The AWM teacher partnership program. In L. Paditz & A. Rogerson (Eds.), *Proceedings of the 10th International Conference on the Mathematics Education into the 21st Century Project, September 11–17, 2009 in Dresden, Germany* (pp. 255–258). Dresden: HTW Dresden. Retrieved June 19, 2016, from http://www.qucosa.de/fileadmin/data/qucosa/documents/7923/Proceedings-636pages-Dresden2009.pdf.

Madison, B. L. (2016). *Mathematics at the school-to-college transition*. CollegeBoard AP Central; Retrieved June 17, 2016, from http://apcentral.collegeboard.com/apc/members/features/9394.html.

Morganson, V., & Jones, M. (2010). Understanding women's underrepresentation in science, technology, engineering, and mathematics: The role of social coping. *Career Development Quarterly, 59*(2), 169–79.

Pólya, G. (1945). *How to solve it: A new aspect of mathematical method*. Princeton, NJ: Princeton University Press.

Seshaiyer, P. (2008). Partnering to make a difference. *AWM Newsletter, 38*(4), 14–15.

Seshaiyer, P. (2012). CUR (Council on Undergraduate Research) focus: Transforming practice through undergraduate researchers. *CUR Quarterly, 33*(1), 8–13.

Chapter 9
Developing Collaborations Among Mathematicians, Teachers, and Mathematics Educators

Kristin Umland and Ashli Black

Abstract Mathematics education is a discipline in the overlap between mathematics and education, and solving problems in mathematics education requires expertise from both domains. Scholarly work in mathematics education is accomplished either by individual scholars who have expertise in both areas or as a collaboration among scholars who collectively have the necessary kinds of expertise. Illustrative Mathematics, an example of an organization that supports such collaborations, is described as one model for supporting scholarly work in mathematics education. Reasons why departments of mathematical sciences should actively support such scholarship are explored.

Keywords Mathematics education • Collaboration • K-12 mathematics • Illustrative Mathematics

9.1 Introduction

The discipline of mathematics education is broad, including work in K-12, undergraduate, and graduate mathematics. Each of these areas requires different types of expertise, both mathematical and pedagogical. In the next section, we describe a collaboration of mathematicians, mathematics educators, and K-12 teachers working to improve K-12 mathematics education in the US. This example serves as the

---

MSC Codes
97Uxx
97 B40
97Axx

K. Umland (✉) • A. Black
Illustrative Mathematics, Oro Valley, AZ 85737, USA
e-mail: kristin@illustrativemathematics.org; ashli.black@gmail.com

J. Dewar et al. (eds.), *Mathematics Education*, Association for Women in Mathematics Series 7, DOI 10.1007/978-3-319-44950-0_9

backdrop for a broader discussion in the following section about how to evaluate and support the work of faculty members in departments of mathematical sciences involved in mathematics education.

There are many examples of mathematicians working on problems in mathematics education; Felix Klein's reforms in the early 1900s in Germany (Chislenko and Tschinkel 2007) and Andrey Kolmogorov's reform in the Soviet Union in the 1970s (Boyko 2013) are two famous examples. Mathematicians were also the architects of the New Math reform movement in the US that ran parallel to the Kolmogorov reforms in the Soviet Union, although the New Math reform was broadly regarded as a failure (Boyko 2013). It is important to learn from these examples and think carefully about how mathematicians can be more effective when working on problems in mathematics education. Issues in school mathematics are complex: they involve children's learning, school systems, and educational policy. These are issues that a professional mathematician may be ill-equipped to deal with. For this reason, many believe that the best way for mathematicians to contribute to K-12 mathematics education is in partnership with others, including K-12 teachers and education experts.

The forces that shape K-12 mathematics education in the United States are unfamiliar to many people who have not worked in this area, so we include some relevant background here.

Mathematics standards describe the mathematics that students should know and be able to do. Sometimes standards are described as *learning standards*.

Mathematics curriculum is used both to name the mathematics that students study and the materials that are used to present mathematics to students.

The curriculum describes the day-to-day mathematical work that students do, while standards describe the end-state we hope students will achieve as a result of their study.

Federal policies in the United States require each state to set learning standards in mathematics and English language arts for K-12 students. For many years, mathematics standards varied greatly from state to state. For example, an analysis of the standards from 42 states indicated that the introduction of addition and subtraction of fractions might happen as early as grade 1 or as late as grade 7 (Reys 2006). In response to this situation, there were various reports and publications aimed at aligning standards and bringing greater coherence to school mathematics: *Adding it Up* (National Research Council 2001), The American Diploma Project (Achieve 2004), the Curriculum Focal Points for Prekindergarten through Grade 8 Mathematics (NCTM 2006), reports by College Board (2006) and ACT (2007), the Report of the National Mathematics Advisory Panel (2008), and Focus in High School Mathematics (NCTM 2009).

In 2009, 48 states signed on to an initiative of the National Governors Association and the Council of Chief State School Officers to write "college-and-career-ready" school standards for mathematics and English language arts. The final draft of the Common Core State Standards for Mathematics (National Governors Association 2010) was first released on June 2, 2010, and 45 of those states have since adopted them or a set of standards very similar to them. These standards were informed by many sources, including mathematics education research, studies of what is necessary for college and career readiness, comparisons to standards from high-performing states and nations, the input of many different stakeholders, and the structure of the discipline itself (McCallum 2012).

These standards have a well-developed structure that sets them apart from the majority of state standards that came before them. The highest level structure consists of domains that typically span multiple years. For example, one of the domains that spans K-5 is "Numbers and Operations Base Ten." Within each grade level, there are clusters of standards within each domain. For example, in grade 4 there is a cluster "Generalize place value understanding for multi-digit whole numbers," and within this cluster there are three standards about understanding the relationship between places in a multi-digit whole number, reading and writing base-ten numbers, and using base-ten structure to round numbers. Each level in this hierarchy has a purpose.

9.2 Illustrative Mathematics

Founded by William McCallum, a mathematician and the lead writer of the Common Core State Standards for Mathematics, Illustrative Mathematics began as a project in the Department of Mathematics at the University of Arizona. The purpose of the project was to enhance public understanding of the standards by providing mathematical tasks that illustrate the expectations for mathematical work that the standards describe. Initially, it appeared to be a simple project to write some K-12 mathematics problems and post them on a website. Later, as more people reviewed more problems, it became clear that mathematics problems are often like Rorschach tests: what different individuals see in a problem varies, and often reflects more about the observer than the problem. Part of the reason for this is that mathematics problems are always used for particular purposes in particular contexts. Consider this problem:

> Fufa had 4 bags with 5 marbles in each. How many marbles did she have altogether?

In grade 1, this is an advanced problem that students could solve by drawing pictures or with addition. In grade 2, it could be used to teach students about skip counting (in preparation for grade 3 work on multiplication). At the beginning of grade 3, it could be used to introduce the concept of multiplication. In the middle of grade 3, it could be used to demonstrate the associative property of multiplication by viewing 4 as 2 times 2. At the end of grade 3, it could be used to assess understanding of basic multiplication facts. In grade 4, it could be used for review. A particular task may be appropriate for one use but not another, so the quality of a task cannot be judged without understanding its intended use.

It also became clear that just posting mathematics problems on a website would do little to clarify the expectations of the standards. Each problem needed to be presented in a context of use and have one or more examples of how students would be expected to solve it in that context. Thus, the team of mathematicians (including the first author, who was then in the Department of Mathematics and Statistics at the University of New Mexico), mathematics educators, and classroom teachers (including the second author, who was then a high school mathematics teacher in Seattle, Washington) working on the project decided that rather than post mathematics problems, they would post *mathematical tasks* composed of a *problem* coupled with an explicitly stated *purpose* (or context of use) along with detailed commentary and solutions for each task. The team consisted of dozens of people from all over the US.

From the inception of the project, the tasks were reviewed by at least two people: one with mathematical expertise, and another with classroom expertise. The team developed a set of eight criteria (http://www.achievethecore.org/page/310/illustrative-mathematics-task-review-tool) to evaluate each task and asked each reviewer to make judgments about whether a particular task met each criterion:

1. The task illustrates the specified standard, cluster, domain, or conceptual category.
   An *illustration* is a set of tasks that clarify the intention, depth, breadth, meaning, or faithful implementation of a standard, cluster, or domain, or conceptual category. No single task can illustrate a standard by itself, so this criterion is met if the task is appropriate to include as part of a complete illustration.
2. The task's purpose is clearly stated in the commentary and is likely to be fulfilled.
   We will call the mathematical idea and/or habit of mind that a task is intending to develop or assess, along with its intended use, *the purpose of the task.*
3. The task has at least one appropriate solution.
   The solutions for tasks should be mathematically correct and reflect the kind of reasoning students could be expected to show. Other possible solutions should be indicated.
4. The mathematics is correct.
   A task must be mathematically correct in the context of the Common Core State Standards, both in the obvious sense of having correct calculations, but also in the use of correct mathematical reasoning and terminology. For example, a task solution should not confuse equations and expressions or congruence and equality.
5. Any diagrams or pictures have a clear mathematical or pedagogical purpose which they are likely to fulfill.
   Tasks should only include diagrams, pictures, or illustrations that support comprehension of or provide mathematical meaning for the problem. If a diagram (such as a tape diagram or number line) is meant to represent a quantity, it should be well-labeled so that its interpretation is not ambiguous.
6. The context supports the purpose of the task.
   Contexts can be either mathematical or real-world, and they can play different roles in different tasks. Contexts can variously:

   (a) Support students in understanding the mathematics,
   (b) Motivate students to work on the mathematics, or
   (c) Provide an opportunity to apply their knowledge in a novel context.

7. The task write-up appropriately addresses units and numerical precision.

   While one might argue that attending to units and numerical precision could fall under the rubric of being mathematically correct, it is such a common problem in tasks and so easily overlooked that it has been explicitly included.
8. The language of the task is unambiguous and grade-appropriate.

   In some cases, the purpose of a task is to require students to take a context that does not have an unambiguous mathematical interpretation and make a choice about what mathematical interpretation would be appropriate (for example, in a very complex modeling task, there may be more than one reasonable mathematical model). However, the language of the task should not be ambiguous in any case.

Not surprisingly, the team found that a group of people with different kinds of expertise were collectively able to provide more detailed, critical analysis related to the different criteria. This confirmed that both mathematical and classroom perspectives were essential for shaping the tasks appropriately. For example, here is the task statement and commentary for a grade 4 task written by one of the authors:

> Historians estimate that there were about 7 million people on the earth in 4,000 BCE. Now there are about 7 billion! We write 7 million as 7,000,000. We write 7 billion as 7,000,000,000. How many times more people are there on the earth now than there were in 4,000 BCE?

The commentary that accompanies the task states:

> The purpose of this task is to help students understand the multiplicative relationship between commonly used large numbers (millions and billions) by using their understanding of place value. This task also builds on students' work on multiplicative comparison from 4th grade. The task "Thousands and Millions of Fourth Graders" is a good task to do before this one as it requires the same kind of reasoning but the numbers are smaller. The population estimates come from Historical Estimates of World Population from the US Census Bureau.

The first review of an earlier version of the task coming from a mathematician states:

> I am a little worried about [the standards alignment] because although it fits this category for the mathematics, it does not for the numbers involved, which are too big. I don't know when students start to see these numbers so don't know if this is more generally problematic. It is definitely multiplicative comparison though. An alternative would be to link it to [some grade 5 standards]. Though not the focus of the problem, I wonder if it is worth showing the associative property of multiplication, which is implicitly being used when you multiply the three factors of 10?
>
> Finally, I understand the sentence, "Every time we move one place to the left we multiply by 10" at the very beginning of the solution, but maybe something a little more precise would be better, for example, "The value of each place is ten times the value of the place immediately to the right."

A later review by a classroom expert suggested additional solutions reflecting how real students might approach the task. Subsequently public comments confirmed that the task should be moved to grade 5, so it was moved to a grade 5 standard soon after publication on the website.

As they worked together, the team realized that the community of mathematicians, teachers, and mathematics educators forming around the work was an outcome of the project equal in importance to the tasks themselves. They applied for a larger grant to enlarge this community and extend its work to a more comprehensive set of tasks. The expanded project included plans to create a behind-the-scenes "automated task delivery system" to move tasks through the review process, assuring that the right kinds of expertise were applied to their development along the way.

The original ambitious project was funded on a declining scale: 80 % for the first year, 60 % for the second year, and 20 % for the third year. However, finding additional foundation funding for the project to develop an elaborate review process that could train hundreds of people and produce many more tasks proved challenging. The team did secure funding to create middle and high school course outlines, but working within the structure of a mathematics department in a university proved to be a challenge. In 2013, the team incorporated into an organization named Illustrative Mathematics, which received its non-profit status the following summer.

As a non-profit organization, it was easier for Illustrative Mathematics to pursue multiple projects in collaboration with a variety of different organizations. In addition to publishing tasks and course outlines, the team worked on a variety of large-scale projects related to both formative and summative assessment, including reviewing and giving guidance to national assessment companies, and provided professional development to teachers all over the United States. In some cases, the materials that were created and reviewed have had an impact on millions of children in the US. One of the advantages of the way the work of the organization is structured is that mathematicians can be involved in small or large ways. Some mathematicians just review material on a project-by-project basis as their schedule allows. Some mathematicians (as with one of the authors) are involved in almost every project.

An important outcome of the collaborative work of the organization is that mathematicians develop greater respect for the work of both teachers and mathematics educators while at the same time bringing their particular expertise to the work in appropriate ways. The culture of mutual respect is most evident with the interaction between teachers, mathematics educators, and mathematicians, which is one of trusted peers.

Both of the authors have been working with Illustrative Mathematics from its earliest days. Umland was involved at the very beginning of the project as the middle school mathematics lead. She has worked on many different projects, including writing and reviewing tasks and other content for the organization. She is now the Vice President for Content Development. Black has been a long-time classroom expert for the project and then for the organization, involved in reviewing, professional development, and now curriculum content development. Having helped build this community from the ground up, both authors are happy to now call the organization their professional home.

How do faculty in departments of mathematical sciences get involved in such work? How do their colleagues regard their participation in this type of work? We have talked about the kinds of work in education to which mathematicians can contribute; next we will talk about how such work is recognized and rewarded in academic departments.

9.3 Evaluating Work in Mathematics Education

Universities have long recognized that the nature of scholarly and professional work varies from discipline to discipline; what counts as scholarly work and how its value and impact is assessed is typically left to individual academic departments to decide. While research articles in peer-reviewed journals have long been accepted as scholarly work in almost every discipline, there has also been a decades-long discussion about what other activities should count as well (Boyer et al. 2015). There are articles and books about the impact of communication technology on the nature and evaluation of scholarly work in different disciplines (e.g., Brown and Simpson 2014; Priem 2013) as well as discussions about the impact of economic changes on institutions of higher education (e.g., Greenstein 2013; McGee 2015). These forces have, for better or worse, affected the way many people both inside and outside of academia view and value scholarship. So such issues in mathematics and education are framed by this larger, long-term discussion about scholarship in higher education.

How should we evaluate scholarly work in mathematics education? There are several common methods for evaluating the quality of a research article. Both peer reviews and the opinions of widely recognized experts in the discipline are considered important. This second approach only works in disciplines where there is consensus about who the experts are; such consensus is harder to find in interdisciplinary domains. Another evaluation option is the use of measures of scale and impact, some of which are called "impact factors" and are based on citation metrics applied both at the article level and the journal level. While many scholars think that using metrics like impact factors to evaluate the quality of research is a flawed methodology (e.g., Amin and Mabe 2004; Seglen 1997), it does point to a desire among scholars to equate "quality" with the importance, influence, and applicability of the work. In the case of scholarly work that does not have its final output format in research journals, this suggests two ways we might measure quality: through an expert or peer-review process, or by identifying metrics of the influence and impact that the work has on the communities that are affected by the work. However it is done, measuring the quality of scholarly work is necessarily complex and should not be taken lightly.

In order to evaluate this kind of work, there first must be a tangible product of the work. Let's examine different categories of scholarly work that are not traditionally considered research and discuss what such artifacts might look like and how they might be evaluated.

9.3.1 Writing and Evaluating Instructional Materials

One possible measure of the quality and impact of instructional materials is how widely adopted they are, although this speaks more to the impact than the quality of such materials. Another potential measure is research that evaluates their quality. To do this well requires the help of experts in education research. It is also possible to set up rigorous peer-review processes for such materials.

The set of tasks on the Illustrative Mathematics website provides an example of both of these kinds of measures. As with a research article, the tasks are reviewed by people with appropriate expertise, so there is an expert peer-review process. The website itself gets over 100,000 visits per month, and is linked to by state departments of education and many other websites, so it is possible to get a sense of the scale and impact of the site. And as is the case with journals, different sources for online materials develop different reputations for overall quality, and the Illustrative Mathematics website is frequently recommended as a general resource for teachers by various experts and organizations around the country.

9.3.2 Teaching Undergraduate and Graduate Courses in Mathematics

Teaching is clearly an important part of scholarly work. To evaluate it, there must be residual artifacts that can be examined in order to determine its quality or impact. In principle, peer observations could be used, but in practice these are often not performed in a systematic way, and are therefore difficult to compare. A more systematic way of observing teaching in college classrooms is needed. There are also peer-reviewed journals dedicated to the scholarship of teaching and learning that provide a mechanism for scholars to document the impact of their teaching in a way that can be more easily evaluated.

9.3.3 Writing and Evaluating K-12 Policy Documents

Examples of the types of policy documents that require substantial commitment and mathematical expertise include K-12 mathematics standards and teacher education and licensure requirements. Working on state or national policy documents necessarily has a great impact because so many teachers and students are affected by them. Research on the effect of such materials can be done (Cross et al. 2004; Schmidt et al. 2005), and such research necessarily requires the input of mathematical experts.

9.3.4 Outreach Projects

Outreach projects are often hard to evaluate on an individual level, but aggregating measures across multiple sites can help show impact. For example, Math Circles and Math Teachers' Circles bring K-12 students or K-12 mathematics teachers together with mathematically sophisticated leaders in an informal setting to work on interesting problems or topics in mathematics (National Association of Math Circles

2016). Because these groups are often small and informal, it is difficult to evaluate the impact of a single one, but the national networks of people involved in Math Circles and Math Teachers' Circles gives access to resources that can help evaluate the impact of such work.

9.4 Supporting the Involvement of Mathematicians in Mathematics Education

The problems in K-12 mathematics education are thornier and more complex than many mathematicians realize (Beckmann 2011; Yong 2012). Couple the complexity of an individual classroom with the complexities of schools, communities, and state and national government, and it can seem impossible to make a difference, or to imagine that the mathematical aspects of the work really matter. Yet we have seen that the mathematical quality of standards, assessments, and curriculum materials depends on mathematics experts getting involved in their creation and review in appropriate ways. So how do we make it possible for mathematicians to get involved in this work, and how can we be sure they are prepared to do it well? First, it is important for both the mathematics community and the broader mathematics education community that mathematicians be involved at varying levels of time and commitment. Even a simple outreach program benefits not only the participants, but also the mathematicians involved. Many mathematicians find they have as much to learn as to impart in their interactions with teachers and mathematics educators. For example, organizing a Math Teachers' Circle is a way to build trust and goodwill between the mathematical community and the teaching community. It is also a way for mathematicians to share their love of mathematics and learn more about teachers, teaching, and school systems. Keeping the lines of communication open makes it possible for mathematicians to get involved in other types of projects at whatever level of commitment interests them.

As much as there is a role for mathematicians to be involved at varying levels in scholarly work in mathematics education, there is an equally important role for their colleagues to support such work, even when they are not involved themselves. Unfortunately, many faculty members in departments of mathematical sciences are unaware of the complexity of the problems, the need for mathematical experts to be involved, and the difficulty experts may have in finding meaningful ways to make contributions to these problems. Still others are overwhelmed by the complexity and are unsure how to contribute. Typical promotion and tenure guidelines do not address (or inadequately address) the types of scholarly work discussed here. Even when such work is recognized as appropriate, adequate methods for evaluating it are often lacking. Of course there are exceptions, and departments and universities that have found ways to recognize and reward this kind of work do exist. For example, University of Arizona College of Science (1992) adopted guidelines for evaluating faculty members who play a substantial role in pre-collegiate mathematics and science education that include the following language:

> Worthy contributions could include scholarly books that make a significant contribution, textbooks that are substantially different from, and better than, previous textbooks (if any) on a worthy subject, articles in refereed respected journals that describe and advocate better practice or that present research results relating to learning science or mathematics, improved methods and instruments for evaluation, computer software, movie or television productions that enhance education, and so on.
>
> No one person, of course, will make contributions in all of these ways, but any of these activities, and many similar ones, should be thought of as legitimate research or creative activities. The quality and impact of the work must be seen as the important issues.

Such guidelines should be the norm rather than the exception. Given that the culture in some mathematics departments is more insular than others, college administrators who see the value in supporting diverse perspectives and activities within departments can help set guidelines that encourage departments to reward such work. There are even examples of institutions that have set up administrative structures that bridge departments, connecting the small number of faculty across different disciplinary departments that work at the intersection of their individual disciplines and education.

Mathematicians should be concerned about these issues both for altruistic and selfish reasons. Not only do mathematicians have a responsibility to support the mathematical well-being of the people in their communities, the mathematical community itself suffers when people in those broader communities do not understand and value mathematics. To put this into perspective, imagine a world where the medical experts made sure that the health care system worked well for them, but didn't concern themselves with how well it worked for anybody else. Mathematicians like Klein and Kolmogorov displayed the intellectual and moral leadership needed to bring these issues to light, but the hard, painstaking work is done by a community of mathematicians, teachers, and mathematics educators working together. This can only happen with the concrete support of departments of mathematical sciences everywhere.

References

Achieve. (2004). *Ready or not: Creating a high school diploma that counts, report of the American Diploma Project*. Retrieved June 25, 2016, from http://www.achieve.org/readyornot.

ACT. (2007). *ACT College and career readiness standards*. Retrieved June 25, 2016, from http://www.act.org/standard/.

Amin, M., & Mabe, M. (2004). Impact factors: use and abuse. *International Journal of Environmental Science and Technology, 1*, 1.

Beckmann, S. (2011). The community of math teachers, from elementary school to graduate school. *Notices of the American Mathematical Society, 58*(3), 368–361.

Boyer, E. L., Moser, D., Ream, T. C., & Braxton, J. M. (2015). *Scholarship reconsidered: Priorities of the professoriate* (Expandedth ed.). San Francisco, CA: Jossey-Bass.

Boyko, M. (2013) *The "New Math" movement in the US vs Kolmogorov's math curriculum reform in the USSR*. Retrieved March 18, 2016, from https://mariyaboyko12.wordpress.com/2013/08/03/the-new-math-movement-in-the-u-s-vs-kolmogorovs-math-curriculum-reform-in-the-u-s-s-r/.

Brown, S., & Simpson, J. (2014). The changing culture of humanities scholarship: Iteration, recursion, and versions in scholarly collaboration environments. *Scholarly and Research Communication, 5*(4), 1–16.

Chislenko, E., & Tschinkel, Y. (2007). The Felix Klein protocols. *Notices of the American Mathematical Society, 54*(8), 960–970.

College Board. (2006). *College Board standards for college success: Mathematics and statistics*. Retrieved June 25, 2016, from http://www.collegeboard.com/prod_downloads/about/association/academic/mathematics-statistics_cbscs.pdf.

Cross, R., Rebarber, T., & Torres, J. (2004). *Grading the systems: The guide to state standards, tests, and accountability policies*. Washington, DC: Thomas B Fordham Foundation and Institute.

Greenstein, D. (2013). Change is coming. Inside Higher Ed. Retrieved March 17, 2016, from http://www.insidehighered.com/views/2013/12/16/essay-arguing-major-changes-are-coming-higher-education.

McCallum, W. (2012). *The common core state standards for mathematics*, presented at ICME 12, Seoul, Korea. Retrieved June 25, 2016, from http://commoncoretools.me/wp-content/uploads/2012/07/2012_07_12_icme_mccallum1.pdf.

McGee, J. (2015). *Breakpoint: The changing marketplace for higher education*. Baltimore, MD: Johns Hopkins University Press.

National Association of Math Circles (2016). Retrieved June 25, 2016, from https://www.mathcircles.org/Wiki_WhatIsAMathCircle.

National Governors Association. (2010). *Common core state standards for mathematics*. Retrieved June 25, 2016, from http://www.corestandards.org/wp-content/uploads/Math_Standards1.pdf.

National Mathematics Advisory Panel. (2008). *Foundations for success: The report of the national mathematics advisory panel Washington*. DC: U.S. Department of Education.

National Research Council. (2001). *Adding it up: Helping children learn mathematics*. Kilpatrick, J., Swafford, J., and Findell B. (Eds.). Mathematics Learning Study Committee, Center for Education, Division of Behavioral and Social Sciences and Education. Washington, DC: National Academy Press.

NCTM. (2006). *Curriculum focal points for prekindergarten through grade 8 mathematics*. Reston, VA: National Council of Teachers of Mathematics.

NCTM. (2009). *Focus in high school mathematics: reasoning and sense making*. Reston, VA: National Council of Teachers of Mathematics.

Priem, J. (2013). Scholarship: Beyond the paper. *Nature, 495*(7442), 437–440.

Reys, B. (Ed.). (2006). *The intended mathematics curriculum as represented in state-level curriculum standards: Consensus or confusion?* Charlotte, NC: IAP-Information Age Publishing.

Schmidt, W., Wang, H., & McKnight, C. (2005). Curriculum coherence: An examination of US mathematics and science content standards from an international perspective. *Journal of Curriculum Studies, 37*(5), 525–559.

Seglen, P. O. (1997). Why the impact factor of journals should not be used for evaluating research. *BMJ: British Medical Journal, 314*(7079), 498.

University of Arizona Faculty of Sciences. (1992). *Promotion and tenure guidelines*. Retrieved June 25, 2016, from http://math.arizona.edu/~mathedconf/willoughby.pdf.

Yong, D. (2012). Adventures in teaching: A professor goes to high school to learn about teaching math. *Notices of the American Mathematical Society, 59*(10), 1408–1415.

Part III
Benefitting STEM Majors

Chapter 10
Finding Synergy Among Research, Teaching, and Service: An Example from Mathematics Education Research

Megan Wawro

Abstract Being a faculty member in higher education involves the balance and integration of various roles and demands. In this chapter I present my own story, as a mathematics education researcher in the teaching and learning of undergraduate mathematics focusing on linear algebra. Using my experience as an example, I describe how synergy among research, teaching, and service can impact career goals and institutional needs.

Keywords Research in undergraduate mathematics education • RUME • Linear algebra • Research • Teaching and service

10.1 Introduction

I am passionate about understanding how undergraduate students learn and participate in mathematics. What are the various ways in which they reason about new mathematics content? How do they grow as learners from lower division to upper division courses? In what ways do they become fluent in the practices that characterize the discipline of mathematics? How do they make sense of and use mathematics in other STEM disciplines? Compatible with these interests are related explorations regarding university faculty: How do we modify instruction so as to

---

MSC Codes
97H60
97U50
97C99
97D40

M. Wawro (✉)
Department of Mathematics, Virginia Tech,
McBryde Hall 438, Virginia Tech, 225 Stanger St., Blacksburg, VA 24061-1026, USA
e-mail: mwawro@vt.edu

J. Dewar et al. (eds.), *Mathematics Education*, Association for Women in Mathematics Series 7, DOI 10.1007/978-3-319-44950-0_10

best assist students in learning new content? How can we model what it means to be a mathematician? What supports do we need as we transition to more student-centered instruction at the university level? How can we assist students in connecting the mathematics content they learn in our mathematics classes with content they learn in other STEM courses? Not only have these curiosities driven my research in the learning and teaching of undergraduate mathematics, they have also become integrated in my teaching and university service.

In this chapter I present my own story, as a mathematics education researcher, as an example of how synergy between research, teaching, and service can impact career goals and institutional needs. After a brief personal introduction, I begin by summarizing my research program in the teaching and learning of linear algebra. Next, I describe some of my current responsibilities as a teacher in a mathematics department and how they influence and are influenced by my research program. I then briefly highlight some of the ways in which this interplay between research and teaching has prepared me to influence my department and institution through service.

My personal involvement in mathematics education has spanned both secondary and university levels. After teaching high school mathematics for 3 years, I earned a master's degree in mathematics and taught university mathematics courses. It was then that I became fascinated with investigating how students learn undergraduate mathematics and how that research could affect teaching practices at the university level. Since earning a Ph.D. in mathematics and science education, I have been a faculty member in the Department of Mathematics at Virginia Tech.

Virginia Tech, a research-intensive land-grant state university with over 30,000 students, provides a productive environment for mathematics education research. The very active mathematics education research community has faculty housed in either the Department of Mathematics in the College of Science or in the School of Education in the College for Liberal Arts and Human Sciences. It also has two Ph.D. options: one through the Mathematics Department and one through the School of Education's Faculty of Teaching and Learning. It is in this supportive setting that my own opportunities for rich interplays between teaching, research, and service have been fostered.

10.2 Research

My research centers on the learning and teaching of undergraduate mathematics, with a specific focus on linear algebra. In general, the research my colleagues and I conduct explores the development of mathematical meaning over time for both individual students and the classroom community at the undergraduate level. A primary assumption we make in our work is that students' mathematical development is a process of individuals actively constructing their own knowledge as they participate in and contribute to the mathematical activities in the classroom community of practice. The theoretical framework upon which these assumptions are based is the

Emergent Perspective, which is further described in Sect. 10.2.1. Consequently, many of our research publications about student reasoning in linear algebra explore how individual students understand specific content (Plaxco and Wawro 2015; Wawro 2015), how communities of learners make mathematical progress (Wawro 2014; Zandieh et al. 2016), or a coordination of the two (Rasmussen et al. 2015). In Sect. 10.2.1, I highlight this body of work through two examples of student reasoning in linear algebra: a study investigating individual students' understanding of the notions of span and linear independence, and a study investigating students' mathematical progress at both the individual and the collective level.

The aforementioned body of work about student learning in linear algebra informs and complements our research in developing instructional materials for student-centered, active learning classrooms in linear algebra. As detailed in Wawro et al. (2013), my colleagues and I rely on the close integration between theory and practice as we engage in the cyclical process of alternating between analyzing student reasoning and creating and modifying task sequences that support student reinvention of key ideas. One of the main products of this work is instructional materials; in Sect. 10.2.2, I highlight our unit on the concepts of span and linear independence as an example of this work.

10.2.1 Student Reasoning in Linear Algebra

In Plaxco and Wawro (2015), we characterized students' conceptions of span and linear independence, framed in light of their mathematical activity, to provide insight into their understanding. Data came from individual interviews with linear algebra students. We organized the wide range of student conceptions of span and linear independence into four categories: travel (described in Sect. 10.2.2.1), geometric, vector algebraic, and matrix algebraic. To further illuminate participants' conceptions of span and linear independence, we classified the participants' engagement into five types of mathematical activity: defining, proving, relating, example generating, and problem solving. By coordinating these two categories, we were able to produce fine-grained analyses of students' conceptions and the potential value or limitations of such conceptions in certain contexts.

The second example from our research documents the mathematical progress, at both the individual and collective levels, of a particular community of learners. A challenge in mathematics education research is to coordinate different analyses to develop a more comprehensive account of teaching and learning. In Rasmussen et al. (2015), we contribute to these efforts by offering four constructs to analyze learning: classroom mathematical practices, disciplinary practices, individual participation, and individual conceptions. The first two constructs document mathematical activity at the collective level. We describe the mathematical concepts and ways of reasoning that come to function as if they are mathematical truths in the classroom (classroom mathematics practices). We also describe how the class's mathematical behaviors align with the practices that typify the common activities of professional mathematicians, such as defining, symbolizing, and conjecturing (disciplinary practices). The second two constructs document mathematical activity at the individual level by describing how individual students participate in the classroom mathematics and disciplinary practices (individual participation) and by describing individual students' understanding of the mathematical concepts being developed (individual conceptions). These four constructs arise from and align with the Emergent Perspective (Cobb and Yackel 1996), a theoretical framework in which learning is viewed as "a process of both active individual construction and enculturation" (p. 186) that occurs as "students participate in and contribute to the practices of the local community" (p. 185). Thus, mathematical development at the collective level and the individual level are reciprocally related in that they are inextricably bound together in their respective developments.

In Rasmussen et al. (2015), we illustrated these four constructs for making sense of students' mathematical progress using data from the same undergraduate mathematics course in linear algebra as in Plaxco and Wawro (2015). In particular, we considered video recordings of whole class discussion, small group work, and individual student interviews to document students' mathematical progress as they engaged in a task sequence on span and linear independence. (This task sequence, referred to as the "Magic Carpet Ride sequence," is described in detail in Sect. 10.2.2.) First, we documented various classroom mathematical practices that *emerged* as the classroom community reasoned about linear independence and span, such as "For a given set of n vectors in R^m, if $m < n$, the set must be linearly dependent." Second, we described analysis of individual mathematical conceptions by summarizing results from Plaxco and Wawro (2015), which focused on how individual students from this same classroom understood span and linear independence. Next, we described individual participation by characterizing students' roles (such as author or relayer) in the production and spread of mathematical ideas at the classroom level. Finally, we highlighted students' work on Task 4 of the task sequence as an example of the disciplinary practice of "theoremizing" (which consists of activity related to both conjecturing and proving) in which students generated their own conjectures related to linear dependence and justified them.

10.2.2 Instructional Materials in Linear Algebra

Because of the importance of the transitional role that linear algebra plays in students' mathematical development, effective instruction at this juncture is paramount. Research has shown that classrooms in which students are central participants lead to learning gains (e.g., Mazur 2009; National Research Council 2012), and inquiry-oriented instruction has been shown to offer more equitable learning opportunities than traditional lecture-based approaches in that it diminishes the achievement gap (Laursen et al. 2014; Tarr et al. 2008). In our *Developing Inquiry-Oriented Instructional Materials for Linear Algebra* project, which is based on and contributes to the research about student thinking, we embrace these recommendations to engage students in learning mathematics through inquiry.

In this project we produce curricular materials, known as Inquiry-Oriented Linear Algebra (IOLA), that promote a student-centered, inquiry-oriented approach to the teaching and learning of introductory linear algebra. This effort is guided by the instructional design theory of Realistic Mathematics Education (RME) (Freudenthal 1991). A central tenet of RME is that mathematics is first and foremost a human activity, as opposed to being a predetermined collection of truths. Two RME heuristics—guided reinvention and emergent models—help navigate our curriculum design efforts. The notion of guided reinvention emphasizes the active role an instructor plays in utilizing student ideas and justifications to move forward the mathematical development of the class. The notion of emergent models emphasizes that classroom endeavors should support students in developing models of their mathematical activity that can in turn be used as models for subsequent mathematical activity. These are both facilitated by drawing on task sequences that are based on realistic starting points and are designed to support students in making progress toward a set of associated mathematical learning goals. As such, students' activity evolves toward the reinvention of formal notions and ways of reasoning about the mathematics initially investigated. In IOLA, this framework facilitates a transition from students' current, informal ways of reasoning about key concepts in linear algebra towards more formal, mathematically mature ways of reasoning.

At present, three units comprise the IOLA materials[1]. The tasks within each unit were informed by our engagement in a cyclical process of alternating between analyzing student reasoning and creating and modifying task sequences that support student reinvention of key ideas. Each unit focuses on a deep conceptual understanding of particular mathematical concepts, as well as how the ideas relate to each other. Unit 1, informally referred to as "The Magic Carpet Ride sequence," supports student reinvention of linear independence and span (Wawro et al. 2012). Unit 2, informally referred to as "The Italicizing N sequence," facilitates student exploration of matrices as linear transformations (Andrews-Larson et al. 2016). Unit 3, referred to as "The Blue to Black sequence," supports students' reinvention of diagonalization, eigenvectors, and eigenvalues (Zandieh et al. 2016). The units are independent of each other in the sense that an instructor could use one without using another; however, if an instructor chose to use all three plus the bridging material on systems of equations and row reduction, the majority of topics that one would expect to address in an introductory level linear algebra course (in R^n) would be explored.

A summary of Unit 1 follows as an example of this body of work. Unit 1 is intended to start on the first day of class, prior to any formal instruction, and consists of four main tasks. The tasks are grounded in a "realistic" scenario (of vectors relating to modes of transportation) that allows students to build rich imagery and formal definitions, both of which students use to reason throughout the semester.

[1] Interested faculty can gain access to the password-protected IOLA materials at http://iola.math.vt.edu by requesting an account.

Throughout, small group work (3–5 students per group) is alternated with whole class discussions in which students explain their tentative progress, listen to and attempt to make sense of others' progress, and finally come to justified conclusions on the various tasks and related questions that arise from their investigations.

10.2.2.1 Task 1: Investigating Vectors and Their Properties

In Task 1, students are asked to imagine they are young travelers leaving home with two modes of transportation at their disposal. They are asked to investigate whether it is possible to reach where Old Man Gauss lives, a location 107 miles east and 64 miles north of their home. One of the modes of transportation is a magic carpet. Its movement, when ridden forward for one hour, is denoted by the vector $\begin{bmatrix}1\\2\end{bmatrix}$ to indicate motion along a "diagonal" path resulting from displacement of 1 mile east and 2 miles north of its starting location. The other mode of transportation, a hoverboard, is defined similarly along the vector $\begin{bmatrix}3\\1\end{bmatrix}$. The goals of Task 1 are to (a) have students present and discuss multiple solution strategies; (b) have the instructor label student work and introduce formal notation for scalar multiplication, linear combinations, vector equations, and system of equations; and (c) coordinate geometric and algebraic views of the problem situation and its solution. This underlying metaphor of "travel" explored with vectors in R^2 often becomes a grounding imagery for students as they learn about linear combination, span, and linear independence not just in R^2 but in R^n as well.

10.2.2.2 Task 2: Reinventing the Notion of Span

The second task in the instructional sequence asks students to determine if there is any location (in the plane) where Old Man Gauss could hide so that they would be unable to reach him using the same two modes of transportation from the previous task. The goal of Task 2 is to help students develop the notion of span in a two-dimensional setting before formalizing the concept with a definition. As students work on this task, they begin to conceptualize movement in the plane using linear combinations of the vectors $\begin{bmatrix}1\\2\end{bmatrix}$ and $\begin{bmatrix}3\\1\end{bmatrix}$ to determine that there is nowhere for Gauss to hide. This involves developing a coherent geometric interpretation for a linear combination of vectors with all possible sign combinations of scalar coefficients, as well as an algebraic interpretation of $a\begin{bmatrix}1\\2\end{bmatrix}+b\begin{bmatrix}3\\1\end{bmatrix}=\begin{bmatrix}x\\y\end{bmatrix}$ for all possible real numbers x and y. Class discussion of this task sets the stage for the instructor to label the students' work with the term "span" and introduce a more formal and general definition of span, such as: The span of a set of vectors is all possible linear combinations of those vectors, or in other words, all places you could reach with those vectors. Furthermore, any vector $\boldsymbol{v}$ that can be written as $\boldsymbol{v} = c_1\boldsymbol{v}_1 + c_2\boldsymbol{v}_2 + \ldots + c_p\boldsymbol{v}_p$ for scalars $c_1, \ldots c_p$ is in the span of $\{\boldsymbol{v}_1, \boldsymbol{v}_2, \ldots, \boldsymbol{v}_p\}$.

After the class experience with span has been aligned with the formal notion of span, the students are positioned to engage in problems and theorems related to span not only in R^2 but also, with the guidance of the instructor, in R^n. Student feedback regarding this introduction to the concept of span is quite positive. For instance, on a written portfolio assignment in which students were asked to explain three encounters in linear algebra (either in class or in homework with problems, examples, proofs, etc.) that helped document their progress in understanding linear algebra, one student wrote:

> I would have to say that the problem for this class that has made the most impact on me was the very first problem that we did. The reason why I think it had the most impact was because of how the problem set us up to understand the underlying principles that the class would be working with in the coming weeks. By thinking about what points the two vectors could and could not reach formed the basis for understanding the span of a set of vectors, which in terms of the carpet ride problem would simply be the set of all points that the vectors could reach.

10.2.2.3 Task 3: Reinventing Linear Dependence and Independence

In this task, students are asked to determine if, using three given vectors $\boldsymbol{v}_1, \boldsymbol{v}_2$ and $\boldsymbol{v}_3$ that represent modes of transportation in a three-dimensional world, they can take a journey that starts and ends at home. They are also given the restriction that the modes of transportation could only be used once for a fixed amount of time (represented by the scalars c_1, c_2, and c_3). The purpose of the task is to provide an opportunity for students to develop geometric imagery for linear dependence and linear independence that can be leveraged in the development of the formal definitions of these concepts. Initial progress on this task is made when the class establishes that a trip that begins and ends at home could be represented by a homogeneous vector equation, which enables connections to previous algebraic activity. Students often encounter obstacles with the notions of linear independence and dependence because of the difficulty in interpreting the formal definitions and using formal systems (Dorier 1998). In the Magic Carpet Ride sequence, students' work provides them with rich geometric and algebraic imagery for linear independence and existence of solutions that is strongly connected to the formal definitions.

10.2.2.4 Task 4: Generating Examples and Generalizing

The handout for Task 4 asks students to generate sets of vectors that satisfy three varying constraints: number of vectors (2, 3, or 4), vector space (R^2 or R^3), and linearly independent or dependent. The main goal of Task 4 is to guide students to develop generalizations and supporting justifications regarding linear independence and dependence for any given set of vectors, shifting away from a reliance on the Magic Carpet Ride scenario. Although worded in terms of R^2 and R^3, the students' reasoning behind the strategies should inform the development of generalizations that extend to any R^n as they work on and discuss Task 4. A couple of examples of conjectures frequently created and justified by students are: "If the zero vector is included in a set of vectors, then the set is linearly dependent" and "If a set of 2 (or 3) vectors in R^2 (or $\mathbb{R}^3$) spans R^2 (or $\mathbb{R}^3$), then the set is linearly independent." A secondary goal is to develop an intellectual need for efficient computational strategies and sophisticated solution techniques; this need arises when students want an efficient way to check the linear independence of one of their generated example sets, such as three vectors in R^3.

In summary, the "Magic Carpet Ride" sequence starts with students' limited experience with vectors in R^2 and R^3 and, through the metaphor of travel, fosters deep student understanding of the concepts of linear combination, span, and linear independence. In addition to students learning about this concepts, the design of the curriculum—around the notion of inquiry—students are exposed to the mindset that mathematics is a human activity in which they can have an active role in knowledge creation. For instance, one student wrote in his portfolio assignment (mentioned in Sect. 10.2.2.2) that after working on Unit 1 Task 4:

> It was at this moment that I realized that … there were generalizations that could apply to this table of sets and linearly independencies and that we, mere students, were practically coming up with theories. It felt just like being a mathematician except with really basic concepts!

10.3 Teaching

For me, teaching and research are interconnected because each informs the other. This reciprocal relationship motivates me to investigate how students learn particular ideas in mathematics, as well as to seek out and develop tools and lessons that meet my pragmatic needs as a teacher. Teaching is never a fully developed static state; it is under constant alteration, being affected by research, by myself and by others, on how people learn. I ground my teaching in the tenet that mathematics is a human activity and, as such, believe that mathematics is more than a body of facts and skills to acquire. Mathematics also involves participation in the mathematical practices that typify the discipline. In line with my desire to promote students' view of themselves as active learners and participants in the practice of mathematics, I aim to assist students in developing from their current or naïve ways of reasoning towards more formal mathematical reasoning via engaging in mathematical practices in the classroom.

My role as a teacher is to start where students are and help guide them, through engaging in mathematical practices, towards the reinvention of mathematical ideas. This process requires two types of inquiry; as the teacher, I learn about where students are by inquiring into their thinking, as the students simultaneously inquire into the mathematics (Rasmussen and Kwon 2007). In the classroom, students often work on tasks in small groups and then present their ideas, defend their solutions, and ask others about their thinking or strategies during whole class discussion. As the teacher, I help move the development of the mathematical content forward by bringing out key ideas and identifying the ways in which the ideas brought forth by students connect to the language, notation, and conventions of the broader mathematical community. In this way, the mathematical content is also moved forward. By looking for opportunities for students to reinvent significant mathematical ideas through their own creative ways of thinking, I aim to promote my students' views of themselves as doers of mathematics. In every new situation, the interrelationship of research and practice changes me in that as I realize more about how students learn mathematics, I am motivated to adjust and improve my teaching practice.

At my current institution, my teaching spans three course types: linear algebra (both an introductory, sophomore-level first course and a proof-based second course) for undergraduate majors from across the STEM disciplines, mathematics content for pre-service secondary mathematics teachers, and mathematics education research for graduate students. With respect to the first category, my investigations into student reasoning in linear algebra have allowed me to know not only when and how students often struggle but also that they are capable of thinking creatively and effectively in solving problems. Locally, our research effort in curriculum design resulted in a recent collaboration with a colleague in my department. My colleague taught a morning section and I taught an afternoon section of the department's introductory course in linear algebra. Each class during the semester, we attended and videotaped each other's class, as well as debriefed twice a day (immediately after each of us had taught). During the debriefs we would discuss what we noticed about student thinking that day, our impressions of how the day went with respect to our learning goals, and our plans for subsequent lessons based on the students' mathematical progress. That intense collaboration, which provided a space for us to jointly reflect on and invest in not only our own teaching but also each other's, has had lasting effects on us as instructors and as colleagues.

With respect to mathematics content for pre-service secondary mathematics teachers, my knowledge of the research corpus has a positive impact on course content. The course goals include developing a deeper understanding of high school mathematics content, which is facilitated by: (a) explicitly drawing connections between mathematics they learn at university and that which they will teach at the high school level; and (b) learning what the research says about student learning in those content areas. Regarding (a), for instance, we use Unit 2 (Matrices as Transformations) from the IOLA materials (Andrews-Larson et al. 2016) as a starting point for exploring the ideas of one-to-one, onto, composition, and inverses of linear transformations—concepts that the students will teach as high school teachers for real-valued functions—and connecting that to research on student understanding of function (Oehrtman et al. 2008). The pairing of these two activities prompts preservice teachers to ponder their own understanding of linear transformations in light of Oehrtman et al.'s explanation of action and process views of function, as well as their role as a teacher in promoting their future students' deep understanding of functions. Regarding (b), I find great value in having preservice teachers read and digest research on student thinking. Within a given course, for instance, they read research on the teaching and learning of graph theory (McDuffie 2001), combinatorics (Lockwood 2014), and probability (Shaughnessy 2003).

10.4 Service

Finally, my role as a teacher is not independent from my responsibilities as a mathematics department faculty member. Furthermore, the aforementioned synergy between research and teaching allows me to have an impact on my department and institution through service that relies on that interplay. For instance, I have had the opportunity to actively participate in shaping undergraduate and graduate curricula: contributing substantially in the development of both the introductory and proof-based linear algebra courses; leading the development of a doctoral course in research on undergraduate mathematics education (RUME). In addition, I view mentoring graduate and undergraduate students as an integral aspect of my role as a faculty member. I enjoy seeing graduate students from both PhD options find the research area they want to be involved in so that they can contribute to and make a difference in our field's collective body of knowledge.

10.5 Reflection

In this chapter I presented my own story, as a mathematics education researcher in a mathematics department, as a case study of integration among research, teaching, and service. After reflecting on my experiences, I believe that finding balance in these academic responsibilities has positively impacted my own professional goals and my local institution. I am fortunate to be able to grow as a faculty member in a department and university that value the information discovered through research in the teaching and learning of undergraduate mathematics. I am inspired by others who continue to seek out opportunities in which their research expertise and passions could have an impact on the growth of their local department, college, and university. I look forward with hopeful anticipation towards the future and to seeing all that can be accomplished in our field through continued balance and integration of the various aspects of our research, teaching, and service.

Acknowledgments The research discussed in this chapter is based upon work supported by the National Science Foundation under Collaborative Grant numbers DUE-1245673. Any opinions, findings, and conclusions or recommendations expressed in this material are those of the author and do not necessarily reflect the views of the National Science Foundation.

References

Andrews-Larson, C., Wawro, M., & Zandieh, M. (2016). *A hypothetical learning trajectory for conceptualizing matrices as linear transformations.* Manuscript submitted for publication.

Cobb, P., & Yackel, E. (1996). Constructivist, emergent, and sociocultural perspectives in the context of development research. *Educational Psychologist, 31*(3/4), 175–190.

Dorier, J.-L. (1998). The role of formalism in the teaching of the theory of vector spaces. *Linear Algebra and Its Applications, 275*(27), 141–160.

Freudenthal, H. (1991). *Revisiting mathematics education.* Dordrecht, The Netherlands: Kluwer Academic Publishers.

Laursen, S. L., Hassi, M. L., Kogan, M., & Weston, T. J. (2014). Benefits for women and men of inquiry-based learning in college mathematics: A multi-institution study. *Journal for Research in Mathematics Education, 45*(4), 406–418.

Lockwood, E. (2014). Both answers make sense! Using sets of outcomes to reconcile differing answers in counting problems. *The Mathematics Teacher, 108*(4), 296–301.

Mazur, E. (2009). Farewell, lecture? *Science, 323*, 50–51.

McDuffie, A. R. (2001). Flying through graphs: An introduction to graph theory. *Mathematics Teacher, 94*(8), 680–683.

National Research Council. (2012). *Discipline-based education research: Understanding and improving learning in undergraduate science and engineering.* In: S. R. Singer, N. R. Nielsen, & H. A. Schweingruber, (Eds.). Committee on the Status, Contributions, and Future Direction of Discipline Based Education Research. Board on Science Education, Division of Behavioral and Social Sciences and Education. Washington, DC: The National Academies Press.

Oehrtman, M., Carlson, M., & Thompson, P. (2008). Foundational reasoning abilities that promote coherence in students' function understanding. In M. Carlson & C. Rasmussen (Eds.), *Making the connection: Research and teaching in undergraduate mathematics education* (pp. 27–42). Washington, DC: The Mathematical Association of America.

Plaxco, D., & Wawro, M. (2015). Analyzing student understanding in linear algebra through mathematical activity. *Journal of Mathematical Behavior, 38*, 87–100.

Rasmussen, C., & Kwon, O. N. (2007). An inquiry oriented approach to undergraduate mathematics. *Journal of Mathematical Behavior, 26*, 189–194.

Rasmussen, C., Wawro, M., & Zandieh, M. (2015). Examining individual and collective level mathematical progress. *Educational Studies in Mathematics, 88*(2), 259–281.

Shaughnessy, J. M. (2003). Research on students' understandings of probability. In J. Kilpatrick, W. G. Martin, & D. Schifter (Eds.), *A research companion to principles and standards for school mathematics* (pp. 216–226). Reston, VA: National Council of Teachers of Mathematics.

Tarr, J. E., Reys, R. E., Reys, B. J., Chavez, O., Shih, J., & Osterlind, S. J. (2008). The impact of middle- grades mathematics curricula and the classroom learning environment on student achievement. *Journal for Research in Mathematics Education, 39*(3), 247–280.

Wawro, M. (2014). Student reasoning about the invertible matrix theorem in linear algebra. *ZDM The International Journal on Mathematics Education, 46*(3), 1–18.

Wawro, M. (2015). Reasoning about solutions in linear algebra: The case of Abraham and the Invertible Matrix Theorem. *International Journal of Research in Undergraduate Mathematics Education, 1*(3), 315–338.

Wawro, M., Rasmussen, C., Zandieh, M., Sweeney, G., & Larson, C. (2012). An inquiry-oriented approach to span and linear independence: The case of the Magic Carpet Ride sequence. *PRIMUS Problems, Resources, and Issues in Mathematics Undergraduate Studies, 22*(8), 577–599.

Wawro, M., Rasmussen, C., Zandieh, M., & Larson, C. (2013). Design research within undergraduate mathematics education: An example from introductory linear algebra. In T. Plomp, & N. Nieveen (Eds.), *Educational design research – Part B: Illustrative cases* (pp. 905–925). Enschede, the Netherlands: SLO.

Zandieh, M., Wawro, M., & Rasmussen, C. (2016). Inquiry as participating in the mathematical practice of symbolizing: An example from linear algebra. *PRIMUS*. doi: 10.1080/10511970.2016.1199618.

Chapter 11
Communicating Mathematics Through Writing and Speaking Assignments

Suzanne Sumner

Abstract This chapter contains examples of writing and speaking assignments in mathematics courses at all levels, from a first-year seminar to a graduate course for teachers of mathematics. Courses in chaos theory, differential equations, history of mathematics, and mathematical modeling are illustrated as case studies for implementing writing and speaking assignments. These assignments are described, along with grading guides and reflections on the impact of these forms of assessment.

Keywords Communication in mathematics • First-year seminar • Speaking in mathematics • Writing in mathematics

11.1 Introduction

The University of Mary Washington (UMW) is a small public liberal arts institution of 4000 undergraduate and 500 graduate students. UMW is committed to developing better researchers, writers, and speakers, starting with our first-year seminars and continuing throughout the students' college experience. Through writing and speaking assignments in mathematics classes, my colleagues and I develop the

MSC Codes
97D40
01A05
34-01
37-01
97A30
97B50
97M10
97U50
97U70

S. Sumner (✉)
Mathematics Department, University of Mary Washington,
1301 College Avenue, Fredericksburg, VA 22401, USA
e-mail: ssumner@umw.edu

J. Dewar et al. (eds.), *Mathematics Education*, Association for Women in Mathematics Series 7, DOI 10.1007/978-3-319-44950-0_11

students' ability to research information, assess its quality, and communicate that information. Communicating quantitative ideas forces students to articulate their mathematical knowledge and compels them to find the meaning in the mathematics and to demonstrate their understanding (Barrass 2006; Montgomery 2003; Walters and Walters 2004).

Written assignments in mathematics classes can serve two purposes: writing to learn and learning to write (i.e., writing in the discipline). As Connolly and Vilardi (1989, p. 4) stated, "'Writing to learn' in science or mathematics classes is most basically about developing students' conceptual understanding of these subjects by developing their capacity to use the languages of these fields fluently," and they added, "The writing-to-learn movement is fundamentally about using words to acquire concepts" (1989, p. 5). If we can teach our students how to become better writers (and speakers) in addition to teaching them mathematics, so much the better.

The same dual purposes hold for speaking assignments in mathematics classes: speaking to learn and learning to speak. Smith (1997, p. 49) wrote, "The use of speaking assignments across the curriculum beginning in the first year of college not only develops the ability to speak coherently and persuasively, but also helps students learn course content." This explanation resonates with my own experience teaching the last 30 years: "we learn by doing…We understand concepts better and retain them longer when we express these concepts in our own words" (Smith 1997, p. 49).

My teaching builds on the two pillars of written and oral communication, along with a third pillar: interdisciplinary applications. I enliven the mathematics by connecting it to applications in art, biology, ecology, education, environmental science, geology, history, and physics. I want my students to link their interests with the mathematics I teach, knowing their engagement is the key to their learning.

Overall, my students appreciate having multiple forms of assessment, especially assignments that demonstrate their creativity. I can tell if students really understand the mathematics when articulating it in writing or speaking, much better than if they memorize rote techniques for test problems. My courses are structured so that 25 % of the grade originates from writing and speaking assignments and 75 % of the grade originates from traditional quizzes, tests, and exams. This mixture allows students who are not good test-takers to demonstrate their knowledge through coursework that plays to their strengths (Weimer 2002). Conversely, for the students who are weak in their communication skills, the writing and speaking assignments require them to strengthen these areas through continued practice (Bean 1996).

11.2 First-Year Seminars: The Mathematics of Chaos

UMW requires all incoming first-year students to take a First-Year Seminar, with these specific learning outcomes: to utilize research techniques for retrieving and synthesizing information and to communicate their results via writing and speaking. To achieve these goals, the first-year seminars have an enrollment cap of 15 students, designed to alleviate the workload associated with grading numerous writing and speaking assignments. The small class size is one of the enticements to teach the course.

Another enticement: UMW faculty members have great freedom in the seminar topics offered; we can explore compelling topics we might not normally teach. Each fall the mathematics department offers three or four sections from the following seminars: The Art of Mathematics; Cryptology; Game Theory; Infographics; Numbers Rule Your World; Pirates, Liars, and Pigeons; and The Mathematics of Chaos.

My first-year seminar, The Mathematics of Chaos, includes the butterfly effect, iterative processes, fractal geometry, chaotic dynamics, and the mathematical definition of chaos. This content has remained fairly constant over time (since first created by my colleague Jeffrey Edmunds), although new research in chaos theory is added as appropriate. Students begin by reading popular articles from primary sources where the mathematics is not too technical, and important ideas are explained in an intuitive manner (May 1976; Smale 1998). By the end of the semester they are reading more technical articles (Costantino et al. 1997; Li and Yorke 1975; Vellekoop and Berglund 1994).

Over the course of the semester, students complete four short writing assignments to investigate different problems using *A First Course in Chaotic Dynamical Systems Software*: *Labs 1-6*™. Then the students explain their conclusions about whether the dynamics lead to patterns or chaos and about what patterns are hidden within chaos, while defining the necessary mathematical terminology. Through these papers, the students have many opportunities to improve their mathematical understanding and their writing mechanics. I assess the papers on the completeness of the response, along with the correctness of the mathematics and the quality of the writing (please see Appendix A.1 for more details).

For another assignment, students create their own fractal design using matrices and the *Fractal Attraction*™ software and present their design in class (Appendix A.2). This project allows the students' creativity to shine. Examples of students' fractal creations are found in Fig. 11.1.

The seminar culminates with a research project consisting of a paper and a presentation (Appendix A.3). With my approval, students choose a topic related to chaos theory and submit an annotated bibliography that performs our library's

Fig. 11.1 "UMW Eagle," David Peworchik and "UMW Runner," Holden Vanderveer, The Mathematics of Chaos

recommended CRAAP Test to evaluate their sources' Currency, Relevance, Authority, Accuracy, and Purpose (Blakeslee 2004). After my comments on their sources and corrections on their writing, students submit a draft of their project paper for more feedback. (To prevent students from interpreting the word "draft" too freely, I tell them they are submitting their final paper, and after receiving my feedback, they can submit a revision.)

11.3 Writing and Mathematics: History of Mathematics

UMW's general education curriculum requires that all students take at least four Writing Intensive (WI) courses, because writing enhances learning across all disciplines and instructors across disciplines share a collective responsibility to help students become skilled writers. A course that is WI must require multiple writing assignments with frequent instructor feedback.

Some sections of introduction to statistics, probability and statistical inference, abstract algebra, and directed study have been taught as WI. History of Mathematics is the only UMW mathematics course designated WI, regardless of instructor. This sophomore-level course runs every semester at one or two sections. I focus this course on great mathematicians' lives and important historical problems.

For example, one assignment in History of Mathematics is to write a biographical paper on a mathematician from an under-represented group and to deliver an in-class poster presentation (Appendix B.1). Students describe the mathematician's life story and accomplishments. Figure 11.2 shows an example of a student's mixed media poster featuring the mathematician Omar Khayyam.

A National Science Foundation (NSF) grant reviewer once attended my class on poster presentation day. The grant reviewer commented that the students really took ownership of "their" mathematicians and made their stories their own.

The majority of the writing in History of Mathematics occurs in daily journals where the students write a paragraph about what they find interesting in that day's reading (Burton 2011). Each class begins with a discussion of the journals, forming a natural segue into that day's material and using the reading as a first exposure to the course material. To hold the students accountable for the reading, I ask each student, each class, to discuss their favorite part of the reading. I write encouraging and specific comments in their journals so they know I really do read their entries. I also use these journals to correct writing mistakes without a grade deduction, in the hope students will subsequently avoid these mistakes on their formal writing assignments (Appendix B.2). One student remarked that the journals "allowed me to find something interesting in each chapter and help evolve my writing style…They were helpful with…correcting grammar and style." Another student wrote, "I liked the comments [on the journals] because it made it seem more like a conversation rather than just a grade."

In a semester-long research project, students reenact Galileo's measurement of Earth's gravity and Eratosthenes' measurement of Earth's radius to ultimately mea-

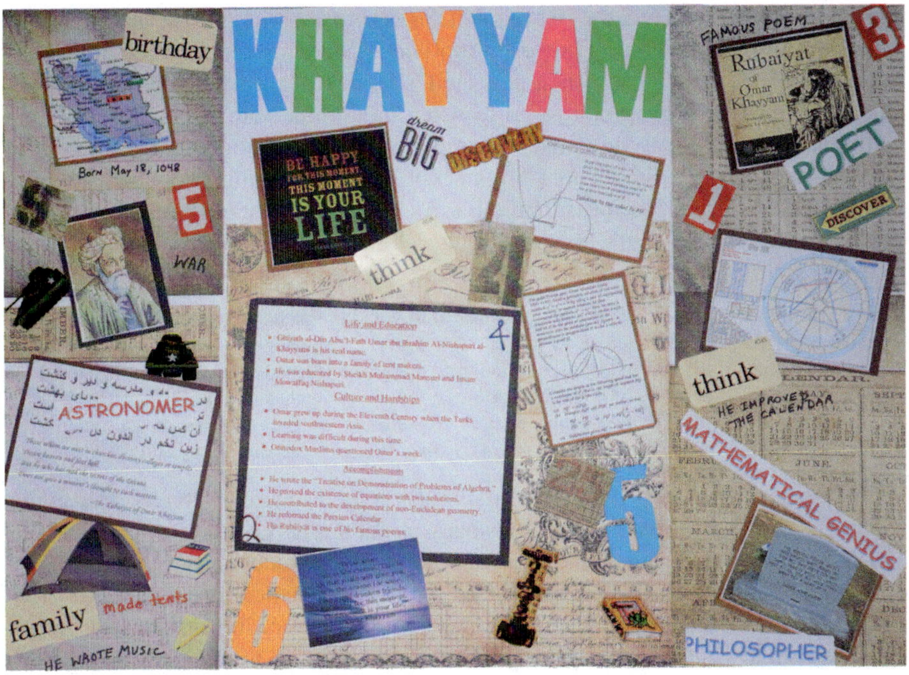

Fig. 11.2 "Omar Khayyam," Kellie Hurley, History of Mathematics

sure the Earth's density (Vacher 2012). Students, in groups of four, design and conduct the necessary experiments and then individually write a laboratory report on the results (Appendix B.3). Many students are nervous about writing a scientific report, claiming they are not familiar with writing in that format. However, they soon discover that scientific writing is not that mysterious; they are merely to explain the experiments and the experimental outcomes. Students are instructed to imagine they are writing this report to guide a fellow UMW student to perform the same experiments.

For the most part, students' grades on the Earth Density research project are higher than or at the same level as their overall course grades. In fact, data collected since Fall 2002 when I first initiated this project shows that 86.5 % of 689 students performed at a higher level or the same level as their overall course grade.

I can offer several likely reasons for students' improved performance on their research projects. First, this assignment is spread throughout the entire semester, starting with group online discussions. This format prevents a last minute rush job, and shows students the beauty of taking a large problem and breaking it up into manageable parts. (My hope is that they carry this idea to their other courses and assignments. One student agreed, "Everybody knows that you want to start your research as early as possible, but this process makes it happen. Our experiment was unsuccessful but the paper was a success.") Second, for students who feel stronger in their writing than in science, the report gives them an opportunity to shine. Third, the online discussions and the experiments are performed in groups. In my experience, working with peers is a powerful motivator. Students appear to be more conscientious on work they know their peers will see, and they work harder if their work affects their peers' grades, compared to when they think I am the only one seeing their work. Last, I like to think that the students find this assignment fun and out of the ordinary–we actually spend a class session outside taking Global Positioning System (GPS) readings to measure the Earth's radius and dropping balls out of windows to measure the Earth's gravitational acceleration–a novelty among mathematics classes. Students have made the following comments on anonymous surveys:

> It was fun to include a topic dealing with astronomy and the physical traits of the Earth, as those have always been interests of mine.
>
> The project itself was helpful for visualizing things we learned in class. It was also nice to break up the monotony of a math class.
>
> We even replicated the famous ancient experiment of Eratosthenes measuring the Earth's circumference (well, slightly tweaked since Eratosthenes did not use GPS).

Students in my History of Mathematics class come from every conceivable major and have widely different mathematics backgrounds. This course counts as a major and minor elective, so mathematics and other science students take the class. The dual general education designations as WI and Human Experience and Society (for courses exploring the forces shaping human activity, relationships, social structures, and intellectual systems) attract the mathematics and science students who are hoping to find an interesting course in a topic they enjoy. And the additional designation of Quantitative Reasoning entices the non-science students wanting to find a friendly mathematics course.

11.4 Speaking and Mathematics: Introduction to Mathematical Modeling

UMW's general education curriculum requires that all students take at least two Speaking Intensive (SI) courses, for active participation in their learning, for increased motivation, and for better understanding of course material. For the SI designation, a course must provide multiple opportunities for speaking assignments and instructor feedback.

Specific sections of finite mathematics with applications, statistical methods, number theory, discrete mathematics, and numerical analysis are taught as SI. Some professors use whole class discussions and small group discussions as ways to satisfy the speaking requirement, in addition to, or as alternatives to, traditional class presentations.

My Introduction to Mathematical Modeling, a first-year course, is SI. This class originated through an NSF grant awarded to our colleague, Marie Sheckels, to revise our curricula to improve how we prepare pre-service teachers in mathematics and science, so that they in turn are better mathematics and science teachers for their own students. We first offered the course in 1998, and we offer several sections each semester.

Colleagues Patricia Dean, Debra Hydorn, and I designed this course with the environment as a focus. Every example and data set used in this course describes some environmental issue, such as overpopulation, endangered species, pollution, limited resources, natural disasters, and epidemiology. The mathematical topics used to understand these problems are linear regression, curve fitting, and difference equations (Gordon et al. 2004). According to course evaluations and informal feedback, students really appreciate such a narrowly focused mathematics course. One semester after the final exam, a history major shook my hand to thank me for presenting the mathematics within a common theme. Apparently every other mathematics class he had ever taken felt like a sequence of disparate topics, and this class had been the first time mathematics had made sense to him.

Speaking opportunities in Introduction to Mathematical Modeling take various forms: group work in class, two group projects with papers and poster presentations, and a formal presentation; group sizes are three to four students. The first project involves finding the best-fit model (linear, exponential, power, or logarithmic) to a data set from Pfaff's Sustainability Math website (Pfaff 2014). I usually select data sets about carbon emissions, global temperature, grain production, oil consumption, ozone depletion, and wind power production. Groups use logarithmic transformations and linear regression to find their best-fit curves, with *Excel*™ and graphing calculators. Then they must decide which of the four models is the best-fit overall to the data and justify their selection by assessing each model's goodness of fit (Appendix C.1). Figure 11.3 shows an example of a group poster for a tree ring data set used in Patricia Dean's class.

For the second project, I provide the student groups with various environmental scenarios and ask them to create difference equations to describe these scenarios. The environmental topics generally are pollution, recycling, overpopulation, and invasive species. In one scenario, the group imagines that it is a wildlife organization studying goose overpopulation. For a summary of that assignment, see Appendix C.2. Figure 11.4 shows an example of a group poster for a goose overpopulation model.

For these projects, the groups work together to analyze the mathematical models and to create posters summarizing the environmental issues and their mathematical work. Then each student presents the group's poster individually in the poster session. The poster sessions follow the conference poster session model, i.e., the classroom is filled with posters, and multiple presentations occur simultaneously. The atmosphere in the room buzzes with activity, a departure from the usual sedate mathematics classroom. Grading all these presentations occurring at the same time is a challenge as well. I enlist friendly faculty and former students to grade the poster presentations (after training them on the grading rubric). My student graders love that I trust them with this responsibility, and I make sure to mention their help in letters of recommendation.

The third formal speaking assignment is an individual presentation to the entire class about an infectious disease, a topic of their choice. The student is given a collection of images about the disease to use in their presentation; at least one image is some type of mathematical graph that they must explain. Other images are micro-

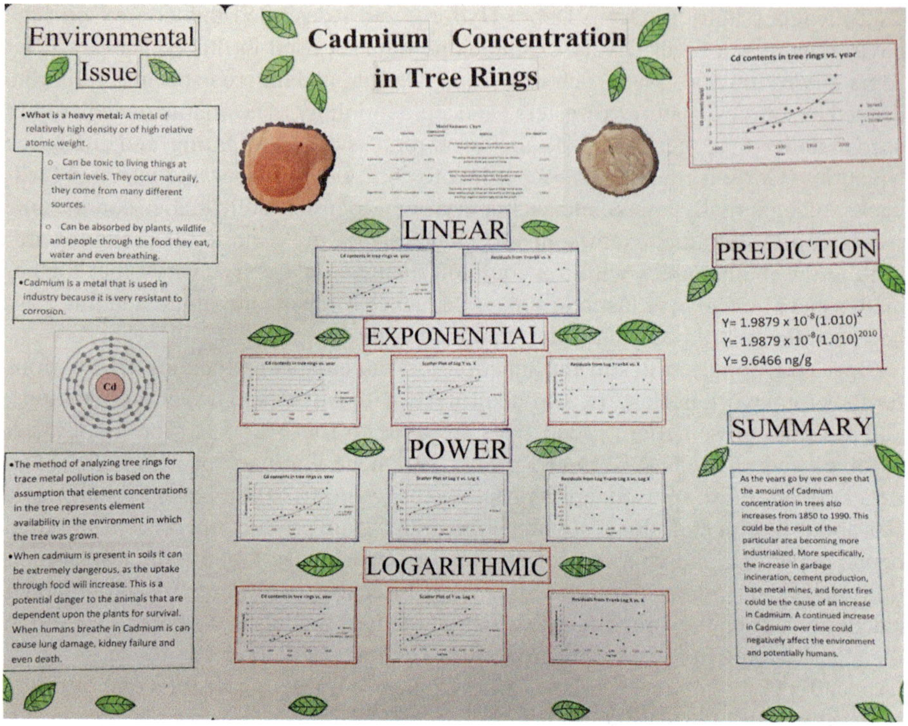

Fig. 11.3 Tree ring data analysis poster, Patricia Dean's Introduction to Mathematical Modeling students

scopic pictures of the disease, geographical maps of affected regions, public health posters, etc. The student creates a summary handout and posts it on the class's online discussion board (Appendix C.3). After the presentations, the class has a better understanding of the biology of infectious diseases, which makes it easier for us to model the spread of disease with a Susceptible-Infective-Removed (SIR) difference equation model.

In Spring 2016 the following comments were made on anonymous surveys about the speaking assignments:

> I feel like the presentations were <u>super</u> helpful for gaining speaking skills. I would want <u>more</u> of them!
>
> Projects were helpful to understand the importance of environmental issues and how the issues can be mathematically expressed.
>
> I think group interactions teach us a lot of the material by working in groups.

When we first created the Introduction to Mathematical Modeling course at UMW, we collected data in two semesters for two classes through pre- and post-course surveys to measure the impact of the course, through its material and pedagogy, on student perception of their confidence and skill in mathematics (Dean et al. 1999). Both classes showed significant improvement (p-value<0.05) in students'

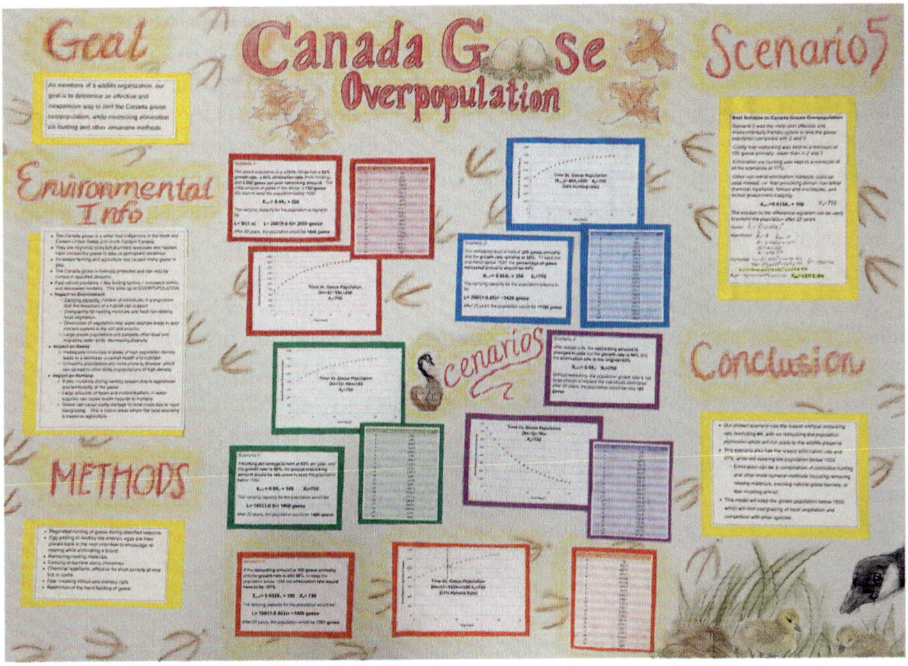

Fig. 11.4 Goose overpopulation poster, Patricia Dean's Introduction to Mathematical Modeling students

confidence in their mathematical abilities. The post-course survey contained two additional questions assessing students' perceptions of greater insight into the integration of mathematics and science and of better understanding of environmental issues. On these questions, the combined classes averaged 4.25 on the former and 4.00 on the latter (where the scale was 5 = strongly agree, 4 = agree, 3 = neutral, 2 = disagree to 1 = strongly disagree).

11.5 At the Upper-Level: Ordinary Differential Equations

Due to the success I have had using writing and speaking assignments in History of Mathematics and Introduction to Mathematical Modeling, I now assign writing or speaking assignments in all of my mathematics classes.

In my junior-level Ordinary Differential Equations course, I assign a paper about problems famously solved by differential equations (e.g. the van Meegeren art forgeries or atomic waste disposal) for students to summarize (Braun 1993). The assignment comes early in the semester before the students have learned the mathematics for solving the differential equations. Instead of assessing students' understanding of particular mathematics topics, I urge the class to concentrate on the real world problem to be solved and how to state that problem in terms of differential equations, being careful to explain the variables and parameters involved and why the equations describe that problem. Then I ask them to omit the mathematics involved in solving the problem and rather explain what the solution tells them about the real world application. My intention is to have the students recognize the rich variety of problems that differential equations can describe, along with gaining an appreciation for the modeling process and giving them a preview of the mathematics to come. Not all assignments need to be in the format of students explaining mathematics that is new to them.

Other writing in Ordinary Differential Equations comes by means of computer laboratory assignments, using *Mathematica*™ software to investigate modeling applications such as Newton's Law of Cooling, population growth, the AIDS epidemic, forced harmonic oscillators, and predator-prey interactions. I chose these topics in consultation with my colleagues in biology, chemistry, and physics for strengthening areas where our students need more reinforcement. In these assignments, students fit different models to data sets, study the effects of changing parameters, and then explain their results in short two-page essays.

11.6 At the Graduate Level: History of Mathematics for Teachers

At the request of UMW's College of Education, I developed a History of Mathematics for Teachers graduate-level course for a master's degree in mathematics education cohort of K-12 mathematics teachers. This course was the last course before the capstone course where the graduate students completed their master's projects. To prepare the graduate students for their upcoming projects, this course emphasized writing, speaking, and research skills, much as the first-year seminar does, but from the opposite end of their college experience. I have also modified this course to act as an independent study for pre-service K-12 teachers.

Students write a (graded) daily journal about a topic from the reading that they could incorporate into their teaching. Another assignment is a "Number Biography," meaning students research a number or class of numbers and answer the following questions: What is this number? When and where was this number born (the times, places, and cultures involved)? What is its life story (the history of its development)? Who were the mathematicians involved? Why is this number special? In addition to a summary paper, the students prepare a poster or bulletin board for use in their classroom.

The final research project for History of Mathematics for Teachers allows the students to investigate a research question from either the history of mathematics or the pedagogy of mathematics, pursuing it in more detail, using primary sources. Students can pick from a list of topics or follow their own interests. The requirements to write an annotated bibliography and submit an initial draft provide opportunities for instructor feedback. At the end of the semester, students write a paper and make a presentation to the class. Throughout the years students have researched topics such as: how the stereotype that mathematicians do their best work before 30 argues against women doing mathematics and how child prodigies support that stereotype; the impact of Hitler and Nazism on mathematics before, during, and after World War II; mathematics ability and gender; and the Four-Color Conjecture and Kepler's Conjecture and whether computer-aided proofs can be considered real proofs.

11.7 Lessons Learned

Over the years I struggled with spending hours reading and commenting on students' writing, only to have my comments ignored and the same mistakes reappear on later assignments. I have arrived at a partial solution to that problem. In the classes where students have done research for a written report to present in class, I tell students that they will have an essay question on the next test where they will discuss another student's topic (of their choosing). So for History of Mathematics, students write an essay on the life and achievements of another student's chosen mathematician from an underrepresented group. For Introduction to Mathematical Modeling, the essay is to explain one of the other environmental issues and how and why that group chose their best-fitting model. To help each other with their respective essays, I ask the class to post their revised papers to the class discussion board.

The advantages to this approach are twofold: students are motivated to read my comments and make my suggested revisions, and they pay better attention to each other's presentations, knowing they are responsible for the material later. Similarly, for the disease presentation in Introduction to Mathematical Modeling, students post their summary handouts to the class discussion board. The final exam contains an essay question asking students to compare and contrast their disease topic with another disease presented in class. This question often leads to interesting essays that showcase students' critical analysis skills. The fact that students tend to write very good essays on tests when they know the topic ahead of time is a reward for both student and instructor.

11.8 Reflections on Departmental Impact

Because First-Year Seminars, WI, and SI courses are requirements in our general education curriculum, my work in these areas has been well received and supported by my colleagues campus-wide since this work addresses learning outcomes the university values. UMW provides excellent resources for both students and faculty through our library, Speaking Center and Writing Center.

Within the mathematics department, however, because the majority of my work with First-Year Seminars and WI and SI courses occurs for lower-level courses, it has been pretty much overlooked by departmental colleagues. The first WI and SI mathematics courses were taught by four women in the department, either for the lower-level non-major courses or for major elective courses. We had a great deal of freedom and flexibility in designing the content and pedagogy of these courses; these courses did not involve the others in the department who mostly taught calculus or upper-level courses. In recent years, to spread out the lower-level teaching load, each department member, each semester, is now assigned at least one course at the lower level, raising interest in different teaching approaches.

Department members also noticed that WI and SI courses and sections had healthy enrollment numbers compared to those without them. In addition, we wanted to make it easier for our mathematics and science majors to meet their general education requirements. Now we have WI and SI courses throughout our mathematics curriculum, albeit on an ad hoc basis.

If we seek to be more intentional in providing WI and SI courses in the mathematics major, the department must consider the following issues carefully. We should not violate academic freedom by forcing an unwilling department member to teach a course in this manner. We also must weigh the risk that a WI or SI designation on a required major course might lower enrollments on major elective courses, with or without one of these designations. UMW's mathematics department currently offers a wide variety of electives in the major, a flexibility that should be kept.

Also, some department chairs value a professor's willingness to teach these WI and SI courses (in merit pay raises), whereas others believe the reduced class sizes are reward enough. A typical lower-level mathematics class at UMW ranges from 25 to 35 students, and we are fortunate our upper-level mathematics classes are capped at 15. Admittedly, our class sizes are enviably small. However, when our lower-level SI and WI classes were capped at 15 students, that size reduction really was sufficient reward for the added workload of grading writing and speaking assignments. But it feels less like parity to teach a WI section of History of Mathematics at 20 students rather than a normal section of Calculus I at 25.

Furthermore, we have seen an increase in the enrollment limits for WI and SI courses. When WI and SI designations were first approved at UMW, faculty members were assured that SI and WI class sizes would not exceed 15. Now the SI and WI class sizes are capped at 20, a 33 % increase in students. Aside from the obvious increase in grading for the written and oral assignments, it is impossible to schedule 20 individual student presentations in one class period for a 75-min class, let alone a 50-minute class. It may seem petty to grumble about class sizes this small, given the large class sizes faculty at other institutions face. But at UMW we have a teaching load of 12 credit hours (typically four 3-credit courses) per semester. Each increase in workload is another lost opportunity to pursue our research interests. And therein lies the rub.

The reality is that research accomplishments are generally more valued than teaching accomplishments, even at a self-proclaimed teaching institution such as UMW. My former chair, Debra Hydorn, decided to change the culture of our department to interweave both pedagogy and research, by expanding our emphasis on undergraduate research. With an increased number of students completing honors theses, the result was more students needing assistance with their writing and presentation skills (D. Hydorn, personal communication, February 20, 2016).

11.9 Reflections on Student Impact

Still, my students are the most important audience of these pedagogical efforts. Mathematics major Aaron Thomas (personal communication, March 9, 2016) noted the courses discussed in this chapter have given him the opportunity to research mathematical topics and communicate quantitative information with models and in words, skills not taught in high school but would be advantageous in his future career as an actuary. Another student, Kellie Hurley (personal communication, May 4, 2016), remarked that these courses have helped her learn about different presentation techniques from watching the other students' presentations. Mathematics graduate Kimberly Hildebrand (personal communication, May 7, 2016) stated, "Taking WI and SI classes helped me to better communicate technical work, both in college and in my career [in government]. Being able to formally explain problems, whether technical in nature or not, is something I do on a daily basis," adding, "I would not be as effective of a communicator in my career whether it be over email, in a paper, or at a presentation or meeting, without having the experiences of taking WI and SI classes."

For these reasons, I doubt I could return to teaching mathematics in the traditional, 100 % test and examination, format. I learn so much more about my students and their abilities through their communication assignments, and their research assignments never fail to teach me something new–what more could a professor want?

Acknowledgements My thanks go to Debra Hydorn, Wyatt Mangum, the editors, and the reviewers for their insightful comments on this chapter, as well as my students for being a continual source of inspiration. I thank Patricia Dean for graciously sharing her students' work. I also thank Marie Sheckels and the Virginia Collaborative for Excellence in the Preparation of Teachers (VCEPT) NSF grant DUE-9553789 for their support in creating the Introduction to Mathematical Modeling course.

Appendix A: Grading Guides for The Mathematics of Chaos Seminar

Short Writing Assignments

These four papers have a length of two to three pages and are worth 10 points each (10 % of the course grade each). For mistakes in mathematics and grammar, my first-year student grading deductions are: 1 to 4 mistakes = 0.5 point, 5 to 8 mistakes = 1 point, 9 to 12 mistakes = 1.5 points, 13 to 16 mistakes = 2 points, etc.

Fractal Design Presentation

This five-minute presentation is graded on a $\surd$+= 1 point, $\surd$=0.5 point, $\surd$−=0 points scale for a total of 5 points (5 % of the course grade) with the following criteria (modified from a UMW Speaking Center rubric):

1. Explanation of how the fractal design is self-similar
2. Description of the fractal design (number of transformations used, their size, placement, rotation, shearing, color)
3. Material is presented clearly and use of visual aids is effective
4. Presenter is well prepared and organized and responds well to questions
5. Mannerisms are appropriate and not distracting (dress, posture, gestures, voice, eye contact)

Research Project

I assess the four-page project paper for correctness in mathematics, writing, and citation. The first draft is worth 5 points (5 % of the course grade) and the revised project paper is worth 10 points (10 % of the course grade). The presentation is graded on a $\surd$+= 0.5 point, $\surd$=0.25 point, $\surd$−=0 points scale for a total of 5 points (5 % of the course grade) with the following criteria (modified from a UMW Speaking Center rubric):

1. Introduction
2. Background information
3. Description of the topic
4. Information from References and Resources
5. Conclusion
6. Material is presented clearly
7. Use of visual aids is effective
8. Presenter is well prepared and organized
9. Presenter responds well to questions
10. Mannerisms are appropriate and not distracting (dress, posture, gestures, voice, eye contact)

Appendix B: Grading Guides for The History of Mathematics

Biography Project

The two-page biography paper is 8 points (4 % of the course grade) and deductions for mistakes in mathematics, writing, and citation are: 1 to 2 mistakes = 0.5 point, 3 to 4 mistakes = 1 point, 5 to 6 mistakes = 1.5 points, etc. The biography presentation is 4 points (2 % of the course grade) and is graded on a $\surd$+= 1 point, $\surd$=0.5 point, $\surd$−=0 points scale with the following criteria:

1. Description of mathematician's personal life and educational training
2. Description of the times and culture
3. Obstacles faced as a minority/if not minority, status within culture
4. Indication of the mathematician's accomplishments

Daily Journal

The journal counts for 13 points total (6.5 % of the course grade), because Burton's text *The History of Mathematics: An Introduction* has 52 sections counted at 1/4 point per section (Burton 2011). I award credit just for completing the assignment.

Earth Density Research Project

This Earth density project counts for 25 points (12.5 % of the course grade), 10 points for five online discussions to design the experiments to measure the Earth's density and 15 points for the scientific report on the experimental results. Each online discussion consists of four questions the group of four students answers to design their experiments (2 points total). Each student earns one point for contributing an answer and then earns the second point if all questions are answered correctly by the deadline. I grade the six-page scientific report using a checklist (title, introduction, hypothesis, methods, experimental protocol, materials, results, data tables, discussion section, conclusion about the hypothesis, error analysis, reference list, and all the formulas needed for the calculations) with my usual grading deduction for writing mistakes (Appendix B.1).

Appendix C: Grading Guides for Introduction to Mathematical Modeling

Project #1: Curve Fitting to a Data Set

The four-page Project #1 paper is 10 points (5 % of the course grade) and is graded as in Appendix B.1. The 5-min poster presentation is 10 points (5 % of the course grade) and is graded on a $\surd$+= 1 point, $\surd$=0.5 point, $\surd$−=0 points scale with the following criteria:

1. Description of environmental issue
2. Description of data set (variables, units, patterns)
3. Results of model fitting
4. Justification of best-fit model

5. Using best-fit model to make a prediction
6. Material is presented clearly
7. Use of visual aids is effective
8. Presenter is well prepared and organized
9. Presenter responds well to questions
10. Mannerisms are appropriate and not distracting (dress, posture, gestures, voice, eye contact)

Project #2: Modeling with Difference Equations

Assessing Project #2 is done in essentially the same way as with Project #1 (Appendix C.1). An example of Project #2 is:

"Suppose the goose population in a wildlife refuge grows exponentially with an annual growth rate of 50 %. Each year hunters eliminate 60 % of the geese. The state's wildlife department moves 200 new geese into the region each year by restocking. Suppose the initial number of geese is 750. The wildlife refuge wants to keep the number of geese below 1500 within 20 years. For each scenario, write the difference equation for the number of geese x_{n+1} in the next year in terms of x_n and graph the solution for 20 years.

1. Find the equilibrium value for your difference equation. What does this value represent?
2. If the restocking is held at 200 geese annually, what percentage of the geese should be harvested annually to keep the number below 1500 within 20 years?
3. If the annual harvesting/hunting rate is held at 60 % annually, at what level should the geese be restocked annually to keep the number below 1500 within 20 years?
4. Suppose budget cuts force the refuge to suspend its restocking program. Your wildlife organization studies the model with no restocking. What long-term prediction does your organization make about the goose population?
5. Your wildlife organization successfully lobbies to reinstate restocking, although at a reduced level of 100 geese annually. What annual percentage of hunting do you recommend to keep the goose population below 1500 within 20 years?
6. Which scenario is better, #3 or #5? Predict the goose population in 20 years using your answer.
7. Solve your best difference equation in #6 to find the formula for the solution. Check your prediction for $n = 20$ using the formula for the solution."

Infectious Disease Presentation

This five-minute presentation counts as 10 points (5 % of the course grade) and is graded on a $\surd$+= 1 point, $\surd$=0.5 point, $\surd$−=0 points scale with the following criteria:

1. Description of the infectious disease (symptoms, transmission, mortality, etc.)
2. Distribution of the disease; epidemic/endemic

3. Increasing, decreasing, or constant problem
4. Cure/treatment/prevention
5. Impact on society/culture/economics/history
6. Material is presented clearly
7. Appropriate mannerisms, gestures, voice, eye contact
8. Presenter is well prepared and organized
9. Presenter responds well to questions
10. Mature approach to the subject matter

References

Barrass, R. (2006). *Scientists must write: A guide to better writing for scientists, engineers and students* (2nd ed.). New York, NY: Routledge.

Bean, J. (1996). *Engaging ideas: The professor's guide to integrating writing, critical thinking, and active learning in the classroom*. San Francisco, CA: Jossey-Bass.

Blakeslee, S. (2004). The CRAAP test. *LOEX Quarterly, 31*(3), 6–7.

Braun, M. (1993). *Differential equations and their applications* (4th ed.). New York, NY: Springer-Verlag.

Burton, D. (2011). *The history of mathematics: An introduction* (7th ed.). New York, NY: McGraw-Hill.

Connolly, P., & Vilardi, T. (1989). *Writing to learn mathematics and science*. New York, NY: Teachers College Press.

Costantino, R., Desharnais, R., Cushing, J., & Dennis, B. (1997). Chaotic dynamics in an insect population. *Science, 275*, 389–391.

Dean, P., Hydorn, D., & Sumner, S. (1999). Impact of a new introductory mathematical modeling course on student confidence in mathematical ability and skills. *The Journal of Mathematics and Science: Collaborative Explorations, 2*(2), 111–115.

Gordon, S., Gordon, F., Tucker, A., & Siegel, M. (2004). *Functioning in the real world: A precalculus experience* (2nd ed.). New York, NY: Pearson/Addison Wesley.

Li, T.-Y., & Yorke, J. (1975). Period three implies chaos. *The American Mathematical Monthly, 82*(10), 985–992.

May, R. (1976). Simple mathematical models with very complicated dynamics. *Nature, 261*, 459–467.

Montgomery, S. (2003). *The Chicago guide to communicating science*. Chicago, IL: The University of Chicago Press.

Pfaff, T. (2014). *Sustainability math*. Retrieved March 1, 2016, from http://www.sustainabilitymath.org.

Smale, S. (1998). Finding a horseshoe on the beaches of Rio. *The Mathematical Intelligencer, 45*(1), 39–44.

Smith, G. (1997). Learning to speak and speaking to learn. *College Teaching, 45*(2), 49–51.

Vacher, L. (2012). *Earth's planetary density: Constraining what we think about the Earth's interior*. Retrieved May 3, 2016, from http://serc.carleton.edu/sp/ssac_home/general/examples/15087.html.

Vellekoop, M., & Berglund, R. (1994). On intervals transitivity = chaos. *The American Mathematical Monthly, 101*(4), 353–355.

Walters, D., & Walters, G. (2004). *Scientists must speak: Bringing presentations to life*. New York, NY: Routledge.

Weimer, M. (2002). *Learner-centered teaching: Five key changes to practice*. San Francisco, CA: Jossey-Bass.

Chapter 12
Real Clients, Real Problems, Real Data: Client-Driven Statistics Education

Talithia D. Williams and Susan E. Martonosi

Abstract In this chapter we describe two client-focused educational experiences at Harvey Mudd College that offer students the opportunity to work on real problems for real clients using real data. The first is the Harvey Mudd College Clinic capstone program, in which teams of students spend an academic year working on a project for an external sponsor. The second is a course project in an upper level statistics elective in which the students analyze data provided by a campus partner. For both of these, we describe their structure, recent projects, as well as student and client feedback. We also offer our reflections on how providing these educational experiences has influenced us personally and professionally.

Keywords Course projects • Capstone • Statistics education • Client projects

12.1 Introduction

Demand has recently surged for data-savvy individuals in organizations ranging from government agencies to start-up businesses to nonprofit organizations. The rise of data science, an interdisciplinary field that combines computer science, statistics, and mathematics to gain insights from large data sets, has led to a surplus of jobs for data scientists, but the supply of workers who are equipped to solve data-driven problems is limited. The skills needed to be a successful data scientist increasingly depend on a combination of technical expertise, effective communication, teamwork, and attention to the client's needs. While the phrase "data-driven

---

MSC Code
97K80

T.D. Williams (✉) • S.E. Martonosi
Department of Mathematics, Harvey Mudd College,
301 Platt Blvd., Claremont, CA 91711, USA
e-mail: twilliams@hmc.edu; martonosi@g.hmc.edu

J. Dewar et al. (eds.), *Mathematics Education*, Association for Women in Mathematics Series 7, DOI 10.1007/978-3-319-44950-0_12

statistics education" might seem redundant, the reality is that statistics is often taught without requiring the analysis of real data and seldom taught with considerations of the needs of a client. At Harvey Mudd College (HMC), we have created client-focused data science experiences both in the classroom and in the senior level capstone course, known as "Clinic."

HMC is an undergraduate liberal arts college of science, engineering and mathematics that is part of the Claremont Consortium consisting of four other undergraduate colleges and two graduate institutions. All HMC students complete a common core curriculum in the technical disciplines represented on campus (mathematics, computer science, chemistry, biology, physics, and engineering) and an extensive sequence in the humanities, social sciences, and the arts. Therefore, we have a student body that is proficient in the STEM fields and for whom data analysis is likely to become an integral part of their future careers.

The purpose of this chapter is twofold. First, we illustrate two types of data analysis course experiences: a client-sponsored field capstone program in Sect. 12.2, and a project run within a statistical linear models elective in Sect. 12.3. We provide the reader with implementation models that they could adapt to their own institution. Second, we reflect on our experiences teaching these courses and describe some of the challenges and opportunities they have yielded. Williams, the first author, has taught the statistical linear models course using client-based projects, so in Sect. 12.3 the narration changes to the first person. Sect. 12.4 concludes the chapter by looking to the future of statistics education. Throughout this chapter, we use the term *deliverable*, commonly used in industry, to denote a report, a presentation, an algorithm, or other work product that is delivered to a client in the course of a project.

12.2 The Harvey Mudd College Clinic Program

The Harvey Mudd College Clinic program started in the HMC engineering department in 1963. Drawing its name from the training of medical students, in which they practice their skills during clinical rotations on real patients under the watchful eye of an experienced physician, the HMC Clinic was developed to give students the opportunity to practice their skills on a real engineering problem sponsored by an external client under the supervision of a faculty member. This experience helps students synthesize their classroom knowledge and bridge the gap between the theory of the classroom and the real world.

In 1973, the HMC mathematics department adopted Clinic in its curriculum, and since then, the departments of physics and computer science have followed suit (Borrelli 2010). Participation in Clinic is a graduation requirement for engineering, computer science, and computer science/mathematics joint majors, while mathematics and physics majors can choose either Clinic or a traditional thesis to fulfill their capstone requirement. (Biology and chemistry majors typically complete a traditional thesis, although some have participated in Clinic when an appropriate project was available.) Our departments work collaboratively to recruit interesting Clinic projects. When a project requires interdisciplinary skills as is increasingly the case, we create cross-departmental Clinic teams.

Coming from applications-driven disciplines (statistics and operations research), both authors were naturally drawn to Clinic. In our disciplines, the word "research" does not always imply proof-based theoretical advances, and often refers to the creative application of existing methodologies in new ways to solve real problems. This is the heart and soul of Clinic. Given our skills and interests, we have been involved in the program as faculty advisors, director, or associate director for six and ten years, respectively.

We often hear that HMC's unique focus on science, engineering and mathematics make it difficult for instructors at other institutions to implement programs like Clinic. However, as surveyed by Martonosi (2012), and by Martonosi and Williams (2016), other institutions have taken notice of the HMC Clinic program and have successfully emulated it. The Olin College website describes their Senior Capstone Program in Engineering, which was modeled very closely after HMC's Clinic (Olin College 2016). Moreover, in recognition of the broad impact the program has had on engineering education worldwide, the professors who cofounded the program in 1963 were awarded the 2012 Bernard M. Gordon Prize for Innovation in Engineering and Technology Education by the National Academy of Engineering (Harvey Mudd College 2012). Our hope is that readers can adapt some of the characteristics described in this chapter to create a Clinic-like program at their own institution.

We outline the structure of the program in Sect. 12.2.1, describe some of the projects that have been completed in Sect. 12.2.2, provide excerpts of student and liaison feedback in Sects. 12.2.3 and 12.2.4, and reflect on our personal experiences in the program in Sect. 12.2.5.

12.2.1 Structure of the Mathematics Clinic Program

We start by listing the cast of characters involved in the Mathematics Clinic: the team, the team's project manager, the team's faculty advisor, the sponsoring organization of the project, the sponsor's liaison, and the director and associate director of the Mathematics Clinic program. We describe each of these in more detail in the subsequent paragraphs.

In Clinic, teams of four to six students, mostly seniors, work for a full academic year on a problem posed by an external sponsoring organization. The students are responsible for determining the appropriate methodology for solving the problem, conducting relevant literature review, managing the project timeline, and preparing all deliverables in a professional manner. The teams are formed by the Clinic director and faculty advisors to balance student project preferences (as determined by a survey), background in skills needed for the project, grade point averages, student interest in serving as a team's project manager, and to avoid known personality conflicts. Clinic strengthens students' technical skills by exposing them to a complex, real problem. Additionally, Clinic builds professional skills by requiring students to communicate clearly both orally and in writing, manage a large project, and work effectively in a team.

One student on the team is appointed to serve as the team's project manager, acting as the primary point of communication between the team and the sponsor's liaison. The project manager also ensures that the team is making steady progress towards deliverables and intermediate deadlines. He or she does this in addition to technical contributions to the project alongside the other team members.

The teams are advised by a faculty member and by a liaison from the sponsoring organization. The faculty member serves primarily as a coach, mentoring the students on their research habits, team dynamics and communication. Because we feel that lessons learned from mistakes often have more impact than those learned from successes, the faculty advisor intervenes in the specifics of a project only when the team is heading far off course. For this reason, the faculty advisor need not have expertise in the mathematical methodology used in the project. The role of the sponsor's liaison is to provide domain expertise and context to the students throughout the year. The team meets weekly with the liaison by teleconference or Skype to share intermediate results and to ensure that the team's direction aligns with the project's goals. The liaison must also provide the team with data and background information in a timely fashion.

The Mathematics Clinic director and, in some years, an associate director oversee the three to five Clinic projects being run in the department each year. Their primary responsibility is to recruit projects of sufficient quantity and quality, working together with the other departmental Clinic directors and HMC's Director of Corporate Relations. The directors leverage the HMC alumni network and make site visits to companies across the west coast and, occasionally, other parts of the country. We charge a substantial fixed fee per project, which covers administrative costs, travel (for recruiting trips and for the teams to visit the sponsors), computing equipment and software, and other supplies. Because of the fee, we seek sponsors who are invested in the outcome of the project and, accordingly, we assign them the intellectual property rights to the completed work. We advise potential sponsors to propose projects whose results are not critically needed in the short-term, but the outcomes of which will be very useful to them in a few years' time. This mitigates some of the risk associated with relying on a team of undergraduates to complete a project that the company cares about. We also require the sponsor to identify a liaison who will be able to dedicate sufficient time and energy to the team to ensure that the team will have consistent access to contextual information needed to produce a useful product.

Clinic counts as a regular three-unit course in each semester, and we expect the students to devote approximately 10 h per week on the project. Of these 10 h, one hour is spent in a weekly meeting with the faculty advisor, one is spent in a weekly teleconference with the liaison, and one is spent in a weekly classroom session. The remaining 7 h are spent working on the project, and we strongly encourage the teams to schedule those 7 h as a team. We have found that the team can work more productively and resolve obstacles more quickly if everyone is in the same room working at the same time. Moreover, we have found fewer issues of students relying on their teammates to do most of the work when they are held accountable to the rest of the team for their work hours.

The weekly classroom sessions involve professional development workshops that we created as director and associate director and which continue to be used. The topics are effective team dynamics, teleconference and site visit etiquette, project management tools, and effective oral presentations. To keep the students engaged and help them retain the information, we structure these workshops as comical, improvisational skits. In these skits, faculty advisors portray students having traits and behaviors we have observed over the years as being detrimental to a team's success, such as interrupting teammates, arriving late to meetings, or not allowing every member of the team to contribute. The class discusses what they saw and what the characters should have done differently. The students enjoy the humor in a full panoply of bad behaviors parodied in a series of two-minute skits. At the start of the year, it is hard for them to realize how true-to-life these skits are; however, over the year, most of these behaviors do indeed arise on some of the teams, and the memory of the skits gives the students a starting point for addressing them. During the spring semester, Clinic teams from all departments (approximately 45 teams per year) assemble together for the weekly classroom sessions, which are dedicated to team presentations. Each team makes one 15–20 min presentation over the semester, allowing the students to review each other's work and practice giving oral presentations.

Imposing intermediate milestones and deliverables helps students manage the large, open-ended project:

- Clinic and Liaison Orientation: In early fall, liaisons are invited to campus to debrief the team on the context of the problem.
- Fall and Spring Site Visits: Each team travels to the sponsor's site early in the fall semester and again late in the spring semester. The teams gain insight into the context of their project and make presentations to the sponsoring organization.
- Marathon Push: In the first month of the project, teams devote time to the project beyond the required ten hours per week to immerse themselves in the project and write the team's Statement of Work (SOW).
- SOW: This document summarizes the team's understanding of the problem, literature review, proposed methodology, and timeline for the year. The liaison reads and approves the SOW in writing.
- Fall and Spring Presentations: In November, each team gives a presentation to the full group of Mathematics Clinic students about the progress they have made on their project. Throughout the spring semester, each Clinic team on campus makes a presentation to all Clinic students, faculty advisors and special visitors.
- Projects Day: In early May, all liaisons are invited to campus for our annual celebration of Clinic. Each team makes several presentations of their final project and presents a poster of their work in a general poster session. Receptions and a celebratory dinner conclude the event.
- Reports and Deliverables: At the end of the fall semester, the team submits a midyear report to the liaison. At the end of the academic year, the team delivers to the sponsor a final report, along with all software, computer code, data and other intellectual property of the project.

12.2.2 Recent Projects

The range of mathematical disciplines represented in the completed Mathematics Clinic projects over the past 40 years is breathtaking, including modeling pollutant transport in the atmosphere, fraud detection, and optimal control of satellite motion. Abstracts for all past Mathematics Clinic projects can be viewed on the Mathematics Clinic website (Harvey Mudd College Mathematics Clinic 2016). Although our Mathematics Clinic program does not recruit solely statistics projects, we have seen a rise in data-focused Clinic projects in recent years. For example, in the past 5 years (2010–2015), exactly half (11 out of 22) of our Mathematics Clinic projects (or those run jointly between mathematics and another department) fall in the category of data analytics, while in the preceding 5 years, that percentage was closer to 20 % (Harvey Mudd College Mathematics Clinic 2016), (Harvey Mudd College Computer Science Clinic 2016).

This shift towards data projects poses some challenges to our department. First, there is a sense of loss for the more traditional applied mathematics projects in the areas of fluid mechanics, differential equations, and linear algebra. While these areas of mathematics are still relevant, current sponsors are more interested in areas pertaining to analytics. This is especially the case for federal sponsors such as national laboratories, where funding that can be used to pay the Clinic fee is more plentiful in cutting-edge research areas like data science. A second problem is that some of our faculty feel ill-equipped to advise data-focused projects. Although the faculty advisor is not responsible for completing the project, many advisors prefer a project in their general research area. In the past, it was easier to match faculty advisors to projects in their areas of expertise. Third, for the authors, who serve as two of only a handful of statistical experts at HMC, the rise in data-related Clinics across campus results in a lot of ad hoc advising of Clinic teams on matters of experimental design and data analysis. This can take up a great deal of time on top of our regular teaching responsibilities, particularly during the spring semester when Clinic teams are completing their analyses.

12.2.3 Student Experience and Feedback

The Clinic experience is often a seminal one for students, motivating them and boosting their confidence before they venture into the "real world." At the end of each semester, students complete peer- and self-evaluations in which they reflect on the Clinic experience. In addition to showing appreciation for the technical knowledge gained in the experience, the evaluations invariably emphasize the professional skills gained and the students' satisfaction in having worked on a real problem for a real client:

> I have learned a lot about machine learning and software development, but also about documentation, working in a team, and research in industry. … It has been fascinating to study mathematical ideas that are applied so readily to an industrial problem.

I enjoyed in particular the aspects of heuristic design and adaptation, as well as the practice doing things like teleconferences, presentations, and reports on our progress, all clearly reflecting the sorts of work we can expect to do in industry.

The evaluations also commonly refer to the challenge of Clinic, and the satisfaction that arises from meeting that challenge:

We've certainly run into our share of issues along the way, but I don't think the clinic experience would have been nearly as valuable if we didn't have those challenges to overcome.

Our advisor held us to a very high standard, which was sometimes stressful but I think it encouraged us to work very hard and resulted in a final product which we are proud of.

Most often, student complaints about the Clinic experience stem from the difficulties in working with a team:

I feel like our clinic team may have had a fair bit of trouble keeping on top of deadlines this year, but that we have produced a good result. I'm not sure why we were so often behind, but if I had to hazard [a guess], I would say communication troubles.

Despite the stress of Clinic, it is an experience that shapes a student's professional trajectory. Moreover, because our Clinic program is well-known, we often hear prospective students speak of Clinic as one of the deciding factors in choosing to attend HMC.

12.2.4 Client Feedback

It is not only the students who find the experience rewarding. Each year, we survey the liaisons, asking them how well the team met the project's goals, managed the project, and communicated with the liaison, and how they rate their overall satisfaction with Clinic. On a five-point scale, five being the best, the College scores higher than four, on average, on these questions.

Additionally, the Clinic Advisory Council, a committee of approximately 20 representatives from industry who have engaged with the Clinic program in the past, conducts phone interviews of all liaisons to discuss their experience with the program. Some comments from recent Mathematics Clinic liaisons are:

I was very impressed by the level of enthusiasm and knowledge of the students. Very well done.

[The liaison] realizes the quality of the "scarce resource" of [HMC students] and is intent upon identifying and hiring the best fits for his company, using the Clinic Program as a way to get to know the team members.

Of course, not every project is completed successfully. In some instances, the liaison and the team realize that the original proposal is infeasible or no longer in the sponsor's interests, so they work collaboratively to redefine the project scope. In other instances, however, the team simply does not meet the project's goals, usually because of poor team communication or project management. Fortunately, truly unsuccessful instances comprise only 5–10 % of all projects. One reason for our high success rate is that Clinic directors thoroughly vet projects in advance to ensure they are of an appropriate scope for the team. Another reason is that teams are encouraged to distinguish in their SOW between achievable baseline goals and "stretch goals" that will be completed if time permits. A third reason is that teams communicate regularly with the liaison so that expectations can be effectively managed if the project scope needs to be adjusted during the year.

12.2.5 Our Experiences

We have both served as faculty advisors of Clinic teams several times. Additionally, the second author served as Clinic director for 5 years, and the first author served as associate Clinic director for 1 year. In this section, we describe some of the personal impacts these roles have had on us.

As faculty advisors, we have found two primary challenges. The first is knowing when to be the "good guy" and when to be the "bad guy" in our team interactions. Generally, we try to observe from a distance to allow the students to direct the flow of the project. At times, however, when a team repeatedly fails to follow our suggestions, misses deadlines, or engages in unproductive habits or behaviors that are disrespectful to the liaison's time, we step in more assertively. The second challenge has been keeping the team and the liaison focused on the project scope. We have occasionally encountered liaisons that push students beyond what is expected from a three-unit course and have had to intervene on the students' behalf.

Being a Clinic faculty advisor sometimes forces us to step out of our comfort zones when the project ventures into an area of mathematical sciences we are less familiar with, but it is rewarding to broaden our knowledge. The opportunity to work directly with many of the companies that hire our students has provided us insight into the future needs and directions for mathematical sciences in industry. This allows us to develop new course materials that are relevant and cutting-edge for our students.

During our time administering Clinic as director and associate director, we developed valuable leadership and administrative skills, balancing the needs of students, faculty advisors, liaisons, and college administrators. The development of these skills was accompanied by a similar development of confidence. Directing the Clinic dramatically reduced time for scholarly activities such as research, but it was enjoyable work, from learning about the fascinating work done at prospective sponsoring organizations, to mentoring the student teams as they developed project management and communication skills.

12.3 Client Projects in Statistics Courses

Clinic is one model for offering statistics students client-driven projects at a departmental scale. On the scale of an individual instructor, final projects in beginning and upper level statistics courses can be used to gauge students' understanding and mastery of the material. One such course that began from this frame of reference was the HMC statistical linear models course (Math 158). I (the first author) based the course on the Guidelines for Assessment and Instruction in Statistics Education (GAISE) (American Statistical Association 2005), whose six broad recommendations include:

1. Emphasize statistical literacy and develop statistical thinking.
2. Use real data.
3. Stress conceptual understanding, rather than mere knowledge of procedures.
4. Foster active learning in the classroom.
5. Use technology for developing concepts and analyzing data.
6. Use assessments to improve and evaluate student learning.

When I first taught the course, I assigned a project based on an interesting dataset I could find and tested students on their ability to correctly analyze the data and present their results before their classmates. While this model proved sufficient in meeting course objectives, the GAISE guidelines, and student assessment, it lacked a connection to a data-driven client experience that would challenge students by placing them in unfamiliar territory. Lazar et al. (2011) have documented the improvement in the statistical analysis and quality of work of students who engage in a consulting-like experience. Much of the noted improvement is a result of incorporating real, messy data into projects and emphasizing the responsibility that students have to clients.

To give the students a more authentic data analysis experience, I restructured the course project to involve a local client with a real problem needing data analytics. In the remainder of this section, I describe that project. Section 12.3.1 describes its structure, Sect. 12.3.2 gives some examples of recent problem statements, Sects. 12.3.3 and 12.3.4 describe the student and client experience, and Sect. 12.3.5 provides some of my personal reflections and advice from having run such a project.

12.3.1 Structure of Community Client Engagement

12.3.1.1 Team Structure

In a typical semester of statistical linear models, students come from several of the Claremont Colleges with varying backgrounds. While an introductory statistics course is a prerequisite, the style of the course depends on the institution in which it was taken. Some courses are full semester and use the open-source statistical programming language R. Others are half semester and teach statistical analysis using Minitab®. This poses a challenge when constructing teams that must coalesce to effectively tackle a project and produce tangible results.

A solution I have found is to intentionally place students in teams based on their strengths. I give a survey to students at the beginning of the semester that asks for their previous computing experience, past statistics courses, and perceived strengths when working in teams. I use this information to create teams that have at least one strong computing person, a strong statistical person, and an effective oral and written communicator. While all students experience all aspects of the project, this distribution helps teams manage more independently by providing them in-house experts in areas of critical need. Teams are typically made up of three to five students, depending on the size of the course.

Once the teams are in place, they are responsible for scheduling additional time outside of class to work on the project together. All students are required to evaluate their teammates' contribution and I take the overall evaluation into account when determining the final project grade. I also try to be especially sensitive to women and underrepresented students by not placing them in groups where they are the only minority. Instead I place them on teams with at least two women, at least two minorities, or some combination of the two.

12.3.1.2 In-Class Lab Experience

I typically reserve two classes per month for an in-class lab where I present a new topic or type of data analysis that needs to be performed on their project dataset. Students bring their personal laptop to class and I reserve additional laptops for students that don't have one. The in-class lab allows me to observe how the teams are working together, gauge their progress so far and answer questions that often arise during their external meetings. I'm also able to observe and intervene if students are having difficulties or becoming disengaged. If the client is local, I often invite them to class to be available to answer questions and provide direction, especially during the first few lab sessions.

The structure of the in-class lab usually begins with stated goals for the session, for instance, to write an R program that will perform an exploratory data analysis and produce various plots. While the analysis of the data occurs over the entire semester, by setting up-front, measurable goals for the lab sessions, students understand that I expect them to have a deliverable by the end of class. To help novice R users get up to speed, it might sound counterintuitive that I tend to pair them together rather than pair a novice with a strong R programmer. However, except in cases where the strong R programmer is also a good mentor and tutor, pairing the novices together often leads to better collaboration.

12.3.1.3 Final Presentation Experience

Final presentations in semester-long statistics courses are commonly used to evaluate students' understanding of the material and their ability to communicate results effectively (Khachatryan 2015). What makes the final presentation experience particularly relevant in a client consulting environment is that students have to explain potentially complicated statistical analysis to people who are not experts in the field. Students learn that they must omit theoretical details and instead deliver results that are clear, concise, and visually appealing. The rich details of the analysis are included in a final written report that each team submits both to me and to the client.

The program and location for the final presentations can be used to create a meaningful closing experience for both the students and the client. Each team, dressed in business casual attire, presents its results and fields any questions that arise. For the past four years, we have had an on-site meeting at the end of the semester where the client invited all staff to the presentation and provided light refreshments.

12.3.1.4 Building Client Partnerships

A key aspect of the success of the course project has been the connection to the community-based client. In response to my solicitation to various constituencies around the Consortium for large-scale data analysis, Sam Kome, the Director of Strategic Initiatives and Information Technology at the Claremont Colleges Library contacted me. The library had large amounts of data and was seeking someone to analyze it.

The steps taken to build the community client relationship and provide a rich experience for students can easily be replicated in other data-driven courses:

1. Send an email to faculty and staff at your campus soliciting data.
   For example, an introductory statistics class could work with the cafeteria staff members to visualize demand by hour of the day, day of the week, or menu items.
2. Involve the community client in developing the project with the students.
   I found that staff members were excited to talk to the class about their data and project goals.
3. Be proactive in getting data at the beginning of the course.
   Students can immediately begin exploratory analysis and apply more advanced techniques as the semester progresses.
4. Create in-class lab days where students work in teams on the final project.
   While this required removing some lecture topics from the syllabus, the hands-on project time was an opportunity for the community client to attend class and answer individual teams' questions.
5. Encourage your client to remain available throughout the semester.
   Our client encouraged the students to contact him by email to review initial results or answer additional questions.
6. Set up a meaningful presentation experience.
   Presentations normally occur at the client facility or in a nice room on campus. Ample time is set aside for informal conversations, and light refreshments are provided.

Although the dedication of class time to the project reduces the number of topics we can treat in the course, the quality of the final project and presentation never ceases to amaze me. Giving students an unknown, unrefined dataset forces them to become researchers, ask their own questions, and go in multiple directions.

12.3.2 *Recent Projects*

In a recent project, the client provided patrons' library Wi-Fi usage data, and each team independently determined the type of analysis required to address the client's needs. Once teams formulated a project direction, they shared among themselves to prevent overlap, and with the client to verify that the analysis would meet their needs. Three approaches to this problem are summarized below as examples:

Team 1: This team established a heuristic for deciding between active and passive connection and defined a metric for wireless usage based on the data transferred and amount of time used. With this metric, the team determined how various factors, such as roles (student, faculty, staff), radio types, and signal quality impacted wireless use.

Team 2: This team performed a comparative analysis specific to Apple products, since they comprise the majority of products connecting to the Wi-Fi. They examined how Macs, iPods, iPads, and iPhones successfully connected to the network and the subsequent signal strength. They compared the performance activity common to each of these device types in terms of data usage and connection time.

Team 3: This team examined measures of connectivity success, including time spent connected, megabytes of data used, number of attempts made to connect, and signal strength, to determine the quality of the Wi-Fi connection.

Through these different approaches, the teams were able to visualize the ways in which various devices connected to the library Wi-Fi and better understand the duration of library visits by campus, day of the week and time of day.

12.3.3 *Student Experience and Feedback*

Through conversations with students and the written course evaluations I have received, students:

- Appreciated the active learning aspect of the in-class labs;
- Felt a sense of purpose by working with real data for clients who valued their solutions;
- Were genuinely surprised to see the excited reactions of the staff to their results; and
- Realized that there are many types of analyses that can be done on a single dataset.

One student sent an email the following semester stating:

> I just wanted to send you a quick note saying thanks for the statistical skills you helped hone in Math 158 last semester. I am doing an experiment for a cognitive science class and I've relied heavily on the techniques and tools you showed us throughout the class in order to analyze the results. It is very useful to be comfortable analyzing data, so thank you for imparting that ability throughout the course.

However, some students were frustrated by the open ended approach, the time it took to clean the dataset, and having to rely on team members who were less dependable.

12.3.4 Client Feedback

We have worked with the Claremont Colleges Library for the past five years and they have been very pleased with the partnership. Below is an excerpt from a letter that Sam Kome sent to the Harvey Mudd College Dean of the Faculty highlighting the experience from the Library's perspective:

> The wireless analysis identified and described a significant authentication issue that actually affected all the Claremont campuses. The library was disproportionately affected, and was able to use the analysis to adjust wireless provision and nearly eliminate complaints. We very much look forward to continuing this fruitful collaboration. … We have found Professor Williams students' work to be thorough and thoughtful, and each presentation yields novel and directly useful information. We also appreciate that the students become deeply engaged with the data and they frequently express gaining a greater understanding of the complexity of today's academic library.

12.3.5 The Instructor's Experience

Creating an authentic data-driven experience required me to rethink my role in the classroom and allow my students to take ownership of the experience. I vividly recall the moment that students took charge of the project with the Claremont Colleges Library. I had invited the staff members to class to talk about the data. After the staff did a very brief presentation, there was an awkward silence as students looked to me for direction. I told them, "I'm not analyzing their data, so I don't have any questions for them. But in three months, you get to stand up and present your results in front of the entire staff. This is your time. We can take as much or as little of it as you need." This was the moment that the students took ownership of their project.

One closing thought on this experience would be the following:

> *Make the mundane meaningful.*

Although necessary, the process of cleaning real-world big data is often tedious and frustrating. Students frequently spend hours writing code to parse the data into a readable format to finally begin performing the statistical analysis. During one particular in-class lab day, two recent HMC alumni, Kyle and Russell, were visiting campus for a recruiting trip and asked to observe the class. The students were cleaning data that day and although I constantly reminded them that this is a necessary process of data analysis, they were obviously frustrated and disappointed with the task.

I asked Kyle and Russell if they would like to look at the data, and the two of them began cleaning the data as I continued helping students. After a short time,

Kyle and Russell had written a script that read in the data, parsed it, corrected the formatting issues, and produced a multiple time series plot. I asked them to display their code on the screen and walk the students through their process. Russell told the class, "I wish I would have had a class like this back when I was at Mudd. In a typical five-day work week, we spend four days just cleaning the data. The stats is pretty easy after that." The mood of the class changed following their presentation.

As a professor, this moment was significant because it provided immediate purpose to a typically mundane classroom lab session by merging it with a real-life career experience. In every lecture, lab and presentation experience, students are learning statistical methodologies to prepare them for their careers. I now intentionally invite former students back to allow them to provide that rich perspective.

12.4 Looking Ahead

What does the future hold for statistics education? The need for computational and analytics skills to mine large data sets is expected to grow (McKinsey Global Institute 2011). Although HMC is unique amongst liberal arts colleges in that its only offered majors are in science, engineering and mathematics, we believe fervently that statistics is itself a liberal art (Moore 1998). It is no longer reserved for specialists, or even just for scientists and engineers. Rather, in the Information Age, it is imperative that all students learn to understand and critically interpret data put before them. Reflecting this trend, data science courses offered at HMC and other Claremont Colleges are routinely filled to capacity with students from many majors, and we are seeking ways to offer more data science electives while still maintaining our existing mathematics curriculum. Client-focused experiences such as Clinic and projects in upper-level statistics electives are one avenue to help prepare students to work effectively with data. We encourage our readers to consider how these types of experiences can be incorporated in their own curricula.

Acknowledgements The authors thank several anonymous reviewers for their thoughtful and detailed feedback that greatly improved this manuscript.

References

American Statistical Association. (2005). *Guidelines for Assessment and Instruction in Statistics Education (GAISE) College Report*. Resource document. American Statistical Association. Retrieved March 31, 2016, from http://www.amstat.org/education/gaise/GaiseCollege_Full.pdf.

Borrelli, R. (2010). The doctor is in. *Notices of the American Mathematical Society, 57*(9), 1127–1130.

Harvey Mudd College. (2012). *Harvey Mudd College faculty members share prestigious 2012 Bernard M. Gordon Prize*. Retrieved March 31, 2016, from https://www.hmc.edu/clinic/gordon-prize/.

Harvey Mudd College Computer Science Clinic. (2016). *Clinic projects*. Retrieved March 31, 2016, from https://www.cs.hmc.edu/clinic/projects/.

Harvey Mudd College Mathematics Clinic. (2016). *Projects indexed by year*. Retrieved March 31, 2016, from https://www.math.hmc.edu/clinic/projects/years/.

Khachatryan, D. (2015). Incorporating statistical consulting case studies in introductory time series courses. *The American Statistician, 69*(4), 387–396. doi:10.1080/00031305.2015.1026611.

Lazar, N. A., Reeves, J., & Franklin, C. (2011). A capstone course for undergraduate statistics majors. *The American Statistician, 65*(3), 183–189. doi:10.1198/tast.2011.10240.

Martonosi, S. E. (2012). Project-based ORMS education. *Wiley Encyclopedia of Operations Research and Management Science* (pp. 1–15). doi:10.1002/9780470400531.eorms1059.

Martonosi, S. E., Williams, T. D. (2016). *A survey of statistical capstone projects*. Manuscript submitted for publication.

McKinsey Global Institute (2011). *Big data: the next frontier for innovation, competition, and productivity*. Retrieved May 12, 2016, from http://www.mckinsey.com/business-functions/business-technology/our-insights/big-data-the-next-frontier-for-innovation.

Moore, D. S. (1998). Statistics among the liberal arts. *Journal of the American Statistical Association, 93*(444), 1253–1259. doi:10.1080/01621459.1998.10473786.

Olin College. (2016). *Senior capstone program in engineering (SCOPE)*. Retrieved March 31, 2016, from http://www.olin.edu/collaborate/scope/.

Chapter 13
A Montessori-Inspired Career in Mathematics Curriculum Development: GeoGebra, Writing-to-Learn, Flipped Learning

Kathy A. Tomlinson

Abstract With an overview of Montessori education, I set the stage for curriculum materials aimed at improving undergraduate mathematics education. I describe four ways to enhance student learning with the dynamical mathematics software GeoGebra: classroom demonstrations, student activities with instructor-created applets, student activities with applets that students create by following podcast instructions, and student-created applets that more advanced students generate independently to solve problems. I discuss two types of writing-to-learn assignments: guided reflection and journaling. I also describe collaborative classroom activities, including associated video lessons that I constructed to implement a flipped or blended learning environment. Connections are made between current mathematics education research findings, Montessori principles and the curriculum materials that I designed. The chapter closes with a reflection on my career path. I discuss my passion for mathematics and social justice, how this led to professional opportunities in mathematics education including a project in the scholarship of teaching and learning, and how my work in mathematics education is useful as I assume leadership as chair of my department.

Keywords Curriculum development • Flipped learning • GeoGebra • Montessori • Writing-to-learn

---

MSC Codes
97D40
97I40
97I50
97I80
97U70

K.A. Tomlinson (✉)
Department of Mathematics, University of Wisconsin-River Falls, 410 South Third Street, River Falls, WI 54022, USA
e-mail: kathy.tomlinson@uwrf.edu

J. Dewar et al. (eds.), *Mathematics Education*, Association for Women in Mathematics Series 7, DOI 10.1007/978-3-319-44950-0_13

13.1 Introduction

In a career that is inspired by Maria Montessori's ideas, I design and implement mathematics curriculum materials that attempt to respond to current mathematics education research. Most of my work seeks to improve instruction in mathematics courses taken by Science, Technology, Engineering and Mathematics (STEM) majors including calculus, differential equations, linear algebra, mathematical modeling, complex variables and discrete mathematics.

As I develop mathematics curriculum I am guided by questions about how students learn and what teaching methods and strategies work for them. How can we help students get a deep conceptual understanding through work with concrete ideas in a way that helps them move to greater abstraction? How can we get students to spend more productive time on task? How can we teach in ways that help students retain knowledge? How can we lower the number of students who withdraw from or fail our classes, while maintaining high learning expectations? How can we help students become engaged with and committed to mathematics?

These questions led me to three forms of curriculum work. The first uses the open source dynamical mathematics software GeoGebra. I have created four types of GeoGebra[1] modules ranging in level of student involvement from the instructor demonstrating in class while students make observations and connections, to students creating their own applets (small computer applications that demonstrate mathematical concepts), making decisions and discoveries along the way. The second focuses on writing-to-learn assignments,[2] encouraging students to reflect and engage with mathematical ideas at many levels. In my third form of curriculum work, I have implemented flipped or blended learning pedagogy, creating collaborative classroom activities supported by video lessons.

13.2 Inspiration from Montessori Mathematics

13.2.1 Principles of Montessori Education

Since Montessori principles have had such a strong influence on the ways I think about teaching and learning, I will outline some key Montessori ideas. While Montessori education is designed for children ages birth through 18, I have found that some Montessori principles translate to the university setting. At the elementary level, Montessori education is characterized by its distinctive classroom environment, teacher role and cognitive goals.

[1] I have created a GeoGebra Book for this chapter: https://www.geogebra.org/book/title/id/RdxKWn2R?doneurl=https%3A%2F%2Fwww.geogebra.org%2Fmaterial%2Fedit%2Fid%2FRdxKWn2R#

[2] Sample writing assignments are available at https://www.uwrf.edu/MATH/SampleMathematicsActivities.cfm

The classroom space is a cross between a cozy living room and a science laboratory. There are open spaces where children spread out their work on rugs, small tables that seat between one and four children, and low shelves where children retrieve beautiful materials that they use in discovery style learning activities. The furniture is arranged to create attractive spaces for each part of the curriculum. Children spend three years in a single room, normally with the same instructor. This enables younger children to benefit from the influence of older children while older children gain leadership experience.

The role of the Montessori teacher is to prepare and organize the learning environment, to provide brief lessons on how to complete learning activities, and above all, to skillfully observe children. Based on these observations, the teacher chooses lessons that capture the child's attention and help each child to make progress at a pace that is appropriate for that child. Using carefully designed hands-on materials, the teacher gives lessons to small groups of children. The children then work autonomously, responsible to practice with the materials over time until mastery is achieved. The teacher serves as a critical link between the child and the prepared learning environment, facilitating the child's construction of his or her own understanding.

Even the cognitive goals in Montessori education are distinctive. They include helping children become self-disciplined, caring, independent, self-motivated, comfortable with error, and able to focus for extended periods. These goals are less tangible than the usual academic content goals and very difficult to measure, especially in a public school setting. Yet giving greater emphasis to these goals often results in higher levels of academic success (Dohrman et al. 2007; Lillard and Else-Quest 2006). One way these goals are attained is through a three-hour uninterrupted work cycle in which children are free to choose what to work on and how much time to devote to it. This promotes problem-solving and concentration by encouraging children to choose challenging work, knowing they will have plenty of time to complete it.

13.2.2 Montessori Principles and College Mathematics

How can these ideas about educating children find relevance in college-level mathematics instruction? While many Montessori practices are specific to the education of children, some of the principles behind the practices are applicable at the college level. Montessori recognizes an important connection between *movement* and *cognition*. Materials are designed to be *self-correcting*. In a college classroom, I have found that hands-on activities using dynamic mathematics software provides students a way to check their work by examining multiple representations. Montessori values *choice* and requires children to *create their own mathematics exercises*. I design mathematics curriculum materials that give students some choices about what mathematical objects to work with and require them to create some of their own exercises. By removing competition and grading systems, Montessori also promotes within the child an *intrinsic motivation* to learn. Inspiring college mathematics students to learn simply because the ideas are so beautiful, important and engaging is one of my greatest challenges. Perhaps the greatest gift of Montessori principles for college level mathematics instruction is its view on the *progression from concrete to abstract.*

As a mathematician, I think about concrete understandings versus abstract understandings in two ways. First, there is the idea of using specific concrete examples to motivate a general abstract principle. We can notice that $23+45$ is even, that $237+841$ is even, and eventually conjecture and then prove, that the sum of any two odd numbers is even. Using pattern recognition to generalize provides one way to progress from the concrete to the abstract.

My second thought about concrete versus abstract relates to the idea of underlying mathematical structure. Mathematicians observe the salient properties of a mathematical object and then generalize to a more abstract version of that object. For example, we notice that Euclidean distance between two points in the Cartesian plane is non-negative, symmetric, zero only when the points are identical, and satisfies a triangle inequality. Based on this observation, we define an abstract metric to be a real-valued function that has these same four properties.

Montessori adds to these understandings of concrete versus abstract in two important ways. First, in the *progression* from concrete to abstract, there are intermediate steps. Montessori mathematics manipulatives are used to guide children gradually from concrete to abstract understanding through a series of small abstractions. Tactile work is associated with the concrete end of the spectrum while purely mental work is on the abstract end. Second, we can teach and learn *a single mathematical concept or process along this progression*, scaffolding student understanding. The mathematics itself is not necessarily getting more abstract, rather the way the student comprehends the mathematics gets progressively more abstract.

In Fig. 13.1 the children are learning to think about place value with a number they chose themselves: 7777. At the most concrete level, they represent 7777 with the "golden beads" (base ten blocks); each golden bead represents one, a bar of ten beads represents 10, a flat of one hundred beads represents 100, a cube of one thousand beads represents 1000. In this representation 7777 is a very tactile concept; there are 7777 beads to touch. Children take the next step in the progression to abstraction with the "stamp game," color-coded tiles with values 1, 10, 100 or 1000 imprinted on them. The stamp game is a more abstract representation of place value than the golden beads because color and numeral, rather than size show the distinction. In another step towards abstraction, the children represent the abstract numeral using color-coded cards whose colors align with the stamp game. The cards with 7000, 700, 70 and 7 are stacked to make 7777. The next material in the progression to abstraction is the "small bead frame," an abacus consisting of four wires each with ten color-coded beads, according to hierarchy. This material is at the abstract end of the spectrum because the numerals imprinted on the stamps are gone and the restriction of ten beads requires the child to do any exchanging between place values immediately. When children work with number operations there are other materials

Fig. 13.1 Progression from concrete to abstract

(not shown in Fig. 13.1) that help students make the transition from concrete understanding of algorithms to abstract paper-and-pencil computations. In all of these representations the child is learning the same concept of place value. The mathematics itself isn't getting any more abstract. However, the child's concept of place value makes a gradual progression on a spectrum from concrete to abstract.

13.3 GeoGebra as a Tool to Improve Conceptual Understanding

GeoGebra, dynamic mathematics open source software, serves as a tool to create Montessori-style activities (think: hands-on, self-correcting) that help students gain abstract understanding through concrete work. Part of the power of GeoGebra is that information may be entered in any one of three ways: symbolically (in the Algebra View using the Input Box), visually (using tools in the Graphics View) or numerically (in the Spreadsheet View). GeoGebra automatically provides the other representations of that same information, cleverly color-coding matching objects in the different representations. Another key aspect is that GeoGebra is dynamic. Once dependent objects have been constructed the user can change one part and the rest of the objects change in a corresponding manner. All of these aspects of GeoGebra work together to provide an experience for students that is hands-on and self-correcting.

13.3.1 Classroom Demonstrations

I began using GeoGebra with classroom demonstrations that I hoped would help students understand the ideas behind the mathematics we were exploring. The GeoGebra software is used to create GeoGebra applets. While there are many such applets available online (GeoGebraTube 2011) for this purpose, I found that writing my own applets gave me better intuition about the power of GeoGebra to support student learning.

One classroom demonstration supports student solutions of an optimization exercise in first-semester calculus. In the exercise students are asked to maximize the area of a rectangle that has its base on the x-axis and its other two vertices above the x-axis and lying on the parabola $y = 8 - x^2$ (Stewart 2008). As with many optimization exercises, the greatest challenge for students is creating the objective function and its domain. Figure13.2 shows one visual result from a GeoGebra applet designed to help students create the objective function, find the domain of the objective function and make sense of their final answer. The applet helps students visualize a concrete sample rectangle. This aids them in understanding the more abstract general rectangle, guiding them next to the discovery that an appropriate expression for the width of the rectangle is $2x$, and then to an appropriate objective function: $A(x) = 2x(8 - x^2)$. As the instructor experiments with the dynamic point (x, y), using the mouse to drag it up and down the parabola, students can discover that an appropriate domain for the objective function is approximately $0 \le x \le 2.8$. The applet also supports the students'

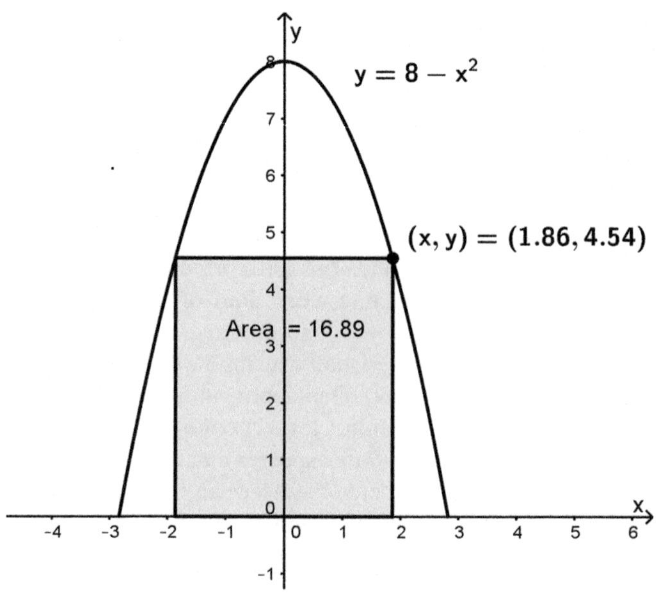

Fig. 13.2 Optimization exercise with GeoGebra

understanding that the upper endpoint of the domain can be found at an x-intercept of the function $y = 8 - x^2$, guiding them to the precise domain, $0 \le x \le \sqrt{8}$. After students have found the critical number of their objective function and tested that their critical number truly maximizes the function, they can make sense of their answer when the instructor experiments with the dynamic point (x, y), observing that the maximum area of the shaded region is approximately 17.4.

This classroom demonstration helps students begin to see how they can create their own objective functions and make sense of their final answers. The dynamic nature of GeoGebra plays an important role in helping students progress from understanding how to think about optimization problems both concretely (a specific rectangle) and more abstractly (a general rectangle).

13.3.2 Student Activities with Instructor-Created Applets

As I created more classroom demonstrations (Tomlinson 2014) and used applets created by others, I realized that it was important for students themselves to interact with the applets. If the research that validates the Montessori principle of movement and cognition (Lillard 2005) is applicable to college students, then students need to get their own hands on the applets. Hence I began to develop *student activities with instructor-created applets* that students access through a learning management system and use to complete exercises both inside and outside of class.

For example, my differential equations students complete exercises outside of class using an applet I developed that illustrates the Euler Approximation Method for first order initial value problems of the form $\frac{dy}{dx} = f(x, y), y(x_0) = y_0$ (see Fig. 13.3). Students complete exercises in which they experiment with different initial value problems, different step sizes and different numbers of steps to get approximate solutions. Their exploration helps them see the connection between the step size and the number of steps. They can also view either an analytic (in the case of an algebraic $f(x, y)$) or a numerical solution, which allows them to make connections between their approximate solution and a more precise solution.

It is fairly easy for students to simply memorize the formulas for Euler's Method: $x_{n+1} = x_n + h, y_{n+1} = y_n + hf(x_n, y_n)$, work exercises from a textbook and correctly solve similar exercises on a test without having even a small clue of what they are doing and what it means. By having students explore Euler's Method exercises with the GeoGebra applet, they begin to make sense of the "$f(x_n, y_n)$" in the formulas and understand that it represents the slope of a line segment. One of the reasons that a unit on Euler's Method is included in a differential equations course is to emphasize that, while many symbolic approaches for solving differential equations are studied, it is important to be able to think about differential equations and their solutions graphically and numerically as well. Thinking in all three modes gives students better intuition about what it means to solve a differential equation. The Euler's Method applet demonstrates these multiple representations very clearly with side-by-side views demonstrating symbolic, visual and numerical representations.

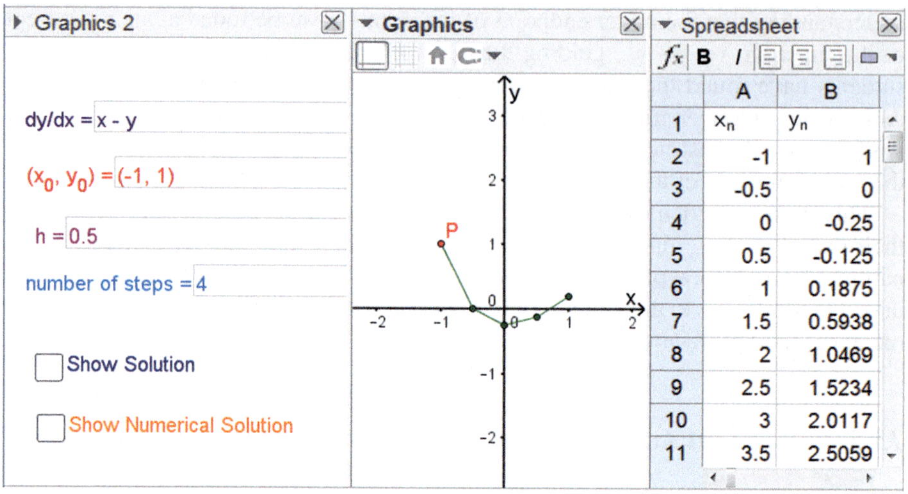

Fig. 13.3 Euler approximation method with GeoGebra

13.3.3 Student-Created Applets Following Detailed Instructions

The third way I use GeoGebra to support student learning is by having *students create their own applets*, outside of class, by following podcast instructions. I provide podcasts instead of live instructions because students can make their applets much more efficiently if they have the ability to pause and re-start my instructions. Once they have created their applets, they bring them to class for exploration to learn mathematics.

Figure 13.4 shows visual output from a student-made applet designed to construct the limit definition of the derivative. In GeoGebra, students graph a function, create a slider, use their slider to draw a dynamic secant line $\overline{PQ}$, and compute the slope of their secant line. Several gains result from having students create the applet themselves. Creating the applet gives students a more concrete understanding of how the coordinates of the points P and Q arise and where the formula for the *slope* of the secant line comes from. It also helps them understand that for a given point P there is a progression of secant lines. So creating an applet helps students gain a deeper understanding of the mathematics. In addition, creating their own applet and assigning colors of their choice gives students a sense of ownership of the knowledge.

In class, students use their sliders to explore the connection between slope of secant line and slope of tangent line. They begin to develop the limit definition of the derivative in a very concrete, visual way. Students explore the results of moving the slider that controls the values of h towards values that are close to 0. This is a tactile version of the abstraction $\lim_{h \to 0}$. By working with the slider students make connections between the slope of a secant line, average rate of change, slope of a

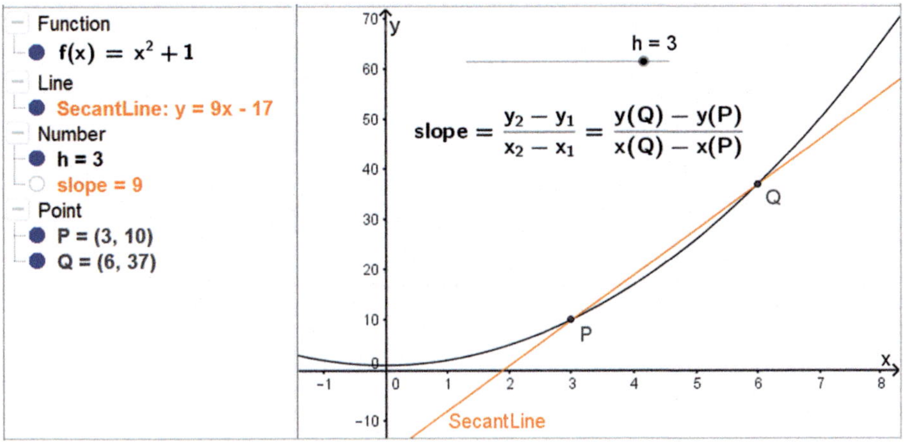

Fig. 13.4 Rate of change with GeoGebra

tangent line and instantaneous rate of change. They come to understand, at a very concrete level, that

$$\text{"slope of tangent line"} = \lim_{h \to 0} \text{"slope of secant line."}$$

The dynamic aspect of GeoGebra next allows students to easily explore the limit definition of the derivative for a variety of functions and points on the functions. Through this exploration students progress from the more concrete formula

$$\text{"slope of secant line"} = \frac{y(Q) - y(P)}{x(Q) - x(P)}$$

to the more abstract formula

$$\text{"slope of secant line"} = \frac{f(x+h) - f(x)}{h}.$$

GeoGebra experiences guide students to combine their rich ideas about the slope of the secant line with their rich ideas about limits to develop the limit definition of the derivative.

13.3.4 Independently-Generated Student Applets

The last type of GeoGebra module I have developed is one for students in more advanced classes to *generate applets independent* of detailed instructions. An example of this type of module is the study of images of lines and circles under

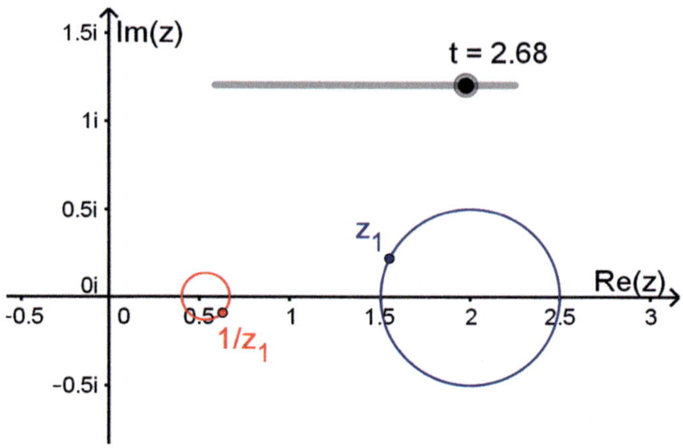

Fig. 13.5 Complex image of circle $|z-2|=\frac{1}{2}$ under inversion with GeoGebra

complex mappings in a complex variables course. Working with students who are adept at parameterizing lines and circles, I demonstrated how to use GeoGebra to make a conjecture about the image of a line or a circle under the complex mapping $f(z)=z^2$ I followed this by showing the students how to prove those conjectures.

Their out-of-class assignment was to explore the images of circles and lines under the inversion mapping: $f(z)=1/z$ by producing their own applets. Students create an appropriate slider to use as a parameter (see Fig. 13.5) and then use this slider to create a complex number, on a circle or a line (in Fig. 13.5, $z_1=2+0.5e^{it}$). Students use the Trace feature in GeoGebra and the slider to create a line or a circle in the complex plane. Next they define the image of the point z_1. in the GeoGebra Input Box, using the function: $z_2=1/z_1$ Using the Trace feature for the point z_2. students make a conjecture about the image of their line or circle. (In Fig. 13.5 this image is the circle centered at $\frac{8}{15}+0i$ with radius $\frac{2}{15}$) In the assignment students find images of lines and circles under inversion through this progression: (1) a line through the origin of their own choice; (2) circle centered at the origin of their own choice; (3) a particular line that doesn't go through the origin; (4) a particular circle not centered at the origin. By experimenting with these lines and circles students conjecture that the image under inversion of any line or circle is another line or a circle. Working with specific concrete lines and circles, students generalize what happens to any line or circle under inversion.

13.4 Writing-to-Learn Mathematics

When I was an undergraduate majoring in mathematics, my very kind physics instructor attempted to engage me in some casual conversation by asking what we were studying in my advanced calculus class. I was flummoxed. The only answer I could seem to provide was something along the lines of "Section 2.5 exercises 3, 11 and 17." I knew there were some epsilons and deltas involved. I could complete those exercises completely to the satisfaction of my advanced calculus instructor. Yet, there was no way I could give a reasonable response like, "We are learning how to think about continuity in a rigorous, symbolic way. This sharpens the notion of 'arbitrarily close' that we used for limits in calculus class and opens the door to proving theorems about continuous functions." I was woefully inarticulate about what I was learning. Furthermore, I didn't even know what I could have been doing that would help me gain this ability to articulate the mathematics.

As a professor, one lesson I could have taken from this conversation is: "Don't try to make pleasant conversation with your students." Joking aside, for me the real lesson is that if I want my students to learn in a robust way that helps them retain knowledge, I need to find ways to encourage them to articulate what they are learning. By developing writing-to-learn mathematics materials, I help students do this. I am also motivated because these activities have the potential to help students spend more productive time on task. Part of the beauty of writing-to-learn activities is that they can simultaneously help weaker students succeed and provide stronger students with a challenge (Meier and Rishel 1998; Sterret 1992).

The writing-to-learn activities that I use require students to reflect on the mathematics they are learning in ways that facilitate construction of their own understandings. Some writing-to-learn assignments encourage students to verbalize their ideas in dialogue with one another and capture that dialogue on paper. Other assignments involve prompts for inner dialogue resulting in deeper mathematics comprehension. These writing activities help scaffold understanding in the same way that Montessori mathematics materials do for children. While the child in a Montessori classroom is progressing from tactile work (concrete) to mental work (abstract), the college student is progressing from working practice exercises (concrete) to constructing mathematical insights (abstract).

Writing-to-learn assignments differ greatly from proof writing that I explicitly teach in some courses (linear algebra, discrete mathematics, etc.) and from report writing done at the culmination of a semester-long research project in other courses (mathematical modeling, senior capstone, etc.) Descriptions of two types of writing-to-learn exercises I developed follow.

13.4.1 Cooperative Guided Reflection

The first involves projects that I call "cooperative guided reflection" (CGR). In these projects, students work in teams solving textbook exercises and then use a list of prompting questions to guide them in a reflection process. Because CGR is time-consuming, typically students will complete only two CGR projects in a semester. Thus, I choose topics for CGR carefully. A CGR topic should be challenging for students, help students synthesize several ideas, and involve either problem solving or strategizing. Here I will describe how CGR has worked for teaching integration strategies in a second-semester calculus course.

After introducing students to a variety of integration techniques (integration by parts, substitution, etc.) we have a classroom discussion about strategies for deciding which techniques to use. Previously, I would assign about 20 exercises for students to practice with and be done with the topic. To employ CGR, I still assign 20 exercises, but I ask them to complete additional activities in assigned groups of two to four students.

The CGR activities begin with teams of students selecting eight of their 20 integral exercises and creating two integral exercises of their own according to definite guidelines. For example, they must make sure that, broadly speaking, their ten integrals show all of the integration techniques we have studied. They must make sure that they have an integral whose solution requires more than one technique. I provide significant support as students create their own integrals. I give them suggestions that include thinking about the inverse relationship between differentiation and integration, deciding on their technique before they create the integral, and making a variation on an integral from their textbook. They are also expected to check their answers to the integrals they create using technology such as GeoGebra or WolframAlpha®.

The next part of the CGR activities is to reflect on and analyze their ten integrals. They complete a grid in which I list the techniques and they supply a corresponding integral with some verbal explanation to help classify their work. In the last part of the guided reflection, they respond to three prompts asking them to reflect on one integral, one technique and one strategy. The prompts include questions about what they find interesting, how making mistakes helps them learn, and how their decision process works. The writing may be considered informal as students are exploring the way they think about the mathematics in addition to analyzing the mathematics itself. Teamwork promotes student dialog that informs the written reflections.

From the point of view of Bloom's Taxonomy and Webb's Depth of Knowledge (DOK), CGR activities provide higher cognitive demand to students than simply working integration exercises (Hess et al. 2009; Webb 2006). With enough practice, a single integration technique requires low cognitive demand, not much more than recalling and organizing (DOK levels one and two, respectively). Strategizing about which technique to use and using multiple techniques to complete an integral requires higher cognition, as students learn to make decisions and revisions in their integration techniques (DOK level three). CGR activities engage students in the highest cognitive demand, because they extend their thinking, by creating their own integrals and analyzing their integral exercises (DOK level four).

13.4.2 Journaling

The other type of writing-to-learn exercise I have implemented is journaling in the introductory differential equations class. Although the value of mathematics journaling has been written about extensively (Meier and Rishel 1998; Sterret 1992), in my experience it is not a common practice at the college level. Undoubtedly, this is because it can be time-consuming for students and instructors alike. So rather than explain journaling in detail, I will describe how I made this activity manageable both for myself and my students.

In my class, students submit ten journal entries with homework sets. I provide overall guidelines along with some general writing ideas for them. General writing ideas include open-ended instructions to summarize a section of the textbook and to connect differential equations with other disciplines. In the guidelines, I describe the purpose of their journaling: to learn by exploring, organizing and synthesizing mathematical ideas. In addition to such general writing ideas, with each homework set I provide two or three content-specific prompts from which students can choose (Farlow 1994). Although some instructors have used mathematics journals to explore the affective realm, I emphasize that their journal is not a place to discuss course mechanics or exam results, that students are expected to confer with me about such concerns.

Scoring for the journals is based on complete, thoughtful entries. Mostly students get full credit, saving instructor time. I take time to provide positive feedback that emphasizes students' best ideas. I also provide corrections for misconceptions or mistakes. When it is clear that students have put thought into their entries, I do not deduct points for such errors.

Both CGR and journaling reveal student thinking and confusion that can become prompts for classroom discourse. The first time I used CGR with the Integration Strategy topic, I had no idea that students conflated integration *technique* with integration *strategy*. Groups of students wrote that *technique* and *strategy* were two words for the same thing. Our subsequent in-class conversation helped them begin to distinguish between when they were using a technique and when they were making a decision about what technique to use next. I often share some of the best student writing with the entire class by displaying it on a document camera. This provides students with models for how to reflect on mathematics and how to articulate their thoughts. One of the most illuminating student journal entries stated, "I can do all of the assigned exercises, but I don't really understand it well enough to journal about it yet." In the absence of the journaling activity, students may equate working routine homework exercises with truly understanding mathematics. This student realized that he needed to do some more thinking, reflecting, or talking to make complete sense of what we were learning.

13.5 Flipped or Blended Learning

13.5.1 Principles and Goals of Blended Learning

Some of my most recent curriculum work involves implementing *flipped* or *blended* learning, as a way to improve *interactive engagement* during class time. *Interactive engagement* teaching methods involve activities that yield immediate feedback through discussion with peers or instructors (Epstein 2013). Primarily, the term *flipped* refers to a flip between learning from lectures in class and then practicing at home to learning from lectures at home and practicing in class with instructor support. In my classes, this means using technology (video lessons) so that students see ideas at home and then work on exercises in groups in class. A second interpretation of the term *flipped* is that it is a flip between the classroom activities being centered on the teacher to being centered on the student activity. The term *blended* means that there is a mix between a flipped classroom and the more traditional approach. For me, blended learning always involves some practice exercises at home.

There are many pointers guiding us in the direction of interactive engagement teaching methods, from Montessori's focus on student-centered learning to work by science colleagues to incorporate more active learning in their classrooms (Freeman et al. 2014). Research on the Calculus Concept Inventory (CCI) is especially compelling. The CCI is a way to measure students' conceptual (but not necessarily procedural) understanding of calculus. Researchers found that students in US colleges did very poorly on CCI and that none of the following had an effect on CCI score: class size, instructor experience, time spent in class, student preparation at entrance. However, *interactive engagement* teaching methods did improve student performance on CCI (Epstein 2013). A national study of calculus instruction also points toward the efficacy of active learning (Bressoud et al. 2015).

Thinking about using flipped learning to help students who were not succeeding in my classes led me to Bergman and Sams' (2012) delineation of three types of students who do poorly in school. There are students whose time is over-extended; many of my students work more than 20 h per week, while taking a full load of challenging coursework. There are students who have an insufficient background. (This is my personal favorite excuse for students doing poorly in my classes, since it takes the onus off of me.) Helping students who need to fill in missing gaps is a significant part of my teaching. The third kind of student Bergman and Sams identified as "playing school." These students come to class, but don't want to learn, aren't trying to learn, and are instead really just trying to figure out how to get a certain grade, by doing the least amount of work possible. It never occurs to them that learning is in their own best interest. Bergmann and Sams argued that they are able to reach all three types of students through flipped learning. As I design materials to implement flipped learning, I keep these three kinds of students in mind along with the students who have great success in more traditional college learning environments.

When I began to implement flipped learning I had already been creating online video content (accessible through a learning management system), in the form of annotated notes, to support my students as they worked on homework exercises. Students appreciated hearing my voice helping them work an assigned exercise, emphasizing the ways I wanted them to think about various aspects of the work. My strongest students used them occasionally; I could see real gains for students who were underprepared. But students who were very busy or who were "playing school" were not benefiting.

13.5.2 Blended Learning in Calculus I

With flipped or blended learning, some of the lessons are provided to students before class. This frees up class time to support students as they interact with the material, helping diverse learners. This is how the flip worked for me in a first-semester calculus unit on areas. Students were assigned four 5-min podcasts to watch before class. In the video lessons, I explained the general ideas and demonstrated two examples. Students were expected to take notes and be prepared to show them to me at the beginning of class.

When students came to class I distributed a packet of exercises consisting of the same examples I had worked on the chalkboard when teaching this unit in a more traditional format. Students worked on the packet with each other using their notes from the podcasts. Students who had not watched the podcasts (or not taken notes) moved next to a classmate with notes. This arrangement worked well because most students had notes (knowing that there were points associated with them) but those who didn't still had a way to be fully engaged. I began to circulate around the room, checking podcast notes and talking to students. Every few minutes I took a break to write down some solutions, projected onto the document camera. This gave students a way to check their work and also kept them on task. A few times, I paused the class work briefly to direct a whole group discussion, addressing an idea that had arisen. As I circulated through the room, I answered individual questions about both the video lessons and the packet of exercises, many of which I would have been unlikely to hear in my more traditional format. I had a personal interaction with each one of my 32 students that day.

There was one student in my class who I knew led a very busy life and who had also missed some class days because of illness. When I checked in with her during class that day, she said, "I'm doing fine with areas, but I am still having problems with integration by substitution," the topic we had been working with the previous week. Normally, I might have asked her to come to my office hours (which was unlikely to actually happen because of her schedule). In the flipped format, I could see that everyone was on task with areas, so I had time to address her questions immediately.

For me, flipped or blended learning is a way to lower the number of students who do poorly in my classes, while maintaining high learning expectations. It is a way to help all students become engaged with and committed to mathematics. There is one major drawback: the amount of instructor time required to prepare video lessons and in-class activities. I did not use materials that others have posted on the internet, because I believe I can create a video lesson in less time than I can find a suitable version online and because of recommendations about students' need to connect with their instructor (Moore et al. 2014). With flipped or blended learning, I have more meaningful interaction with my students, the most pleasurable part of teaching.

13.6 Career Trajectory

My career has been fueled by passions for mathematics and for social justice. I started as a researcher in partial differential equations, investigating questions about the heat equation with space variables that are complex. While I have had a lifelong interest in political, social and economic equality for all, I didn't originally see this as part of my career. However, when my campus was looking for someone to lead our Women's Studies Program, I saw a way to realize my passion for social justice and I took the opportunity to take my career on a brief excursion. As Director of Women's Studies, I taught women's studies courses and coordinated women's studies programming with faculty from a wide array of disciplines. This work led me to the epiphany that mathematics education is a social justice issue. By creating high-quality mathematics classrooms that spark curiosity and foster long-term interest in mathematics, we are helping to create equal access to our economy (Halpern et al. 2007).

The next detour in my career path was motivated by the birth of my children, leading me to a study of Maria Montessori's idea that the most effective education is supported by materials and activities that are hands-on, self-directed, self-correcting and self-chosen. I became an advocate for Montessori education, presenting to community groups and serving on school committees. Through grant-funded work with the College of Education on my campus and volunteering in local schools, I work to bring Montessori mathematics into mainstream classrooms. Eventually, I began using Montessori's ideas in my own college mathematics classrooms.

The opportunity to do mathematics education research presented itself when I participated in a regional scholarship of teaching and learning (SoTL) in mathematics workshop. I completed a project addressing the question of how a cooperative guided reflection (CGR) activity in first-semester calculus improved problem-solving skills by doing a literature review, a quasi-experiment, student surveys about problem-solving, and an analysis of student work. I found that there was a positive impact, qualitatively, on students' mathematical belief systems, as well as quantitatively, on students' ability to solve optimization problems (Tomlinson 2008). This gave me impetus to continue experimenting with CGR.

Another important aspect of my career has been coaching 19 successful mathematical modeling teams. It is a joy to help these students develop skills in mathematics, internet research, teamwork, mathematical technology, and technical writing. Working closely with these students informs my thinking about how to create instructional materials for students at their level.

When GeoGebra, became available, it was a perfect fit, providing another way to implement Montessori principles. The fact that GeoGebra is open source was especially attractive to me because it means that everyone has access. As we are learning more about the importance of interactive engagement classrooms and flipped learning, I am creating curriculum materials to implement these pedagogies as well.

Regular exchange of ideas with colleagues has been key to my success and my continued energy for creating instructional materials. This takes many forms: informal comparison of topic treatment with colleagues in my department who teach the same courses I am teaching, formal discussions with faculty in science departments about how mathematics courses support their work, participation in grant-funded work with colleagues in Teacher Education, and discussions with colleagues outside of my campus at conferences.

My work has been well received on my campus. Any work that improves student engagement usually results in improved retention and recruitment—priorities at most colleges and universities. I have given faculty development presentations on GeoGebra for adjunct and regular faculty, and served as the contact person for people with technology and pedagogy questions about this software. This has been appreciated by faculty and administrators alike. I have accepted invitations to lead GeoGebra workshops for faculty at a local high school and a regional two-year college. Survey responses of students on my work with flipped pedagogy are very positive.

My career has gradually shifted from esoteric, but definitive, questions about partial differential equations to broad-reaching, but nebulous questions about better ways to teach mathematics. It is a comfort to start with a mathematical premise, logically arrive at a conclusion and know that this work is entirely repeatable. On the other hand, it is exciting to create materials that help at least some students become committed and engaged in mathematics, even if we cannot always be certain that the same materials will work for a different instructor or a different set of students.

In the next phase of my career, I am learning how to provide leadership to an academic department of nine tenure-track faculty and 11 other instructional staff members as I begin to serve as chair. While I continue my work developing instructional materials, I am taking on a greater role promoting high-impact mathematics education practices in my department. I believe that this focus will, over time, strengthen my department by making our graduates more employable and attracting more students to our programs. I have started this emphasis by using our curriculum review process as a way to share research results from mathematics education among my faculty. I am learning that through internal grants and personnel processes, I have new opportunities to encourage my faculty to pursue work that improves mathematics education in all of our classrooms.

Acknowledgements I would like to express my gratitude to Gay Ward and Melina Papadimitriou for their guidance in Montessori education. I appreciate the assistance of all the editors for this volume, especially "Editor D," who provided many detailed recommendations. I am grateful for the support of my colleagues at UW-River Falls and in the UW-System. Above all, a very special thank-you to my wonderful colleague Laurel Langford, who regularly supplies me with insightful observations and who provided tireless editing for this chapter.

References

Bergmann, J., & Sams, A. (2012). *Flip your classroom: Reach every student in every class every day*. Eugene, OR: International Society for Technology in Education.

Bressoud, D. M., Mesa, V., & Rasmussen, C. (2015). *Insights and recommendations from the MAA national study of college calculus*. Washington DC: MAA.

Dohrman, K. R., Nishida, T. K., Gartner, A., Lipsky, D. K., & Grimm, K. J. (2007). High school outcomes for students in a public Montessori program. *Journal of Research in Childhood Education, 22*, 205–217. doi:10.1080/02568540709594622.

Epstein, J. (2013). The calculus concept inventory—measurement of the effect of teaching methodology in mathematics. *Notices of the American Mathematical Society, 60*(8), 1018–1026. doi:10.1090/noti1033.

Farlow, S. J. (1994). *An introduction to differential equations and their applications*. New York: McGraw-Hill.

Freeman, S., Eddy, S. L., McDonough, M., Smith, M. K., Okoroafor, N., Jordt, H., et al. (2014). Active learning increases student performance in science, engineering, and mathematics. *Proceedings of the National Academy of Sciences of the United States of America, 111*(23), 8410–8415. doi:10.1073/pnas.1319030111.

GeoGebraTube (2011). Retrieved May 24, 2016, from https://www.geogebra.org/materials/.

Halpern, D. F., Aronson, J., Reimer, N., Simpkins, S., Star, J. R., & Wentzel, K. (2007). *Encouraging girls in math and science*. National Center for Education Research, Institute of Education Sciences, U.S. Department of Education. Retrieved February 17, 2016, from http://ies.ed.gov/ncee/wwc/PracticeGuide/5.

Hess, K., Jones, B., Carlock, D., & Walkup, J. (2009). *Cognitive rigor: Blending the strengths of Bloom's taxonomy and Webb's depth of knowledge to enhance classroom-level processes*. Online Submission.

Lillard, A. (2005). *Montessori: the science behind the genius*. New York: Oxford University Press.

Lillard, A., & Else-Quest, N. (2006). Evaluating Montessori education. *Science, 313*, 1893–1894. doi:10.1126/science.1132362.

Meier, J., & Rishel, T. (1998). *Writing in the teaching and learning of mathematics*. Washington, DC: MAA.

Moore, A. J., Gillett, M. R., & Steele, M. D. (2014). Fostering student engagement with the flip. *The Mathematics Teacher, 107*(6), 420–425. doi:10.5951/mathteacher.107.6.0420.

Sterret, A. (Ed.). (1992). *Using writing to teach mathematics*. Washington, DC: MAA.

Stewart, J. (2008). *Calculus*. Belmont, CA: Thomson Brooks/Cole.

Tomlinson, K. (2008). *The impact of cooperative guided reflection on student learning: The case of optimization problem solving. in calculus I* (Proceedings of the Eleventh SIGMAA Conference on Research in Undergraduate Mathematics Education). http://sigmaa.maa.org/rume/crume2008/Proceedings/Tomlinson%20PRE%20LONG.pdf.

Tomlinson, K. (2014). Using GeoGebra to enhance conceptual understanding in the calculus sequence. In *Proceedings of the Twenty-sixth Annual International Conference on Technology in Collegiate Mathematics* (pp. 308–312). http://archives.math.utk.edu/ICTCM/VOL26/C029/paper.pdf.

Webb, N. (2006). *Mathematics DOK levels*. Retrieved March 11, 2016, from http://www.education.ne.gov/assessment/pdfs/Math_DOK.pdf.

Chapter 14
"The Wild Side of Math": Experimenting with Group Theory

Ellen J. Maycock

Abstract Group theory has traditionally been taught to mathematics majors using the "theorem-proof-example" format. Although this method of presentation is satisfying to a mathematician, many students have difficulty learning the concepts of group theory this way. More than 20 years ago, I transformed my abstract algebra classroom into an active learning environment by using the software package *Exploring Small Groups*. In this chapter, I describe my approach, illustrating it with a specific example, and discuss its impact on my students, my career and other mathematicians.

Keywords Group theory • Teaching with technology • Inquiry-based learning • Computer laboratories

14.1 Introduction

In the spring of 1990, I obtained the software package *Exploring Small Groups (ESG)* (Geissinger, 1989) to use for demonstration purposes in my abstract algebra course at DePauw University. One day, in the midst of trying to explain the commutator subgroup to my class, I stopped and sent the students to the computer lab to try to generate some conjectures about commutator subgroups using *ESG*. The next day, they brought their conjectures to class. I picked three promising ones and asked my students to prove or disprove them for homework. One student raised her hand and said, "How can we do this? We don't know whether they are true or not!" I realized then that there was something terribly wrong—not with the student, but with

MSC Codes
20-01
97D40
97H40

E.J. Maycock (✉)
Bristol, RI, USA
e-mail: ellen.maycock@gmail.com

J. Dewar et al. (eds.), *Mathematics Education*, Association for Women in Mathematics Series 7, DOI 10.1007/978-3-319-44950-0_14

how I was teaching the material. After that experience, I changed how I structured the class: I incorporated a weekly computer lab so that students could construct some of the basic structures of group theory using *ESG*. The addition of a lab component changed the dynamics of the classroom, allowing students to participate more fully in the course.

This chapter details my experiences teaching group theory with a computer laboratory component. The materials that I developed were eventually compiled into a lab manual that I published with the Mathematical Association of America (Parker 1996). I explain the rationale for using such an approach and describe specific details of one lab session. I also discuss the success of this approach, share some comments made by students, assess the broader impact of the materials, and describe the reactions of colleagues in the mathematics community to teaching abstract algebra this way.

14.2 Software

A major factor for the success of my laboratory component was the software, *ESG*, which was designed as a tool to teach group theory. Key aspects of the software were that (1) it was exceptionally easy to use; (2) it had a significant visual component; and (3) the answers to queries, such as "what are the subgroups of this group" were not given immediately but had to be generated by the students. Other faculty members have been successful using a variety of software—ISETL© (Dubinsky and Leron 1994), *Mathematica*™ (Hibbard and Levasseur 1999), *Group Explorer*© (Carter and Emmons 2005), and GAP (Rainbolt and Gallian 2013), for example—to teach this course. However, I found *ESG,* and subsequently *Finite Group Behavior* (*FGB*) (Webb and Keppelmann, 2000), to be especially appropriate for beginning students[1].

14.3 Basic Approach

The idea of teaching abstract mathematical concepts by beginning with concrete material is not new. In February 1894, a professor at Dickinson College wrote in "Application of the New Education to Differential and Integral Calculus" in the *American Mathematical Monthly* (Durell 1894):

> The method proceeds from the concrete to the abstract. It gives the student at the outset something which he can see, make and count, and hence, develops his self activity. He proceeds to other ideas not mechanically or under dictation but for the sake of clear realized advantages.

[1]Although the software *ESG* is now obsolete, Bayard Webb, a student of Edward Keppelmann of the University of Nevada, Reno, wrote a Windows version of *ESG* entitled *Finite Group Behavior* (Webb and Keppelmann 2000). This software is still available and can be used effectively with my laboratory materials. My remarks about *ESG* also apply to *FGB* in this chapter.

A century later, I embraced the same approach in my abstract algebra class: I wanted students to investigate concrete examples and then move to the abstract theory. Of course, my ultimate goal was to introduce the basic concepts and constructions of group theory. With the laboratories, students could develop an intuitive sense of the concepts as they investigated examples generated by hand and with the software. Conjecture-posing was an excellent way for students to explore the ideas. To make conjectures, students needed some understanding of the nature of the mathematical objects under consideration. Developing conceptual understanding and formalizing concepts were done simultaneously in my abstract algebra course; technology was an aid in the process.

Each weekly computer lab was discovery-oriented and had three parts. "Before the Lab" contained instructions for some paper-and-pencil work so that the computer's output would not seem so mysterious. "In the Lab" directed students to work in pairs in the computer laboratory. One student typed in the *ESG* commands and the other recorded the data, discussing the questions together as they worked. The final step was for each student to individually write a lab report and answer the questions posed in "Further Work." These labs appear in detail in Parker (1996).

I now believe that the real learning takes place when students write up their lab reports. The process of carefully writing their results in complete sentences forces students to think in an organized way about what they observed in the computer laboratory. At the end of their lab reports, students were asked to make conjectures based on the patterns they perceived. Their conjectures sometimes anticipated material in the text, often with different phrasing and notation. In the classroom session that followed the computer lab, I made sure that the students understood the concepts that they had encountered. Students wrote formal proofs as well as the less formal, expository paragraphs in their lab reports. Often, in addition to the standard material for the course, we worked through proofs of their conjectures.

14.4 Example: A Laboratory on Quotient Groups

The concept of quotient or factor groups is a difficult idea for undergraduate students to grasp. How can one explain the idea of creating new groups (distinct from subgroups) from the original group? One of my favorite labs took advantage of *ESG*'s use of color to illustrate this concept.

The lab on quotient groups usually fell in the middle of the semester. By then, my students knew the structure of many small groups of orders 3–16 (see Appendix) and had constructed the subgroup lattices of these groups. They had also identified the center and commutator subgroup of each group.

In the classroom, we worked together on some of the questions found in "In the Lab." We worked out the example of the group D_3, or symmetries of an equilateral triangle, by hand before the lab. First, we named the elements of D_3 according to their geometrical actions: Rotations were named *r1* and *r2*; reflections (mirrors) were named *m1, m2* and *m3*. The identity was denoted by 1. The Cayley or group table for D_3 appears in Table 14.1 and the subgroup lattice for D_3 appears in Fig. 14.1.

Table 14.1 Cayley table for D_3

*	**1**	**r1**	**r2**	**m1**	**m2**	**m3**
1	1	r1	r2	m1	m2	m3
r1	r1	r2	1	m3	m1	m2
r2	r2	1	r1	m2	m3	m1
m1	m1	m2	m3	1	r1	r2
m2	m2	m3	m1	r2	1	r1
m3	m3	m1	m2	r1	r2	1

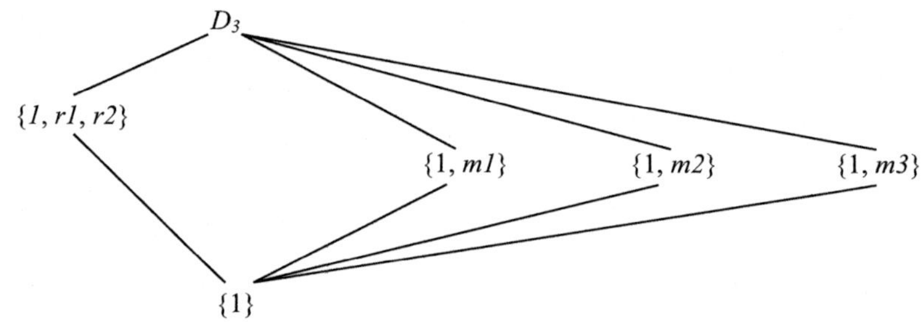

Fig. 14.1 Subgroup lattice for D_3

Table 14.2 Cayley table for the cosets of the subgroup {*1*, *r1*, *r2*} of D_3

*	**{1, r1, r2}**	**{m1, m2, m3}**
{1, r1, r2}	{1, r1, r2}	{m1, m2, m3}
{m1, m2, m3}	{m1, m2, m3}	{1, r1, r2}

Next, we created the right and left cosets of the subgroup of order three consisting of the rotations of the triangle, and observed that the set of left cosets is the same as the set of right cosets: {*1*, *r1*, *r2*} and {*m1*, *m2*, *m3*}. I then explained the coset operation used to multiply two cosets together, and created a Cayley table for the set of cosets, shown in Table 14.2.

Finally, we tried to do the same construction with the subgroup of order two generated by one of the reflections of the triangle, {*1*, *m1*}. The left cosets of {*1*, *m1*} are {*1*, *m1*}, {*r1*, *m3*}, and {*r2*, *m2*}. In this case, the operation on cosets is not well-defined. For example, {*r1*, *m3*} * {*r2*, *m2*} could be either the coset {*1*, *m1*} or the coset {*r2*, *m2*}, depending on the choice of the coset representative. So, of course, we were not able to construct the Cayley table for the set of left cosets of {*1*, *m1*}. Alternatively, I showed that the left and right cosets for this subgroup do not agree. But it was not easy for the students to understand the concept of a group created from cosets from this presentation, especially if a student was not comfortable with abstract symbolic notation.

Table 14.3 The left cosets of {1, E} in D_4 with assigned colors

Left coset	Color
{1, E}	*green*
{A, F}	*red*
{B, G}	*yellow*
{C, D}	*blue*

Table 14.4 The left cosets of {1, B} in D_4 with assigned colors

Left coset	Color
{1, B}	*green*
{A, C}	*red*
{D, F}	*yellow*
{E, G}	*blue*

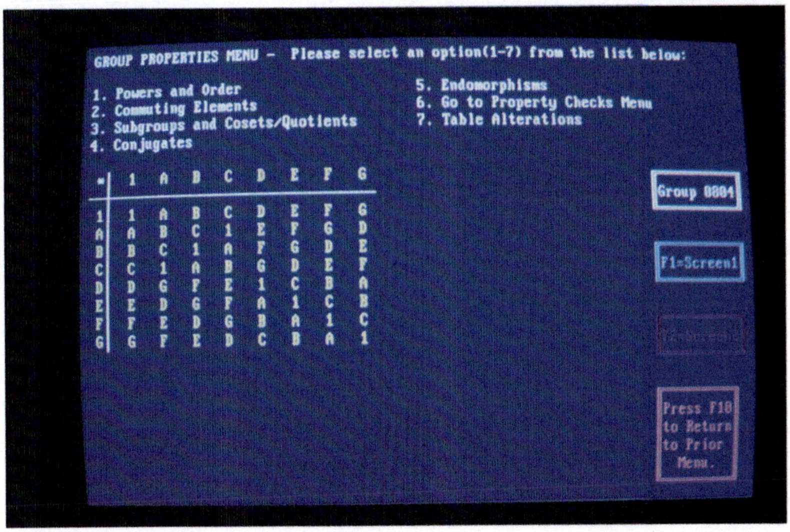

Fig. 14.2 Cayley table for D_4 (*ESG* image)

Due to the visual nature of *ESG*, all of this became much clearer in the computer lab. Together, the students and I used the software to work out the example of D_4, the symmetries of a square (group 0804 in the Group Library of *ESG*). In Tables 14.3 and 14.4 and Figs. 14.4, 14.5 and 14.6, element A is the 90° rotation of the square; B is the 180° rotation; C is the 270° rotation; elements D, E, F, and G are reflections of the square. Figure 14.2 shows the Cayley table of D_4 and Fig. 14.3 shows the subgroup lattice of D_4. The screen shots for this example (Figs. 14.2 and 14.4) have been obtained from *ESG*.

Using the computer, we constructed the left and right cosets for several subgroups. In *ESG*, for each subgroup, the elements along the top and the left-hand side of the Cayley table are rearranged by cosets, and the elements of each coset are identified by a color.

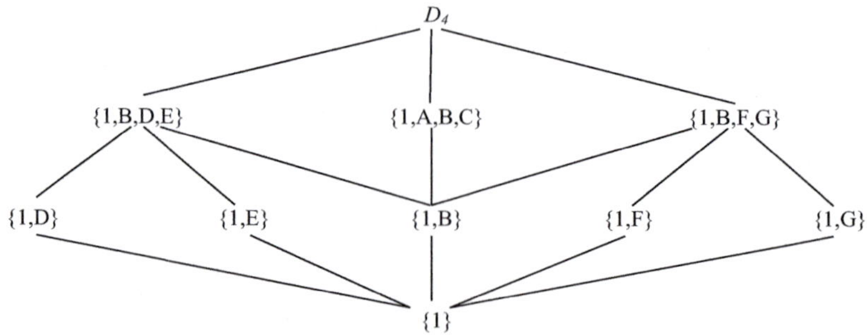

Fig. 14.3 Subgroup lattice for D_4

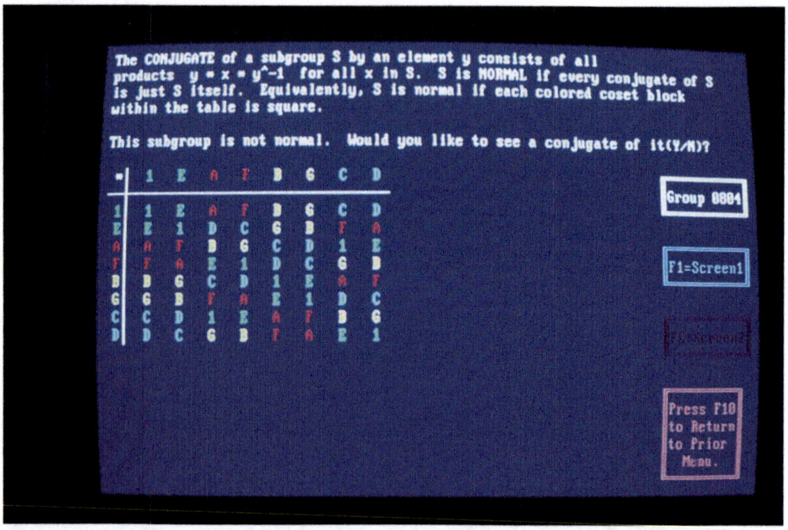

Fig. 14.4 Cayley table for the left cosets of the subgroup {1, E} of D_4 (*ESG* image)

Figure 14.4 shows the Cayley table of D_4 rearranged by the left cosets of the subgroup {1, E} and colored by the assignments in Table 14.3. The subgroup {1, E} is not normal. This can be seen, for example, by noting that *yellow* * *red*=*blue*, if we choose B to be the representative of *yellow*. But *yellow* * *red*=*red* if we choose G to be the representative of *yellow*. Thus, the coset operation is not well-defined. The non-square blocks of color in Fig. 14.4 illustrate this fact.

In the case of a normal subgroup, however, the coset operation is well-defined. We can see this in the group table for the subgroup {1, B} (the identity and the 180° rotation), shown in Fig. 14.5, with the coloring described in Table 14.4. The blocks of colored letters form 2×2 squares and show the normality of the subgroup {1, B}.

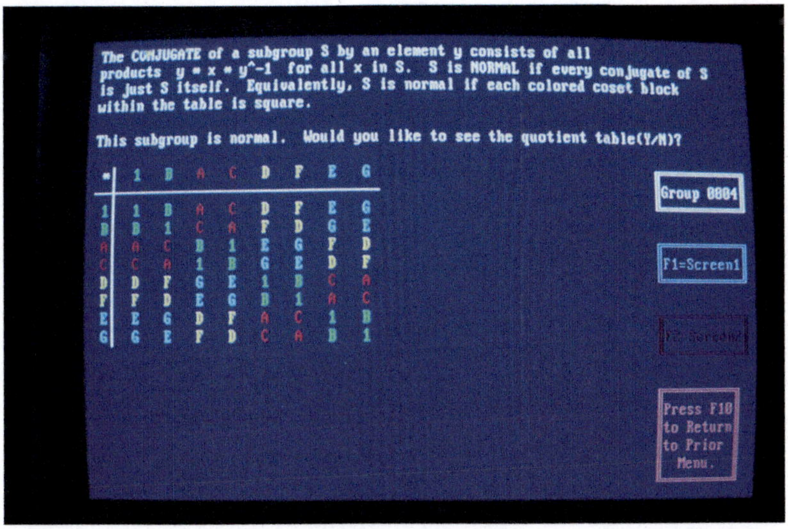

Fig. 14.5 Cayley table for the left cosets of the subgroup {1, B} of D_4 (*ESG* image)

When "Y" was typed in response to the question "Would you like to see the quotient table (Y/N)?" *ESG* displayed the colorful new group created from the cosets, as seen in Fig. 14.6.

By the time my students were working on this lab, they were familiar with the patterns of most of the groups of low order. So it was easy for them to recognize that Fig.14.6 gives the pattern of $Z_2 \times Z_2$, or V, the Klein Four-Group.

The "In the Lab" section then instructed the students to work through a number of different groups, identifying which subgroups of those groups were normal based on the evidence of the computer screen. For each normal subgroup, they were asked to examine the table for its cosets and identify the familiar group that has the same Cayley table. Frequently, student pairs split up the work so that they could generate a larger number of examples.

Finally, students had to write up the computer lab in a lab report individually. The report always included a summary of the data obtained from the computer lab and the responses to questions in the "Further Work" section of my lab manual. For the lab on quotient groups, students were directed to come up with several conjectures as follows:

1. Make at least two conjectures about the kinds of subgroups that always seem to be normal in a finite group.
2. Make at least two conjectures about the factor groups D_n/H, where D_n is the group of symmetries of a regular n-gon and H is either the commutator subgroup or center of D_n.

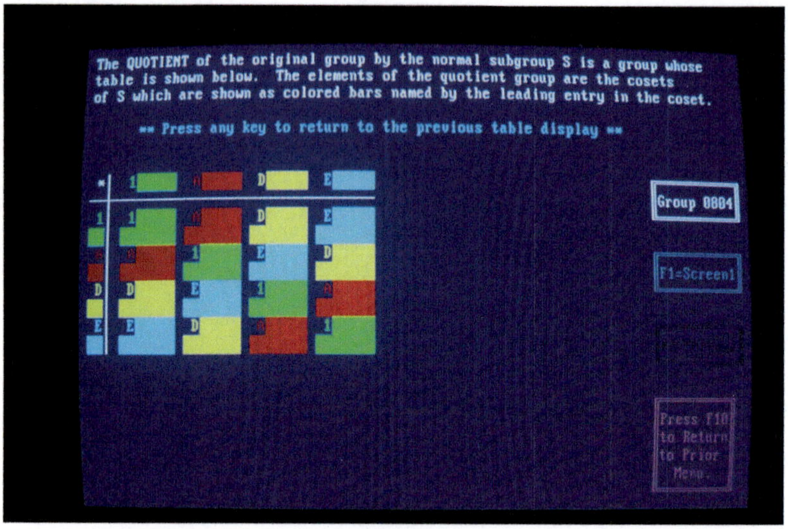

Fig. 14.6 The quotient group $D_4/\{1, \mathrm{B}\}$ (*ESG* image)

In response to these prompts, students in one of my classes generated these conjectures:

- Every subgroup H of a finite group G which has index 2 in G is a normal subgroup of G.
- The commutator subgroup of any group is normal.
- The center of any group is normal.
- Let n be odd and let H be the commutator subgroup of D_n. Then $D_n/H \approx Z_2$.
- Let n be even and let K be the commutator subgroup of D_n. Then $D_n/K \approx Z_2 \times Z_2$.
- Let n be even and let $C = \{r_0, r_{n/2}\}$ (the identity and the 180° rotation) be a subgroup of D_n. Then C is normal in D_n and $D_n/C \approx D_{n/2}$.

Most of these conjectures were actually proposed by more than one student, and all were correct. In some cases, the students had enough background to prove the conjectures. In other cases, they would have to wait until later in the semester to complete the proofs or provide a counterexample. Sometimes we would only be able to partially prove a conjecture at the time of the classroom session following the lab. As we added to the class's knowledge base, we would return to the proof to complete it.

14.5 Impact on Students

I observed that *ESG* allowed my students to investigate examples more easily, because it facilitated computation and enhanced visualization. Instead of working through one example with paper and pencil, a student could generate many in an hour with the computer and subsequently detect patterns based on the examples. In comparison to students who had learned abstract algebra in my classes before I began to use technology, the students whose class had a laboratory component were much more able to understand the concepts behind the formalism and the theory.

Significantly, the laboratories changed the dynamics of my classrooms. The unpredictability of the lab experience meant that I had to be prepared to discard my lesson plans for the day and respond to my students' comments and questions. Students felt empowered by their own discoveries and were pleased to discern that they had more control of the learning environment.

At the end of the first semester of incorporating a substantial laboratory component for the abstract algebra course at DePauw University, I asked each student to respond to the following question: "Please give your overall impressions of the laboratory component of this course." The responses below, from about one-third of the students in the class, were representative:

> The laboratories were a gigantic help with my understanding of the subject matter. The thought process involved helped me see the material in a more concrete way. By first writing the computer's action out ourselves we could understand what the computer was doing. The computer then allowed us to understand higher functions in the group theory and most importantly recognize patterns in group theory.
>
> I believe the lab was very helpful for letting us see the parallels between many different groups. It was something tangible, that did not seem as abstract. This was especially true of the cosets lab. Being able to see the blocks of color provided an immediate link to the group tables that we had already seen. We did discover that our work in the lab made a lot more sense if we worked out a few examples by hand first. Otherwise, the computer was just spitting out random combinations of numbers and we did not understand where they were coming from. I strongly believe that the lab has greatly increased my understanding of this class.
>
> Overall, I definitely think that the lab was a success. It enabled us to examine lots of groups through the use of the computer. If we would have had to look at all of these examples by hand, we would not have had as much evidence to base our many conjectures on. But with the quick-working computers, we were able to look at numerous groups. We were able to see many patterns that would help to solidify our newfound knowledge. More importantly, the labs allowed us to break up the normal classroom routine of lecturing and notetaking.
>
> I'm not quite sure how to react to the labs. In one sense, the topics discussed on the labs helped enable me to better deal with the concepts of class. On the other hand, the labs sometimes confused and frustrated me. I sometimes could not distinguish between pertinent info and busy-work.
>
> I myself am truly grateful for the laboratory component. The labs were a tremendous aid in helping to understand and gain a grasp of the material. Work on the computers helped to make the abstract theory of the course more concrete. It was also a nice alternative to straight lecture. One of the best things about the labs was that we formed our own conjectures about the patterns we saw. Even though the labs were somewhat "directed", we still had the opportunity to exercise creativity. I believe that the progression of 1) lab, 2) conjecture, 3) class discussion, 4) proof was highly beneficial in gaining understanding of the abstract material of the course.

A few years later, a group of my students (those who had taken the abstract algebra course as well as a geometry course with a laboratory component) participated in a taped discussion about creativity in mathematics. They described the frustrating and exhilarating experiences of coming up with conjectures and then trying to prove them. As one student commented, these experiences were a challenge and an adventure—and made the rest of the mathematics curriculum seem "really boring." As he said, this was the "wild side of math."

14.6 Impact on My Career

This project provided an excellent opportunity for me to combine classroom teaching with activity in the area of the scholarship of teaching and learning—essential for me while at an institution with a heavy teaching load but also with expectations of scholarly work. I gave at least ten talks on the project in a variety of venues, and taught mini-courses using my materials and *ESG* at the 1996 Joint Mathematics Meetings (JMM) and at the International Conference on Technology in Collegiate Mathematics (1995, 1997, 1998, 1999). Clifton Corzatt of St. Olaf College and I received an NSF grant (DUE-9554636) to run a conference on using technology to teach algebra and geometry, held at DePauw University in June, 1996. Allen Hibbard of Central College and I co-organized contributed paper sessions on using technology to teach abstract algebra at the 1997 and 1999 JMM. We co-edited the volume *Innovations in Teaching Abstract Algebra* (Hibbard and Maycock 2002) based on those two paper sessions. Edward Keppelmann of the University of Nevada, Reno, and I taught a mini-course at the 2006 JMM using my laboratory materials and *FGB*.

This project was only one part of my efforts to use technology to teach mathematics at the undergraduate level. I had previously participated in the Associated Colleges of the Midwest and the Great Lakes College Association calculus reform project in the late 1980s and early 1990s, and I was eager to see if some of these ideas could be transported to upper-level courses. After the lab manual was completed, I taught real analysis and modern geometry using technology, using *Mathematica*™ in the former and *Geometer's Sketchpad*® in the latter. I summarized my experiences in the online paper "Technology in the Upper-Level Curriculum" (Maycock 2002). Students who were not mathematics majors in the interdisciplinary Honors Scholars program at DePauw University and enrolled in my sophomore-level honors seminar were especially grateful for the insights that technology gave them in trying to understand abstract mathematical concepts.

14.7 Reception of this Approach

DePauw University was quite supportive of my efforts: I was awarded two course remissions during my time at DePauw to work on the laboratories; the completion of the lab manual was my sabbatical project during the academic year 1994–1995. Since faculty members in my department were always allowed significant independence in course design, especially for upper-level courses, I could develop and test the course materials without objections. No colleagues in my department, however, used my materials.

In April, 1994, I was invited to give a talk on my approach to teaching group theory at Purdue University, where I received my doctorate. I was unprepared for the negative reception that I received from several senior faculty members. One particularly hostile questioner asked, "What do you do if the students look in the book?" It wasn't until after the talk that I realized I should have said, "I'd be delighted if my students looked in the book!" In fact, it was usually quite difficult for students to match their conjectures to the formal theorems presented in the textbook. Afterwards, I realized that I was, of course, proposing a dramatically new approach to teaching a very abstract and structured course. Another aspect of my approach that some faculty members found threatening was that students often came up with surprising and unfamiliar conjectures. In those cases, I'd occasionally have to say, "I don't know if they are true," something that can be difficult for a faculty member to say to a class. Ultimately, there is a power shift in the classroom when student discovery is the focus. Because the feedback that I received from students was so positive, I was undeterred by those negative reactions to my laboratory approach.

14.8 Conclusion

By the mid-twentieth century, group theory was being presented very abstractly in the classroom—a satisfying method for the mathematics professor but challenging for students. Many students in a college classroom are not able to learn abstract concepts via lectures that follow the "theorem-proof-example" format—they need to begin with concrete examples before they can grasp theory. I began to realize that my own approach to learning was not the norm and I changed my pedagogy to better reach my students. Beginning with examples and then moving to theory, with students actively involved in discovery learning, was more successful in my own classroom. Not only did my students find that this approach facilitated learning, they also had a glimpse of the excitement of mathematics research. Although using technology is not necessary for a discovery approach, computer labs made the class much more engaging and dynamic.

Now, more than two decades after I began experimenting with a laboratory component in an abstract algebra course, discovery learning or Inquiry-Based Learning (IBL) has become less threatening and more acceptable to faculty members and departments. There are now significant support structures, such as the Academy of Inquiry Based Learning (http://www.inquirybasedlearning.org), to aid faculty members in the transition to this approach. While IBL may never be universally adopted, it is gaining acceptance in mathematics classrooms. I hope that the positive reception that students had to my laboratory approach will encourage other faculty members to experiment with a laboratory component in a group theory course.

Appendix: Group Library of *ESG*, Orders 3–16

Table A.1 lists all groups in the Group Library of *Exploring Small Groups*. The four-digit codes refer to group tables contained in the software. The first two digits of the code give the size of the group. Common names are given with most of the groups.

Table A.1 Group Library of *ESG*, orders 3–16

0301—Z_3 Cyclic	1204—A_4 Alternating Subgroup of S_4
0401—Z_4 Cyclic	1205—Q_6 Dicyclic
0402—$Z_2 \times Z_2$, Klein Four-Group V	1301—Z_{13} Cyclic
0501—Z_5 Cyclic	1401—Z_{14} Cyclic
0601—Z_6 Cyclic	1402—D_7 Dihedral
0602—D_3 Dihedral/S_3 Symmetric	1501—Z_{15} Cyclic
0701—Z_7 Cyclic	1601—Z_{16} Cyclic
0801—Z_8 Cyclic	1602—$Z_8 \times Z_2$
0802—$Z_4 \times Z_2$	1603—$Z_4 \times Z_4$
0803—$Z_2 \times Z_2 \times Z_2$ Elementary	1604—$Z_4 \times Z_2 \times Z_2$
0804—D_4 Dihedral/Octic	1605—$Z_2 \times Z_2 \times Z_2 \times Z_2$ Elementary
0805—Q_4 Dicyclic/Quaternion	1606—$D_4 \times Z_2$
0901—Z_9 Cyclic	1607—$Q_4 \times Z_2$
0902—$Z_3 \times Z_3$ Elementary	1608—A subgroup of $GL_2(Z_5)$
1001—Z_{10} Cyclic	1609—Sylow 2-Sg of $SL_2(Z_4)$
1002—D_5 Dihedral	1610—Semidirect product of Z_4 by Z_4
1101—Z_{11} Cyclic	1611—A subgroup of $GL_2(Z_5)$, M
1201—Z_{12} Cyclic	1612—D_8 Dihedral
1202—$Z_6 \times Z_2$	1613—Sylow 2-Sg of $GL_2(Z_3)$
1203—D_6 Dihedral	1614—Q_8 Dicyclic

References

Carter, N., & Emmons, B. (2005). Group theory visualization with Group Explorer. *Journal of Online Mathematics and its Applications*. Retrieved June 7, 2016, from http://www.maa.org/node/115894.

Dubinsky, E., & Leron, U. (1994). *Learning abstract algebra with ISETL*. New York: Springer.

Durell, F. (1894). Application of the new education to differential and integral calculus. *American Mathematical Monthly, 1*(2), 37–41.

Geissinger, L. (1989). *Exploring Small Groups: a tool for abstract algebra*. [computer software]. San Diego, CA: Harcourt Brace Jovanovich.

Hibbard, A. C., & Levasseur, K. M. (1999). *Exploring abstract algebra with Mathematica*. New York: Springer.

Hibbard, A. C., & Maycock, E. J. (Eds.). (2002). *Innovations in teaching abstract algebra.* Washington, DC: Mathematical Association of America.

Maycock, E. J. (2002). Technology in the upper-level curriculum. *Journal of Online Mathematics.* Retrieved June 7, 2016, from http://www.maa.org/press/periodicals/loci/joma/technology-in-the-upper-level-curriculum-introduction.

Parker, E. M. (1996). *Laboratory experiences in group theory.* Washington, DC: Mathematical Association of America.

Rainbolt, J. G., & Gallian, J. A. (2013). *Abstract algebra with GAP, 2013 version.* Retrieved June7, 2016, from http://math.slu.edu/~rainbolt/manual8th.htm.

Webb, B., & Keppelmann, E. (2000). Finite Group Behavior, version 3.0. [computer software]. Retrieved from http://wolfweb.unr.edu/homepage/keppelma/fgb.html.

Chapter 15
A Departmental Change: Professional Development Through Curricular Innovation

Steve Cohen, Bárbara González-Arévalo, and Melanie Pivarski

Abstract Roosevelt University is a private, comprehensive master's institution with a social justice mission. Faculty are expected to cultivate excellent teaching, create significant amounts of research, and perform much service work. It is challenging to find time for all three when they are approached as distinct tasks. The mathematics department has developed creative activities that integrate all of these professional responsibilities resulting in a significant change in departmental culture. Faculty have added large- and small-scale projects to courses. They are now working with more students on undergraduate research and have created a venue, the Math x-Position, to showcase student work. This was all accomplished by ongoing faculty mentoring, combining teaching and research, and by fostering a safe environment for innovation.

Keywords Departmental change • Mentoring faculty • Projects • Faculty development

---

MSC Codes
97A99
97B99

S. Cohen • M. Pivarski (✉)
Department of Mathematics and Actuarial Sciences, Roosevelt University,
Mailstop AUD 402, 439 S. Michigan Avenue, Chicago, IL 60605, USA
e-mail: scohen@roosevelt.edu; mpivarski@roosevelt.edu

B. González-Arévalo
Department of Mathematics, Hofstra University,
315B Roosevelt Hall, 130 Hofstra University, Hempstead, NY 11549, USA
e-mail: barbara.p.gonzalez@hofstra.edu

J. Dewar et al. (eds.), *Mathematics Education*, Association for Women in Mathematics Series 7, DOI 10.1007/978-3-319-44950-0_15

15.1 The Institution and the Department

15.1.1 Roosevelt University: Social Justice

Roosevelt University's social justice mission began at its founding in 1945, when faculty and administrators walked out of Central YMCA College after refusing to implement a racial and religious quota system (Weiner 2005). Roosevelt is located on two campuses, one in downtown Chicago and the other in Schaumburg, Illinois. It has a diverse student population in terms of ethnicity and age, including many first-generation students, students from underperforming high schools, ones with economic hardships, working students, and students with family responsibilities. We work to serve them all.

The university aims to inspire students to be active citizens who effect positive social change. The Department of Mathematics and Actuarial Science wants students to see how mathematical knowledge can help them understand quantitative aspects of social issues. We motivate mathematical topics through specific problems that provide a convincing answer to the question, "Why are we learning this?" Projects in our courses provide a communication component in which students explain the problem they are solving as well as how to model it. These projects satisfy Roosevelt's three university-wide learning goals: (1) effective communication, (2) knowledge of discipline-focused content, and (3) awareness of social justice and engagement in civic life.

Roosevelt has many institutes and centers dedicated to studying societal issues such as the Illinois Consortium on Drug Policy and the Mansfield Institute for Social Justice and Transformation. They connect with students and faculty across the university, enhance research, and provide resources for transformational service-learning experiences. Historically, our department was not involved with these institutes, even though many social issues have a quantitative aspect to them and thus provide a context for connecting mathematics to the community. Recently, our Preparation for Industrial Careers in Mathematical Sciences[1] (PIC Math) class has started to work with both the university's sustainability committee and the Consortium on Drug Policy.

Roosevelt supports incorporating social justice and civic engagement into classes. There are workshops on service learning, a day-long internal conference on teaching, and support for local conferences. Roosevelt regularly sends faculty to the Science Education for New Civic Engagements and Responsibilities

[1]PIC Math is "a Mathematical Association of America and Society for Industrial and Applied Mathematics program with support provided by the National Science Foundation (DMS-1345499). It aims to prepare mathematical sciences students for industrial careers by engaging them in research problems that come directly from industry. A strong component of PIC Math involves students working as a group on a semester-long undergraduate research problem from business, industry, or government" (PIC Math n.d.).

(SENCER)[2] conferences, acts as a local host for some of the regional and national conferences, and has many faculty serving as SENCER Leadership Fellows. Roosevelt provides internal funds for semester-long research leaves and summer grants.

The expectation that Roosevelt faculty do research is relatively new. In 1990, the university reduced the annual teaching load from eight to seven classes and began to require scholarly work for tenure and promotion. Mathematics research at Roosevelt includes peer-reviewed works in the areas of pure and applied mathematics, as well as pedagogical research. In 2004 the teaching load decreased to six classes while research expectations increased.

15.1.2 The Mathematics Department: Background and Change

For the past several years, the Roosevelt mathematics department has consisted of six to eight tenured and tenure-track faculty, two to four lecturers, and a varying number of adjunct faculty. The tenure-track faculty consist of one full professor, three to five associate professors and one to three assistant professors with backgrounds in statistics, financial mathematics, analysis and probability, algebra, and mathematical engineering. The department offers Bachelor's and Master's degrees in mathematics and actuarial science. The actuarial courses are designed and regularly updated to conform to the guidelines of the Society of Actuaries (https://www.soa.org/) and the Casualty Actuarial Society (http://www.casact.org/). Mathematics courses have more flexibility in content compared to actuarial ones, but in the past they have tended to be taught traditionally. A decade ago, there was little faculty collaboration on teaching innovations. A couple of senior mathematics faculty were on the planning committee for the Chicago Symposium Series (Chicago 2016) and worked independently on active learning in their classrooms. They also collaborated on a project in which pre-service mathematics teachers taught a lesson to a college developmental mathematics class. Although collaboration was limited, there was a culture of support for new initiatives. Instructors did not experience any significant opposition from either the department or the university when trying new things.

---

[2] SENCER is the signature program of the National Center for Science and Civic Engagement (NCSCE). "SENCER applies the science of learning to the learning of science, all to expand civic capacity. SENCER courses and programs connect science, technology, engineering, and mathematics content to critical local, national, and global challenges" (SENCER n.d.). NCSCE is a "national organization that supports a community of teachers and learners. Through grant funding, we help educators in and outside the classroom make connections between the content they teach and real world issues of civic importance" (NCSCE n.d.).

Starting in 2006, there were several retirements and new hires; only three current faculty members were in the department a decade ago. In addition to replacing the retirees we hired experienced full-time lecturers to concentrate on developmental and general education courses. These non-tenure-track faculty have contracts that are renewable every 4–5 years. They teach eight courses per year and have significant service expectations. Although they are not expected to do research, at Roosevelt they participate in pedagogical research activities by giving talks at conferences, publishing non-peer-reviewed papers, and writing grant proposals.

As the department changed and grew, faculty brought new ideas. The department was open to discussions, allowing instructors to experiment with their teaching. As different people began to teach a course, there were natural opportunities for discussions. One of our lecturers introduced an innovative way of teaching quantitative literacy with applied examples and a group project. An NSF-STEP grant funded several science and mathematics faculty to attend the annual SENCER Summer Institute (SSI) (SENCER n.d.). The institute provided an orientation for new faculty on active learning, course development, and civic issues that motivate course content, which reinforced the importance of teaching excellence, research-based practices, and creative work in the classroom. The institute gives small implementation awards for course development that we used to incorporate semester-long projects into integral calculus and later into financial mathematics.

The idea of creating projects spread to other courses. We added mini-projects to finite mathematics and the honors statistics course. Two senior faculty members developed a semester-long problem-motivated industrial modeling course through the PIC Math project (PIC 2016). We joined the Engaging Mathematics project,[3] redesigning our college algebra course into a flipped format[4] using questions that make a connection to Chicago (Engaging Mathematics n.d.). Our faculty supported each new development, sharing their work and learning from each other.

With each new experience of adding projects to courses, we gained more information about student capabilities and the time required to implement the projects. This helped us know when and how to ask for external support and to see how incremental change can lead to significant work. The high level of activity offered a

[3]"Funded through the National Science Foundation's TUES-II program (DUE-1322883), the Engaging Mathematics 3-year initiative aims to significantly increase the use of the SENCER model, and other reformative pedagogies, by a national community of mathematics scholars capable of creating, implementing, and sustaining reforms in mathematics education. Engaging Mathematics is an initiative of the NCSCE" (Engaging Mathematics n.d.).

[4]"*Flipping the classroom* is a pedagogical strategy that replaces the standard lecture-in-class format with opportunities for students to review, discuss, and investigate course content with the instructor in class. There are many ways in which a classroom can be flipped, but the underlying premise is that students review lecture materials outside of class and then come to class prepared to participate in instructor-guided learning activities" (Hughes 2014, p. 137).

Table 15.1 Timeline of mathematics courses with an additional pure or applied project component. Courses that are primarily in computer science are not included. Unless noted, all listed courses continue to incorporate the additional component

Pre-2009	History of Mathematics: Motivated by a university writing requirement, groups create papers and in-class presentations on historical topics.
	Baseball Statistics: Groups create papers and in-class presentations analyzing an aspect of baseball, such as designing a team with a cost constraint, based on predictions of statistical performance.
2009	Quantitative Literacy: Groups create a poster interpreting data on a social issue such as education, homicide, or obesity.
2010	Calculus II: Groups create a semester-long project with a paper and poster on topics such as AIDS, the oil spill, population growth, and the Gini index.
	Math x-Position begins.
	Differential Equations and Modeling: Individuals or groups create projects on topics such as predator-prey systems, writing papers or creating posters.
	Number Theory: Groups present posters on topics such as the golden ratio and the Fibonacci sequence.
2011	Combinatorics: Individuals write papers on the Enigma machine and scratch lotto games.
	Financial Mathematics: Groups complete a semester-long project modeling the mortgage crisis, writing a group paper and poster. (Some semesters.)
	Geometry: Individuals or groups write short papers or posters on a geometric application, or create YouTube videos on geometric constructions and map projections. (Varies by semester.)
2012	Real Analysis: Individuals give oral presentations on advanced analysis topics.
2013	Abstract Algebra: Individuals create a portion of a poster on a concrete application of a group, ring, field, or monoid such as solving the Rubik's cube, quaternions in computer graphics, card shuffling, and Boolean algebra in programming.
	Finite Mathematics: Individuals write a short paper on their choice of topic, such as analyzing poker hands, calculating odds for a game, or applying Venn diagrams or least square lines.
	ANOVA/Experimental Design: Individuals write papers about their experience designing, running, and analyzing an experiment. (Some semesters.)
2014	College Algebra: Individuals or small groups apply algebra studying Chicago-themed problems on population, homicide rates, temperature, and skyway tolls.
	Honors Elementary Statistics: The entire class collaborates to analyze a dataset, presenting their results in a group poster. This course is primarily for non-majors and is cross-listed with economics and sociology.
2015	Industrial Applications of Mathematics (PIC Math course): Groups complete a semester-long research project for an industrial partner, creating a video and a written project report.
Future	Models for Life Contingencies/Actuarial Mathematics I: Project in development.

real advantage. If a faculty member was off for a semester or left, a robust collection of student experiences still occurred. Table 15.1 lists chronologically all the courses (from before 2009) into which we have introduced projects and gives some information about the project topics and implementation.

15.2 Our Curricular Work Over the Past Decade

15.2.1 First Significant Project Experience: Integral Calculus

In 2009, SENCER granted us a small sub-award to develop semester-long integral calculus projects (González-Arévalo and Pivarski 2013). The grant provided structure for our work, a need for accountability, and visibility to different university offices. We originally planned a partnership with a local museum, but we could not make that work. Instead of creating a new project, we modified an existing one on the spread of HIV/AIDS (Janke 1993) to match the backgrounds of our audience of mathematics, actuarial science, chemistry, and biology majors. The HIV problem requires that students model the HIV problem with a separable differential equation; solve the differential equation using partial fractions; and derive the logistic function. They find data sources, fit real data to the curve, make predictions, and assess the strengths and weaknesses of their model.

Modifying an existing project was a good choice for our first experience, given the number of logistical challenges we were to encounter. The planning phase went smoothly. We considered the course topics at that time, their ordering, and in-class time needed for the project. We divided the project into parts, distributed them throughout the semester and removed physics applications to create time for the project. We developed and tested *Maple*™ (version 12) worksheets on numerical integration and curve fitting. We arranged for university librarians to talk with our class about background research. We felt prepared for the new semester.

Unfortunately, a newly installed version of *Maple* (version 13) was incompatible with the worksheets we had prepared. As a result, our first computer lab was a disaster. Not only did *Maple* commands that had worked perfectly in the fall not work in the spring, but also water literally rained into the classroom from the upstairs biology lab. Our labs needed to be revised and run again. We updated the commands and made a list of common programming errors. Students still had problems in programming, but it was a good learning experience, and the list of common errors helped them troubleshoot each other's work.

Students collected data from online sources, such as the Centers for Disease Control and Prevention. Reporting techniques changed over the years, and so the number of people with HIV/AIDS reported was not consistent; sometimes the cumulative number of cases would decrease, even though mathematically it should be an increasing function. Although we did not anticipate this, it led to a stronger experience for the students. They learned about the difficulties involved in obtaining a consistent dataset. When we created the *Maple* worksheets, we used a small dataset to fit a logistic curve. When our students tried it, *Maple* was unable to fit their much larger dataset. Pressed for time, we fit the curve ourselves and told students that this is what an additional collaborator would do. In future semesters, students estimated the inflection point and carrying capacity from the data to fit the curve. At the end of the semester, students presented group posters at the university-wide Science Research Day and wrote group papers.

Students left the class with a substantial modeling experience, which acted as a capstone mathematics experience for many of the science majors. We expected students to learn about differential equations, models, and integration techniques. We did not anticipate they would gain so much insight into the limitations involved in datasets. There is an expectation that computers can magically solve everything, and our students saw that this was not the case. As a condition of our sub-award, students self-reported their learning gains via an anonymous online survey, the Student Assessment of Learning Gains (SALG) instrument (SALG n.d.). Only seven out of about 30 responded, but their description of their learning gains mirrored our impressions. They enjoyed the integration of biology and mathematics, and they appreciated the need to go beyond mathematical knowledge to create a model. All of the survey respondents felt that the use of a real-world issue significantly helped their learning.

We left with many ideas about how to modify the class and improve the experience. Students were capable of more than we anticipated; their work demonstrated a significant level of understanding. It was energizing!

15.2.2 Developing a Problem-Based Course

After implementing successful project experiences in a number of courses (Table 15.1), our department wanted to create a capstone experience for our majors that incorporated the university's interest in transformational learning. The PIC Math project was ideal. Teams of students in the class analyze real data from an industrial partner and create a useful solution to a problem. The problem drives the course content, with students learning mathematical and statistical techniques to model their specific situation. Students communicate mathematics by writing a project report and creating a video explaining their work.

We found two willing faculty members with complementary interests to apply for training and team-teach the class. One is a senior faculty member and an engineer who teaches advanced statistics courses; he was a natural fit for the potential statistical content. The other had experience working with students in history of mathematics, geometry, and calculus group projects, as well as grant experience. We were accepted into the PIC Math program.

The grant supported one instructor's travel to a 3-day program orientation and one student's travel to present at the Mathematical Association of America's (MAA) summer meeting (MathFest). It also provided an external structure for the course, with deadlines for papers and videos. In applying for the grant, it was important that our institution made a commitment to offering the course, so we could plan the class without fear of a last-minute cancellation. As the semester began, the class had 13 students teleconferencing between the university's two campuses.

One group of students worked with a PIC Math-provided industrial partner to model Ebola and the strategic location of treatment centers. We found a second industrial partner when an alumnus connected us with a biology professor who knew a Field Museum scientist with an interesting dataset. The Field Museum has a Zooniverse project that crowd-sources measurements of microscopic plants (MicroPlants 2016). Our PIC Math students analyzed the data's reliability. One student presented a poster on his group's microplant work at MathFest. His project experience made him feel more confident at his summer internship at a market research company.

In Spring 2016 three projects occurred in the class, each with a compelling social component: preservation of species, responsible energy use, and access to drug treatment. Each had a challenging mathematical component as well: data mining, weather and energy use modeling, and regression analysis.

15.2.3 Student Research

Historically, undergraduate research was rare in our department. Our work with the calculus projects helped to stimulate it. We discussed how to employ students to research calculus projects and independently wrote two different job descriptions for research positions. In one, students would investigate qualitative research methods for measuring the effectiveness of new course components. In the other, students would study the different types of calculus projects, identify mathematical concepts and how they fit into the calculus curriculum, and, time-permitting, create a project for use in the calculus course. These two different job descriptions paved the way for students to do a literature search from educational and mathematical perspectives during the spring semester. During the summer students began to develop their own calculus projects for use in our classes. Each new batch of student researchers learned from the previous student researchers' work and projects used in class. They drew upon their recent student experiences for insights that eluded us. This helped us to improve as teachers. We were able to fund student research through an NSF-STEP grant during the summer and the honors program throughout the entire year.

We gained the personal experience and confidence we needed to expand into other forms of undergraduate research. Some students who began designing calculus projects then branched out into projects with more mathematical depth. Our main difficulty was that some students prioritized outside work commitments over their research time; this could be mitigated by paying students by the hour rather than in a lump sum. Regardless, the overall outcome was good; most student projects progressed to a reasonable point. We used two student projects in class, and others became honors theses and internal and external student talks. The students who designed the calculus projects were excited to see other students make direct use of their work. Our work has affected about a dozen student researchers, approximately 100 students in courses with projects, and around 200 audience members in student talks and poster sessions. Although these student research experiences differed from a traditional form, they allowed students to gain communication and organizational skills, self-motivation, and mathematical breadth and depth.

15.2.4 Student Research Day: Math x-Position

We wanted an internal venue for students to present mathematical work. We had worked with the science faculty on a combined research day and learned the planning process. In 2010, we created the Math x-Position; this pun emphasizes that it is a conference for everyone, majors and non-majors alike. The day-long event consists of a student poster session, student research talks, a career panel, a keynote address, and some mathematical games and puzzles. Having applied projects in our courses helped to build both content and an audience. The bulk of the poster session comes from the projects in integral calculus and quantitative literacy. Our student speakers come from independent studies, student research, summer Research Experiences for Undergraduates, the PIC Math course, and internships. When students see others present their work, they start thinking about trying to do their own research project.

Faculty who teach advanced courses, such as number theory and differential equations, took advantage of the dedicated mathematics venue by adding a poster component to their courses. This event required a significant amount of faculty service activity; they arranged for speakers, catering, marketing, attendee registration, room arrangements, funding, and poster printing.

Math x-Position makes our students' work public, demonstrating the broad appeal of mathematics to university leaders. When we initiated this event, our dean, provost, and president all came from the humanities and the arts and had little personal experience with mathematics. They were pleasantly surprised by how well the students communicated about mathematics and that non-majors came willingly to a math-themed event. Seeing the students explain their work publicly gave our administrators an increased appreciation for our students' and our department's efforts.

15.3 Mentoring Faculty in a Changing Environment

15.3.1 Faculty Challenges

At Roosevelt, new faculty members have a range of responsibilities: prepare new classes, advise students, perform service activities, and develop a research program beyond their thesis. Over a 6-year period, faculty must demonstrate growth in all areas. The transition from a doctoral program to one focused on teaching undergraduates and master's students can be challenging. In graduate school, one develops the ability to explain high-level mathematics to other mathematicians, who already know the importance of mathematical research and that conference attendance is essential for developing collaborations and new research ideas. At a teaching university, colleagues from other departments, typically from outside the STEM fields, decide who receives internal grants. Mathematics faculty must be able to explain the utility of mathematics and the importance of their own research.

Experienced faculty members have a different set of challenges. After earning tenure, faculty no longer have an urgent, external motivator. Until recently, in our own institution, full professorship was difficult to attain and provided little reward. It can be easy to develop a sense of inertia, continuing to teach courses without updating them. It was common for faculty members to always teach the same courses, with no opportunity to collaborate with peers teaching the same course. This practice was suitable for our pre-1990s institutional culture, when mathematics faculty had no research expectation. With new demands, senior faculty needed some kind of supportive structure to restart their research programs. Most external funding programs are geared towards early career faculty. Internally, travel was only funded once there were results to present. Summer research funding and research leaves required an already active research program.

15.3.2 Integrating Teaching and Research

While improving student experiences is our main motivation for integrating projects into the mathematics curriculum, these projects also provide natural opportunities for scholarly work in the scholarship of teaching and learning (SoTL).[5] During the transition period when there were no senior research mentors and no support, faculty relied on themselves or connections with colleagues from other universities. At the university level, faculty decided that pedagogical research was as valuable as content-based research. Faculty also explored research as a way to contribute to the broader metropolitan community, and SENCER presented a natural connection by bringing civic engagement into science and mathematics classes. The sciences sent several faculty members to SSI, resulting in the creation of new and redesigned courses for their students (Kim and Szpunar 2010; Wentz-Hunter 2009). This inspired the mathematics department to use SSI as an orientation program, engaging faculty in curricular innovation work from the outset.

SSI provided a structure for faculty who had not previously done pedagogical research. The institute held workshops on how to use, modify, and access data from the SALG survey. Our science peers shared their successful grant application and Institutional Review Board (IRB) paperwork, which helped us to complete our own. This was our first direct experience with securing IRB approval. Using the SALG survey gave us a sense of what we could expect from a survey; even in a good semester, the response rate was about 20 %. Survey information was different from mathematics research; it did not tell us about the entire class, but we could get some information about student impressions. We began to think about what we wanted to study, and how we could get reasonable data for it.

[5] SoTL is the intellectual work that faculty members do when they use their disciplinary knowledge (in our case, mathematics) to investigate a question about their students' learning (and their teaching), submit their findings to peer review, and make them public for others to build upon (Dewar and Bennett 2015).

We submitted a project report about our integral calculus redesign to a SENCER journal (González-Arévalo and Pivarski 2013), and we learned from the refereeing process. Next, we studied the impact of projects on the students who were project designers as well as on students who were tutors for the calculus class. Given the small number of these students, a qualitative study was most appropriate. We learned about qualitative research from a workshop at SSI and discussions with psychology colleagues. We learned about coding, designed a series of interview questions, and received summer support from our provost's office for a student worker to transcribe our interviews. We debated coding categories and worked individually, cross-checking for consistency. We presented our work several times before writing it up (Cohen et al. 2016); the audience questions gave us a sense of what interested others. We particularly enjoyed collaborating on these SENCER projects within our university and learning in a more substantive way about the outcomes of our work.

As we integrated more projects and student research into the curriculum, more faculty became involved with pedagogical research: between 2010 and 2016 six tenured and tenure-track faculty and two lecturers presented at conferences and wrote papers. Recently, five faculty members submitted three papers on calculus and financial mathematics projects to an MAA volume with a theme of social justice in mathematics in the Classroom Resource Materials book series. Three worked on a grant with a larger group of mathematics educators in the Engaging Mathematics project to redesign our college algebra course (Engaging 2016). About half of our mathematics faculty, including tenured, tenure-track, and lecturers, now teach college algebra in a flipped format. They are creating a course manual for wider dissemination. As faculty produced SoTL research, the amount of content-based research required for tenure and promotion decreased to a level that was more in line with their teaching load.

The most satisfying aspect of pedagogical research is the potential to bring immediate value to students. By adding applied components to courses, we generated examples of how mathematics is used at all levels. This helped us provide memorable experiences, tie course content to broader societal issues, and give beneficial career advice to mathematics majors. As an added benefit, these applied projects help us justify internal funding for mathematics research.

15.3.3 Creating a Safe Environment for Innovation

Much has been written about how student attitudes can affect their learning. They benefit when they realize that skills can be developed and occasional failure can help them grow (Kooken et al. 2015). We found that this also applies to teaching innovation. We overcome challenges, in part, by mentoring one another. We rotate courses, so both new and experienced faculty can collaborate with a teaching partner. We share project and computer activities. Monthly departmental seminars provide a low-stakes opportunity to speak about SoTL or pure and applied mathematics research; having an audience provides motivation.

Our department encourages groups of faculty to attend professional development workshops such as the regional SENCER meetings and the Chicago Symposium Series. Groups of faculty attend, hear and present talks, gain new ideas, and refine old ones. We discuss with one another how we can improve our courses, and we discuss our curriculum and teaching practices in our department meetings. We go to longer conferences such as SSI, MathFest, and the Joint Mathematics Meetings; these act as teaching retreats. New faculty members are encouraged to apply to the national and regional New Experiences in Teaching project (Project NExT); this experience provides a valuable disciplinary peer network for them and fosters their professional development.

The senior lecturers lead our developmental and general education mathematics programs. They present at conferences dedicated to these areas and to helping underprepared students. Their work is supported by our department, the dean's office and the provost's office, enabling them to act confidently and decisively even without the protection of tenure. We work as peers with them to improve our curriculum. Our non-tenure-track faculty hold a meeting each semester where all faculty, including adjuncts, discuss our developmental mathematics courses and how to improve them. As a result, we regularly update the topics in our developmental mathematics course. We chat informally throughout the semester about teaching, and all full- time faculty take turns observing each other's classes as formative evaluation.

As a department, we consider many factors when evaluating teaching. We do not rely exclusively on numerical course evaluations to determine the quality of an instructor. Evaluation numbers can be influenced by student bias (Boring et al. 2016). When a faculty member has low ratings, we do peer observations to see how the course is going. In these, the observers comment on what happened in the classroom, what aspects they would like to emulate, and their suggestions for improvement. We also look at student comments to learn the nature of their concerns. Sometimes, students are uncomfortable with a new format, but significant learning is observed to be taking place in the classroom. Their discomfort can lead to low evaluation numbers, while the teaching quality is high (Stark and Freishtat 2014).

By talking with our peers about what works and what does not, we create a space where risk-taking is allowed. Everyone understands that there will be bad days. Mathematicians are professional problem solvers, so we are able to help find solutions to teaching issues. For example, when faculty felt that there was too little time in the integral calculus course to treat the material and do a project, we spent time discussing the issues involved, such as scheduling, faculty work load, and major requirements. Ultimately, we increased the credit hours from four to five. We feel comfortable asking each other for assistance, and this openness helps us improve our teaching and strengthen our courses.

15.3.4 *Reflections: Best Practices for Seeking Departmental Change*

All these experiences and pedagogical experiments promoted a culture where SoTL infused all our academic activities: several faculty started publishing mathematics education papers. All started incorporating best practices and pedagogical innovations into their teaching, and many of these activities added to the service provided to the department, the university, and the community. While reflecting on our work, we created a list of our best practices for others seeking change in their own department:

1. Send groups of faculty to local professional development conferences. These are generally inexpensive and easy to travel to. Designate a time, either at the conference or soon after for faculty to share their experiences with the department.
2. Be aware of the university's objectives and mission and find ways to align goals with them.
3. Start small. Short activities and projects are easier to add.
4. Go big. Projects where students have an opportunity to revise after feedback can yield strong student work.
5. Make mistakes. Learning from mistakes makes better teachers.
6. Ask for advice. Articulating experiences can help to clarify them. Most people like to share expertise, and asking for help can put the focus on problem solving rather than criticism.
7. Share examples of things that did not work. One of the best ways to mentor peers is to share experiences. When more experienced faculty share their troubles, it can help to put junior ones at ease.
8. Do not assume agreement. Taking the time to express and listen to others can lead to more robust ideas.
9. Use classroom work to motivate student activities outside of class. Collaborative projects and classroom activities help students develop friendships, making them more likely to attend events and fieldtrips.
10. Let students help improve teaching. They are useful as researchers and can provide a fresh perspective on classes.
11. Make teaching public by giving talks about teaching projects at conferences. This can help stimulate new ideas, connecting teaching to scholarship.

References

Boring, A., Ottoboni, K., & Stark, P. B. (2016). Teaching evaluations (mostly) do not measure teaching effectiveness. Science Open. doi: 10.14293/S2199-1006.1.SOR-EDU.AETBZC.v1.

Chicago Symposium Series: Excellence in Teaching Mathematics and Science. Research and Practice. (2016). Retrieved March 8, 2016 from https://www.math.uic.edu/chicagosymposium.

Cohen, S., González-Arévalo, B. & Pivarski, M. (2016) Students as partners in curricular design: Creation of student-generated calculus projects. arXiv:1605.06333.

Dewar, J., & Bennett, C. (Eds.). (2015). *Doing the scholarship of teaching and learning in mathematics*. Washington, DC: Mathematical Association of America.
Engaging Mathematics. (n.d.). Retrieved May 3, 2016 from http://engagingmathematics.ipower.com/.
González-Arévalo, B. & Pivarski, M. (Winter 2013) The real-world connection: Incorporating semester-long projects into integral calculus. *Science Education and Civic Engagement: An International Journal.* http://seceij.net/seceij/winter13/real_world_conn.html.
Hughes, H. (2014). Flipping the college classroom: Participatory learning, technology, and design. In L. Kyei-Blankson & E. Ntuli (Eds.), *Practical applications and experiences in K-20 blended learning environments* (pp. 137–152). Pennsylvania: Information Science Reference (IGI Global).
Janke, S. (1993). Modeling the AIDS epidemic. In P. Straffin (Ed.), *Applications of calculus. MAA Notes No. 29* (pp. 210–221). Washington, DC: Mathematical Association of America.
Kim, B., & Szpunar, D. (2010). Integrating teaching experience into an introductory chemistry course: The chemistry of global warming. *Science Education & Civic Engagement: An International Journal*. Retrieved June 19, 2016 from http://seceij.net/seceij/summer10/integrating_tea.html/.
Kooken, J., Welsh, M., McCoach, D., Johnston-Wilder, S., & Lee C. (2015). Development and Validation of the mathematical resilience scale. *Measurement and Evaluation in Counseling and Development*. doi: 10.1177/0748175615596782.
MicroPlants, A Zooniverse Project. (2016). Retrieved March 8, 2016 from http://microplants.fieldmuseum.org/.
NCSCE. (n.d.). National Center for Science and Civic Engagement. Retrieved May 3, 2016 from http://ncsce.net/about/.
PIC Math. (n.d.). Preparation for industrial careers in mathematical sciences. Retrieved May 3, 2016 from http://www.maa.org/pic-math.
SALG. (n.d.) Student assessment of learning gains. Retrieved March 7, 2016 from http://www.salgsite.org.
SENCER. (n.d.). Science education for new civic engagements and responsibilities. Retrieved May 3, 2016 from http://www.sencer.net/.
Stark, P. B., & Freishtat, R. (2014). An evaluation of course evaluations. Science Open. doi:10.14293/S2199-1006.1.-.AOFRQA.v1.
Wentz-Hunter, K. (2009). Cellular and molecular biology: Cancer. SENCER. Retrieved March 8, 2016 from http://serc.carleton.edu/sencer/cellularbiology_cancer/index.html.
Weiner, L. (2005). Roosevelt University. The Electronic Encyclopedia of Chicago. Chicago Historical Society. Retrieved March 9, 2016 from http://www.encyclopedia.chicagohistory.org/pages/1093.html.

Chapter 16
SMP: Building a Community of Women in Mathematics

Pamela A. Richardson

Abstract On the surface, the Carleton College Summer Mathematics Program for Women Undergraduates (SMP) looks like many other summer programs: students spend several weeks engaging with mathematics with peers from a variety of institutions. However, SMP offers both formal and informal mentoring activities that go far beyond one summer, providing support to participants through critical stages of their mathematical careers. The result is a strong community of successful women in mathematics.

Keywords SMP • Community • Mentoring

16.1 Introduction

In January of 1999, I was a sophomore mathematics major at Bowling Green State University. I knew that I loved mathematics, but I had absolutely no idea what I wanted to *do* with it. Like many young mathematicians, I didn't know much about mathematical career opportunities outside of high school teaching. I applied to several summer programs and was thrilled to be accepted to the Carleton College Summer Mathematics Program (SMP). At the time, I didn't understand the need for a women's program in mathematics; I was mostly just excited to get out of Ohio for a few weeks. However, deciding to accept the invitation to SMP was the best decision that I ever made. Since 1999, I have been involved in SMP in many ways: as a mentor, a workshop leader, and an instructor. Each role provided me with experiences that shaped my career.

MSC Code
97B40

P.A. Richardson (✉)
Division of Cognitive and Quantitative Sciences, Westminster College,
Box 40, New Wilmington, PA 16172, USA
e-mail: richardpa@westminster.edu

J. Dewar et al. (eds.), *Mathematics Education*, Association for Women in Mathematics Series 7, DOI 10.1007/978-3-319-44950-0_16

16.2 The Summer Mathematics Program

The Carleton College Summer Mathematics Program was founded in 1995 by Deanna Haunsperger and Stephen Kennedy, faculty members of Carleton College. This National-Science-Foundation-funded program typically runs from mid-June to mid-July and is held on the Carleton College campus in Northfield, MN. SMP is designed as an intensive mathematical experience: participants take two rigorous mathematics courses; attend colloquia, panel discussions and problem-solving sessions; participate in a conference with alumnae of the program; and engage in numerous social activities.

Each year, the program directors select 18 or 19 young women who have only finished one or two years of college. To be eligible for the program, students typically must have taken calculus and linear algebra, and they must be "US citizens, nationals, or permanent residents" (Summer Mathematics Program 2016). The program is quite selective, as usually around 120 women apply. Academic merit as well as the program's potential benefit to the applicant are taken into account. For example, SMP participants are frequently chosen from small colleges that do not have broad mathematical offerings. The directors hire a staff consisting of two instructors, two teaching assistants, and a program assistant. The instructors are female mathematicians selected for their teaching and mentoring abilities. The teaching assistants are typically alumnae of the program who are currently in graduate school in the mathematical sciences, and the program assistants are often alumnae of Carleton College.

The SMP schedule keeps the participants immersed in mathematics. Class sessions are held Monday through Friday mornings from 8:30 a.m. to 12:00 p.m. Courses are designed by the instructors to be accessible to sophomores on a topic not usually introduced in the undergraduate curriculum. For example, I taught a course on Lie theory that introduces the concept of a Lie group and its associated Lie algebra (developed via the tangent space) and exposes students to the quaternions and octonions, matrix-valued functions, symmetry, and even Jordan algebras. Other SMP instructors taught courses on coding theory, geometric topology, low-dimensional dynamics, Morse theory, p-adic analysis, and supermetric spaces. Most courses have an emphasis on mathematical abstraction, preparing students to bridge the gap between more computational mathematics courses like calculus and more theoretical courses like algebra and analysis.

Other formal academic events include semi-weekly colloquia given by dynamic female mathematicians with a wide range of specialties. . When I was a student participant, SMP colloquia introduced me to exciting mathematical topics like ordinal numbers, knot theory, group automorphisms, cryptography, and the history of mathematics. At least three panel discussions are held on these topics: opportunities for undergraduate mathematics majors (research experiences, study-abroad programs, conferences, etc.), applying to and surviving graduate school, and career options for mathematics majors. Each week, an optional but well-attended recreational problem solving session is offered in which participants tackle fun but challenging problems. Most afternoons and evenings are spent working on homework and class projects, seeing instructors during their office hours, and otherwise thinking about mathematics. The program has a collaborative nature: the participants are encouraged (and often even required) to work together in and out of class.

In 2005, the directors added SMPosium, a 3-day reunion conference, to the summer program schedule. Alumnae of the program who earned doctorates (PhD's) in the mathematical sciences are invited to Carleton to interact with each other and with the current participants. The conference program includes research presentations given by the PhD's as well as discussions on graduate school and career issues. More details on SMPosium are provided in Sect. 16.3.3.

SMP integrates scholarly activities and social opportunities. Group dinners are held after each colloquium. Friday evenings and Saturdays have scheduled social events, including dinners and game nights, as well as outings to area parks, Minneapolis, and the Mall of America. All of the students, staff, and visitors engage in these events, and by the end of SMP, the program participants know each other very well and form a tight-knit community.

16.3 Mentoring in SMP

From the beginning, Haunsperger and Kennedy designed the Carleton Program to emphasize support for young women in mathematics. During the summer program, staff members immerse themselves in program activities along with the students, and significant time is devoted to mentoring.

Being an instructor for SMP is an enormous responsibility. Before the program begins, each instructor must design a course that introduces specialized mathematics at a level far below the usual approach to the topic. I first encountered Lie theory in my second year of graduate school, when I took several courses on Lie algebras and Lie groups. At the graduate level, these courses required firm knowledge of algebra and differential topology that went far beyond anything I learned as an undergraduate (and I was lucky enough to take four semesters of algebra and one semester of topology at Bowling Green). Since most SMP students have taken neither abstract algebra nor topology, my SMP course clearly required a different approach. With assistance from texts by Pollatsek (2009) and Stillwell (2008), I developed a course that introduced the ideas of Lie theory through methods that require only basic knowledge of calculus, linear algebra, and complex number systems. Instructors also have to maintain a delicate balance between challenging the students and building their confidence: an SMP course needs to be challenging, but not so difficult that the participants leave the program feeling defeated.

Teaching is just a small part of the SMP instructor's job. Outside of the classroom, the instructors spend a significant amount of time having *conversations* with the participants. As an instructor at SMP, I spent every weekday afternoon holding office hours. The participants made good use of my office hours, but a typical visit did not just focus on class material. The students wanted to chat about practical things like graduate school, courses they might want to take at their home institutions, career options, and research areas. They also stopped by to talk about private struggles and triumphs or just to tell me about things that were happening in their lives. SMP instructors get to know their students on a deep personal level, which allows instructors to give participants individual advice. When asked on the end-of-program survey what she liked about SMP, a recent SMPer wrote, "Pam and Erica [Flapan] were *fantastic*, not just as teachers but as life coaches."

The teaching assistants, program assistants, and directors are additional "life coaches" to the SMP participants. The teaching assistants probably spend the most time with the students, as they live in a dorm with them in addition to helping in class and holding office hours. They, too, develop close personal relationships with the students. Each week, Haunsperger holds a "Deanna Chat," a discussion with the participants and assistants (instructors and visitors are not invited) focusing on the purpose of SMP and on how the program is going. When I was a student, I recall Deanna Chat being a welcoming and open discussion of anything from frustrations with courses to the state of women in mathematics. The numerous social events also provide opportunities for staff members to mentor the participants through informal conversations.

In the early years of the program, the program directors, instructors, and teaching assistants were the primary source of mentoring in SMP. As the SMP community grew, Haunsperger and Kennedy realized the potential for other mentoring activities and implemented new formal structures for interaction among alumnae.

16.3.1 Community Resources

SMP excels at keeping alumnae connected. Haunsperger sends a semi-annual newsletter to all members of the SMP community that includes news from all cohorts. When an alumna joins a research program, goes to graduate school, gets a job, gets married, etc., it is usually recorded in the newsletter. The program also has a Facebook group for members of the SMP community. Through this venue, alumnae can post questions for the group, suggest programs or conferences that might be of interest, or announce personal news. The newsletter and the Facebook group keep the community connected and provide mechanisms for tracking the progress of alumnae.

The directors gather data on other programs in which the alumnae participate. For example, lists are maintained of undergraduate research and graduate programs that members of the community have attended, including names and email addresses of the women who attended each program. These program lists are annually updated and distributed to the SMP participants of that year, and they are available to other members of the SMP community by request. If someone from SMP is considering a particular program, she can contact an appropriate SMPer for advice or information.

The SMP community also periodically surveys alumnae to collect data on life and career events. For example, a document lists alumnae in academic jobs who have experience with pre-tenure or tenure reviews, publishing their research, getting grants, creative teaching methods, etc. A similar document exists for women in industry jobs, detailing an alumna's employer and type of position as well as listing courses, programs, or training that she found relevant to her job. All alumnae can volunteer information on life events like solving a "two-body problem," having children, work-life balance, etc. The survey results are stored in Google Docs so that members of the community can update their responses as needed. The motivation for collecting this information is so that members of the SMP community can share their experiences. For example, an alumna who is approaching her tenure review can contact other SMPers for advice.

Maintaining these resources requires absolutely no funding. However, the success of these resources relies on the willingness of the community members to participate. When an SMPer contacts another to discuss a program or problem, even if the two have never met, it is likely that an enthusiastic response will be received. It is a testament to the strength of the SMP community that the program alumnae are so passionate about supporting their SMP sisters.

16.3.2 Reunions

The SMP family is fortunate to have several opportunities for reunions each year. National conferences, notably the Joint Mathematics Meetings (JMM) and MathFest, provide a natural vehicle for these reunions. Each year at the JMM, Haunsperger organizes a reunion for any SMP community members who are attending the conference. My first SMP reunion at the JMM was in New Orleans in 2001. That year, Haunsperger, Kennedy, and a handful of alumnae gathered in a hotel lounge to chat. A few years later, there were enough alumnae present at the conference that this kind of informal gathering was not practical, so the directors sponsored a dinner at a local restaurant. In 2016, nearly 50 people attended this reunion dinner. The reunion dinners allow members of the same SMP class to reconnect and provide an opportunity for networking across SMP cohorts. In 2005, for example, I was finishing my PhD and applying for jobs. At the SMP reunion, I met several other SMPers who were either currently on the job market or who had recently completed a successful job search. We spent most of the dinner discussing strategies for interviewing and for stress management. Exchanging ideas with other women was very valuable, and I left the dinner feeling much more confident about my upcoming interviews.

In recent years, members of the SMP community held smaller, regional reunions in locations with many SMP alumnae and instructors, such as Los Angeles, Minneapolis, and even Budapest. Regional reunions are not funded by the SMP grant. Individuals pay for their own meals and any necessary travel expenses, but participants value interaction with their SMP family enough to fund themselves.

The reunions held by SMP truly strengthen the bond among its members. Conversations at SMP dinners do not just focus on career topics; the women share all aspects of their lives. Frequently, a dinner attendee can see alumnae passing around pictures of their children or hear a passionate discussion about sports. I seek advice from my SMP friends about my interests outside of mathematics, like knitting, dog training, and film. These personal interactions help SMPers feel like they are indeed part of a supportive family.

16.3.3 SMPosium

The 2005 SMP reunion at JMM inspired the next program innovation. Haunsperger and Kennedy described their revelation in an article in *Math Horizons* (2007).

> When we arrived, most of the SMPers were already there, and the natural thing was happening. They were talking to each other. … We didn't need to do any introductions; in fact, we couldn't have because no one could talk over the roar of conversation. … We had not fully realized, until this dinner in Atlanta, that over the years what we had actually built was an incipient community ready to interact, ready to support itself, and it just needed a tiny push to get it going full strength.

To give the SMP community this "tiny push," the directors created the SMPosium, an annual 3-day reunion conference that occurs during the summer program.

The first SMPosium was held in the summer of 2005. That year, the 12 SMP alumnae and teaching assistants with doctorates were invited, and nine were able to attend. The PhDs gave short research talks at a level that was accessible to the SMP students, served on panel discussions on graduate school and careers, and had informal conversations with the participants and with each other.

I have attended every SMPosium since the initial conference in 2005, and the event has grown significantly. Of course, the number of SMP alumnae with PhDs increases each year, so attendance swells. In 2014, 68 alumnae were invited, and 25 were able to attend. The formal program for the conference also evolved. A recent addition is a session only for PhD's that includes higher-level research talks and structured conversations about career and life issues. The current participants and alumnae have many opportunities for discussion. Group dinners were held each evening, and the entire third day of the conference is devoted to informal social interaction. Everyone goes to a local park or beach to canoe, swim, play games, and talk to each other. Significant mentoring occurs during the outing: an observer would find participants talking to alumnae about what graduate school is like, alumnae talking to each other about job issues or favorite teaching methods, and everyone sharing stories about their mathematical experiences.

The students consistently report that SMPosium is inspiring. In fact, many mention the conference as their favorite part of the program. When asked for feedback about the summer program overall, one student responded as follows:

> SMPosium was absolutely my favorite part. Talking to alumnae about what their careers and lives look like had the most impact for me. It helped me to be able to imagine myself in their career and see if it would be a good fit for me.

The participants learn a lot of new, exciting mathematics. They learn about the struggles the PhD's faced and know they are not alone in their own struggles. They hear multiple perspectives on surviving graduate school, choosing a career, and study abroad programs. Most importantly, though, the participants get to interact with women who were once in their shoes and succeeded.

The PhD's also express that SMPosium is an invigorating experience. They, too, learn a lot of new, exciting mathematics. They get to meet more members of their SMP clan and reunite with some that they already know well. They are enthusiastic about sharing their stories and mentoring both the participants and each other.

16.3.4 Mathematicians in Residence

SMPosium is amazing, but it is also very short. The participants and PhD's have only 3 days together, which often means that the students are just starting to get to know their role models when it is time to leave. Extending SMPosium is not feasible, as feeding and housing many guests would be prohibitively expensive. However, the directors saw the positive interactions that were occurring during SMPosium and sought a way to prolong them.

In 2009, Haunsperger and Kennedy invited a few alumnae to try out a new position in SMP: the Mathematician in Residence (otherwise known as "MiR"). That summer, Jennifer Bowen, Alissa Crans, Katherine Crowley, and I each visited the program for 2–3 weeks to act as role models for the participants and show them what a mathematician actually *does*. During the day, we worked on our research or other professional projects while the students were in class or doing homework. We went to every program colloquium, panel, and social event, and we worked in the same places that the participants did, making ourselves visible and accessible.

The participants were a bit shy at first but soon warmed up to us. We found ourselves answering their questions about graduate school, giving them advice about classes to take or programs in which to participate, telling stories from our first jobs, and discussing our research with them. We learned about their families, their interests outside of mathematics, and their favorite parts of their classes. Our presence gave the participants an additional friend to talk to when they felt overwhelmed or had other concerns. The same interactions occur at SMPosium, but the extended time allowed us to form deeper bonds with the participants and put less pressure on the students to ask all of their questions within a 3-day window.

The experiment was a success, so in the subsequent years, five additional SMP alumnae were invited to be Mathematicians in Residence. Each alumna visits for 1–3 weeks of the program. Mathematicians in Residence are not compensated financially outside of housing and travel; alumnae serve as MiRs because they believe in the program mission and enjoy the mentoring opportunity it provides.

16.3.5 Graduate Education Mentoring (GEM) Workshop

Before 2010, SMP alumnae interested in graduate school could benefit from mentoring events before they entered graduate school and after they completed a

PhD. However, no formal mentoring activities existed for arguably the most critical time: *during* graduate school. To fill this gap, the annual Graduate Education Mentoring (GEM) Workshop was created. The inaugural 2010 program was organized by Bowen, Crans, and Crowley, and I became a workshop leader in 2012.

The GEM Workshop is a full-day event held every January, on the day before the start of the Joint Mathematics Meetings. SMP alumnae who are currently in graduate programs in the mathematical sciences are invited, and about 14 alumnae attend each year. Approximately seven SMPers with PhD's are invited to participate as mentors, in addition to the workshop leaders. The GEM program features talks, panels, and small-group discussions.

The GEM schedule typically includes six or seven talks given by the graduate student participants. Each speaker has 20 min to give a general-audience talk on her research and is assigned two mentors to provide feedback. The mentors complete a written evaluation of the presentation, commenting on organization, clarity, and delivery, and the speakers and mentors discuss the results. Mentors are specifically asked to give supportive advice and to mention strengths as well as weaknesses of the presentations. Most of the GEM participants have not given many professional talks, and the workshop evaluations indicate that the speakers appreciate having a friendly audience for practicing their presentation skills.

Themes for panel discussions are determined by the graduate student participants. Each year, the women attending GEM are surveyed about topics they want to discuss, and the workshop leaders use this information to choose panelists and guide the session. Previous panels covered choosing an advisor, preparing for qualifying exams, changing programs or advisors, navigating the job market, finding research collaborators, publishing research, career options outside of academia, dealing with imposter syndrome (feeling that one's accomplishments are not deserved and fear of being exposed as an imposter) and stereotype threat (concern that one's behavior will confirm negative stereotypes), and finding a successful work/life balance. The mentors are always happy to share their perspectives and advice, and the panels are consistently noted as a highlight of the workshop.

The GEM attendees are often in different stages of their graduate school experience. Recent workshops included informal small-group discussions organized by year in graduate school: first- and second-years in one group, third- and fourth-years in another, and fifth-years and beyond in the last group. The mentors are distributed among the subsets. Separating the women this way allows for more targeted mentoring. The women in their first few years of graduate studies can discuss issues like finding an advisor and choosing a research area, while the women nearing the end of their graduate studies can concentrate on career issues and the job market. The small-group discussions have been extremely popular, and feedback collected from the workshop attendees indicates an interest in more small-group discussions, perhaps based on research area or geographic location.

When the formal workshop is over, all GEM attendees are invited to a dinner to allow time for continued discussions among the participants, the mentors, the workshop leaders, and the directors of SMP. The participants appreciate these informal interactions as much as the official workshop events, as the dinner strengthens the friendships formed during the day. All GEM meals are funded by the SMP grant. The graduate student participants and workshop leaders also receive modest amounts of travel support to attend the event, while the mentors volunteer their time with no travel support from SMP.

At the end of each GEM workshop, the participants and mentors evaluate the program, and the responses are unanimously positive. All indicate enthusiastically that they would attend the event again and would recommend the workshop to other SMP alumnae. Many of the graduate students say that they feel less “alone” after the workshop and that knowing that other women are going through similar circumstances is comforting. Even the mentors note that they learn from the workshop. On a recent evaluation, a mentor had the following comment:

> The small-group discussions helped remind me how different are the issues faced by 1st year grad students compared to 4th and 5th years (the time that’s fresher in my mind!). It will be helpful to me as I advise my students who are deciding whether to attend graduate school.

In all years since the first workshop, the GEM feedback demonstrates an appreciation for the support of the SMP community and for the program’s spirit of collaboration.

16.4 Evidence of Success[1]

SMP has many objectives: introducing students to new areas of mathematics; honing students' mathematical reasoning, proof writing, problem solving, and presentation skills; building self-confidence, encouraging enthusiasm for mathematics, and increasing awareness of opportunities for continued study in the mathematical sciences; and connecting students into a supportive network of other female college math majors, graduates and professionals to support them through their graduate studies in mathematics (Summer Mathematics Program 2016). Underlying all of these goals is an aim to help women *persist* in graduate school and mathematical careers.

To date, the SMP family includes 337 alumnae, teaching assistants, and program assistants who have completed their undergraduate degrees. Of these, 91 have doctorates, 62 have terminal master’s degrees, and 55 are currently in graduate school in the mathematical sciences. Five have doctorates and eight are in graduate school in cognate disciplines. Thus, approximately 47 % of SMPers with bachelor’s degrees already have an advanced degree in mathematics or a cognate discipline, and we expect that number to grow to 65 %. Among those who did not pursue a graduate degree, 19 are secondary teachers, and 51 have mathematics-related industry or

[1]All SMP statistics in this section are from raw unpublished data compiled by the program directors.

government jobs. We anticipate nearly 87 % of the SMP alumnae, teaching assistants, and program assistants will have a graduate degree or a mathematical career.

The success of SMP alumnae in mathematics doctoral programs is particularly remarkable. Nationally, the participation of women in degree programs in mathematics decreases dramatically at the PhD level. In 2014, approximately 41 % of undergraduate degrees and about 41 % of master's degrees in mathematics were earned by women (Vélez et al. 2016). However, only approximately 26 % of PhD's (Vélez et al. 2015) were earned by women. These data imply that mathematics graduate programs are "losing" women before they earn their PhD's.

Calculating a persistence rate for women in PhD programs is difficult, as only aggregate data is reported in national surveys. However, we can estimate a rate using data from the American Mathematical Society (AMS) *Annual Survey of the Mathematical Sciences*. Between academic years 1995–1996 and 2008–2009, the surveys report that 21,858 women were first-year graduate students in the mathematical sciences (Annual Survey 2016). Assuming approximately six years to complete a PhD, we consider academic years 2000–2001 through 2013–2014, during which 6040 women earned a PhD (Annual Survey 2015). See the Appendix for these data. This gives us a persistence rate of about 28 %.

The SMP directors track alumnae progress and have individual data that provides a more accurate persistence rate for members of the SMP community. Assuming approximately 8 years between the summer program and a PhD, we consider the 181 alumnae from SMP 1995–2005; from these years, 72 of the 118 alumnae who entered PhD programs in the mathematical sciences persisted to a PhD, giving SMP a persistence rate of 61 %. In the findings of SMP's program assessments, many of these women specifically cite that SMP had a direct influence on her success in graduate school.

In addition to tracking the progress of the participants, the directors assess the components of SMP. During the summer program, the participants take both pre- and post-program surveys to gauge their perception of their mathematical skills and self-confidence, their knowledge of graduate school and career opportunities, and their experience overall in SMP. Feedback forms are also used to assess the GEM Workshop and the SMPosium. Many of the survey questions are qualitative rather than quantitative, and the responses are overwhelmingly positive. When asked, "What did you gain from participating in the SMP program?" most participants specifically mention an increased confidence in their mathematical abilities and in their likelihood to pursue graduate school, without being prompted to address those goals. A recent SMPer's response to this question included the statement, "I gained a lot of new knowledge, a lot of new friends, and most importantly a lot of confidence. I came in knowing little about graduate school and feeling inadequate, but my confidence has skyrocketed." The feedback collected from all events related to SMP is evidence that the program is meeting all of its goals.

The mathematics profession frequently recognizes the success of both SMP and the women involved in the program. Most notably, in 2014, the American Mathematical Society named SMP one of its "Programs that Make a Difference," an honor that is awarded to programs that succeed in bringing "more persons from underrepresented backgrounds into some portion of the pipeline beginning at the undergraduate level and leading to advanced degrees in mathematics and professional success, or retain[ing] them once in the pipeline" (AMS Committee 2016). The award citation states, "Through all of their activities, the organizers of SMP have been able to form an impressive vertically-integrated network of support and mentoring for and by the members of the SMP community" (AMS Mathematics Programs 2014). In 2012, Haunsperger was awarded the Association for Women in Mathematics Second Annual M. Gweneth Humphreys Award, given each year to a mathematics teacher in recognition of superb mentoring of female undergraduates in mathematics. The award citation ends with the following remark (Second Annual M. Gweneth 2012):

> The AWM is pleased to honor Deanna Haunsperger for her wonderful achievements and unwavering efforts over decades in the mentoring of undergraduate women in mathematics, in particular in attracting them into the study of mathematics and creating a thriving community which supports them throughout their mathematical careers.

Erica Flapan and Margaret Robinson, two regular instructors in SMP, have been awarded the Mathematical Association of America's (MAA) Deborah and Franklin Tepper Haimo Award, given to college or university faculty "who have been widely recognized as extraordinarily successful and whose teaching effectiveness has been shown to have had influence beyond their own institutions" (Deborah and Franklin n.d.). Flapan, a mathematics professor at Pomona College and twelve-time instructor at SMP, was a 2011 recipient of the Haimo Award. Robinson, a mathematics professor at Mount Holyoke College and three-time instructor at SMP, was a 2013 recipient of the Haimo Award. The citations for both awards explicitly mention involvement in SMP as evidence of their excellent teaching (January 2011 Prizes 2011; January 2013 Prizes 2013).

16.5 Career Impact

Involvement in SMP had a profound effect on my career. When I attended as a student, I was certain that I loved mathematics, but I didn't know how I could turn that interest into a career. Through the panel discussions at the program, I learned about careers in industry and government that I didn't know existed. My home institution had dozens of mathematics faculty, but they were *all* men. Until I met the instructors and directors of SMP, I didn't really consider being a professor as a realistic option. I learned about study abroad programs, talked to people who had done undergraduate research programs, and learned about conferences and professional organizations that would welcome me even while I was an undergraduate. SMP opened my eyes to a world of possibilities, and I left the program with both confidence that I had a future in mathematics and a plan to enact that future.

I went to graduate school knowing that I had the support of the SMP community. During the program in 1999, I became very close friends with some of the participants and teaching assistants, all of whom were also in graduate school, albeit at different schools. My closest friend, Emily (Gamber) Burkhead, was a graduate student at the University of North Carolina while I was studying at the University of Virginia. Our geographic proximity allowed us to visit each other often, and we went on vacations together, roomed together at conferences, and frequently talked about our lives. Emily and I could discuss things that were awkward to discuss with the other graduate students in our own institutions, like how much funding we were getting, how we performed on major examinations, and how supported we felt by our departments. Hearing her perspective on these issues was comforting and helped me cope with the stress of earning a PhD. I also attended national conferences more often during graduate school, which introduced me to members of the SMP community who were not in my cohort. I knew that I could reach out to these women when I had a question or concern, and knowing this resource existed increased my confidence even if I didn't use it often.

I later began a tenure-track position at Westminster College in Pennsylvania, where my participation in SMP certainly had an impact on my tenure process. By the time I went up for tenure, I had been to SMPosium six times, been a MiR once, and been an instructor twice. My tenure portfolio required letters of support from faculty outside of Westminster, and I was able to get glowing references from Haunsperger and Kennedy. Being chosen to be an instructor in SMP is a great honor, and being invited to teach four times is more than I ever imagined. My fellow instructors are phenomenally talented mathematicians and teachers, and my inclusion in their ranks did not go unnoticed by my home institution. In 2014, Westminster College nominated me for the US Professor of the Year program, a national award that "salutes the most outstanding undergraduate instructors in the country—those who excel in teaching and positively influence the lives and careers of students" (About the Program 2015). My work in the SMP community was a significant part of my nomination for this award. While I was not a recipient of the national award that year, my nomination is a testament to how much my home institution values the work that I do in the Carleton program.

Being part of the SMP community resulted in several publications and research collaborations for me. In 2009, while we were both Mathematicians in Residence at SMP, Bowen and I began a multi-year research project in non-associative algebra. In 2015, I collaborated with Robinson, with whom I taught in SMP 2011, on research in number theory. Both of these collaborations will likely lead to research publications soon. In 2013, I was solicited to write an article about the program for *Math Horizons* (Richardson 2013). In 2015, Haunsperger and I collaborated on a short piece for the MAA Centennial, variations of which were published on the MAA website, *Focus*, and *Math Horizons* (Haunsperger and Richardson 2015a, b). These publications and research partnerships would not have occurred without SMP.

My work in SMP provides me with less tangible benefits that impact my life but are not evidenced through publications or awards. My involvement in the program increased my visibility in the mathematical community. My professional network expands each time I attend a conference, and most of the new connections I make are through an SMPer (often through Haunsperger and Kennedy, who seem to know *everyone*). For example, my first *Math Horizons* article resulted from conversations with one of the editors, whom I met through Haunsperger and Kennedy. I was appointed to two terms on the MAA's national Committee on Undergraduate Student Activities and Chapters, likely due to the influence of SMP. Being an instructor in SMP was by far the most rewarding teaching experience of my career; to interact with young women who are so in love with mathematics is marvelous. I get to watch the students I have mentored or taught in the program "grow up" and succeed in mathematics. I feel an almost parental pride when SMP alumnae get into an undergraduate research or graduate program, earn a PhD, get a fantastic job, or publish a paper. As a mentor in SMP, I am involved in the development of hundreds of young women in mathematics, which is a reward on its own. The greatest benefit is that I genuinely have a mathematical *family* in SMP—an extended family for me. When I have good news to share, when I face adversity, or when I need support, I turn to SMP.

16.6 Conclusion

In the decades since its creation, SMP has grown to be a strong, supportive community for its alumnae, instructors, assistants, and directors. The Carleton College Summer Mathematics Program for Women is absolutely a program that impacts the lives of its participants.

SMP represents a replicable model for community building. While the program enjoyed significant grants from the National Science Foundation, many of the events that support the SMP community require little funding. All of the online resources mentioned in Sect. 16.3.1 are completely free to implement, and events like the GEM Workshop could be held locally (e.g., at a single graduate institution or in a city with several local graduate schools) with minimal cost. What is required is a long-term commitment to building and maintaining a community.

I am frequently asked if being a *women's* program is part of what makes SMP successful. I do think that, in the often male-dominated mathematical world, we need programs that support women. In my career, I have found that my female mathematics colleagues are more open to discussing their experiences, which is the backbone of SMP mentoring activities. In these ways, it is important that SMP is a program for women. However, I believe that a support community like SMP can be constructed for almost any group. What really bonds the women of SMP together is a shared experience. Even if we attended SMP in different years, we all went through a similar, intense program that inspired us, challenged us, and increased our confidence.

Acknowledgements I would like to thank Deanna Haunsperger and Stephen Kennedy for creating this marvelous community. SMP was supported by National Science Foundation grants DMS-9531237, DMS-9817967 as The Carleton and St. Olaf Colleges' Summer Mathematics Program, DMS-0244538 and DMS-0632713 as The Carleton College Summer Mathematics Program, and DMS-0943597 as The Carleton College Summer Mathematics Program for Women.

Appendix

These data were compiled from many years of data at the AMS Department Profile website. The data for years prior to 2009 can be found in the corresponding "Graduate Student Profile" link on the website referenced in Sect. 16.4.

Table A.1 AMS annual survey data

	Female first-year graduate students	Female PhD recipients
1995–1996	1181	
1996–1997	1208	
1997–1998	1271	
1998–1999	1462	
1999–2000	1549	
2000–2001	1792	311
2001–2002	1820	295
2002–2003	2020	307
2003–2004	1653	331
2004–2005	1511	363
2005–2006	1710	394
2006–2007	1559	365
2007–2008	1558	435
2008–2009	1564	462
2009–2010		514
2010–2011		524
2011–2012		554
2012–2013		577
2013–2014		608
Total	21,858	6040

References

About the Program. (2015). U.S. Professor of the Year Awards Program. Retrieved February 28, 2016 from http://www.usprofessorsoftheyear.org/About_POY.html#.VtMIsZwrIU0.

AMS Committee on the Profession Award for Mathematics Programs that Make a Difference. (2016). The American Mathematical Society. Retrieved February 22, 2016 from http://www.ams.org/profession/prizes-awards/ams-supported/make-a-diff-award.

AMS Programs that Make a Difference. (2014). The American Mathematical Society. Retrieved February 22, 2016 from http://www.ams.org/programs/diversity/citation2014.

Annual Survey of the Mathematical Sciences: Departmental Profile. (2016). The American Mathematical Society. Retrieved January 30, 2016 from http://www.ams.org/profession/data/annual-survey/deptprof.

Annual Survey of the Mathematical Sciences: Doctorates Granted (2015). The American Mathematical Society. Retrieved January 30, 2016 from http://www.ams.org/profession/data/annual-survey/docsgrtd.

Deborah and Franklin Tepper Haimo Award. (n.d.). The Mathematical Association of America. Retrieved February 24, 2016 from http://www.maa.org/programs/maa-awards/teaching-awards/haimo-award-distinguished-teaching.

Haunsperger, D., & Richardson, P. (2015a). What were they thinking? A look at life in 1915. *Math Horizons, 22*(4), 14.

Haunsperger, D., & Richardson, P. (2015b). What were they thinking? A snap-shot of life in 1915. *MAA Focus, 5*.

January 2011 Prizes and Awards. (2011). Retrieved February 24, 2016 from http://www.ams.org/profession/prize-booklet-2011.pdf.

January 2013 Prizes and Awards. (2013). Retrieved February 24, 2016 from http://www.ams.org/profession/prizebooklet-2013.pdf.

Pollatsek, H. (2009). *Lie groups: A problem-oriented introduction via matrix groups*. Washington, DC: Mathematical Association of America.

Richardson, P. (2013). Divine secrets of the mathematical sisterhood. *Math Horizons, 22*(2), 22–24.

Second Annual M. Gweneth Humphreys Award. (2012). The association for women in mathematics. Retrieved February 23, 2016 from https://sites.google.com/site/awmmath/programs/humphreys-award/past-recipients/humphreys-award-recipients-announcements/second-annual-m-gweneth-humphreys-award.

Stillwell, J. (2008). *Naïve Lie theory*. New York, NY: Springer Science+Business Media, LLC.

Summer Mathematics Program for Women Undergraduates (2016). Retrieved January 11, 2016 from http://www.math.carleton.edu/smp/.

Vélez, W. Y., Barr, T. H., & Rose, C. A. (2016). Fall 2014 departmental profile report. *Notices of the AMS, 63*(2), 163–173.

Vélez, W. Y., Maxwell, J. W., & Rose, C. (2015). Report on the 2013-2014 new doctoral recipients. *Notices of the AMS, 62*(7), 771–781.

Part IV
Benefitting Students in General Education Courses

Chapter 17
Creating and Sustaining a First-Year Course in Quantitative Reasoning

Kathleen Lopez, Melissa Myers, Christy Sue Langley, and Diane Fisher

Abstract In 2009, the Louisiana Board of Regents ceased to require that all undergraduate students receive credit for either a version of college algebra or calculus. By that time, many universities across the nation had well-established quantitative reasoning (QR) courses. Since the Department of Mathematics at the University of Louisiana at Lafayette has been active in mathematics reform, it was natural for the department to expand first-year course offerings by creating a QR course. This chapter describes the development and implementation of this course, which was first offered in Spring 2013. It also gives details about continuing challenges and the resources created to support course instructors and students. Of particular concern was whether students in the QR course would progress through their mathematics courses at the same rate as students with similar background who take college algebra. Data collected on student progress suggest that this is the case. The chapter discusses campus reactions to the new course and closes with a brief reflection on how working to develop the QR course fit into the careers of the developers.

Keywords Critical thinking • Financial literacy • Numeracy • Quantitative literacy • Quantitative reasoning

MSC Codes
97A40
97B10
97B40
97D40

K. Lopez (✉) • M. Myers • C.S. Langley • D. Fisher
Department of Mathematics, University of Louisiana at Lafayette,
P.O. Box 43568, Lafayette, LA 70504, USA
e-mail: klopez@louisiana.edu; mgmyers@louisiana.edu;
clangley@louisiana.edu; dfisher@louisiana.edu

J. Dewar et al. (eds.), *Mathematics Education*, Association for Women in Mathematics Series 7, DOI 10.1007/978-3-319-44950-0_17

17.1 Introduction

For several years, faculty and administrators at the University of Louisiana at Lafayette (UL Lafayette) had contemplated the creation of a first-year Quantitative Reasoning (QR) course. This became possible in 2009, when the Louisiana Board of Regents (LBOR) ceased to require all undergraduate students to receive credit for either a version of college algebra (CA) or calculus. Creating a QR course that met not only the initial faculty vision but also that of the administration posed several challenges. Careful consideration was given to creating a course that would benefit students both before and after graduation. Employing a pedagogy that engaged students was also important for the QR course. At the same time, it was important that students taking QR were completing their mathematics requirement at a rate comparable to those taking CA. By preparing the course instructors, utilizing class activities, encouraging student discussion, and prompting students to take ownership of their learning, these objectives were achieved.

17.2 Why Create a Quantitative Reasoning Course?

UL Lafayette, a member of the University of Louisiana System, is a public institution designated "R2: Doctoral University—Higher Research Activity" in the 2015 Carnegie Classification of Institutions of Higher Education (http://carnegieclassifications.iu.edu). UL Lafayette is accredited by the Commission on Colleges of the Southern Association of Colleges and Schools. Enrollment is nearly 19,000, including about 1500 graduate students. The Department of Mathematics offers Bachelor of Science, Master of Science, and Doctor of Philosophy degrees. With over 80 undergraduate degree programs at the university, a large portion of the department's teaching load consists of service courses. The department is committed to providing quality instruction and has a rich history of education reform in its mathematical preparation of teachers, calculus courses, and college algebra courses. In 1995, with financial support from LBOR, the department was the first in the state to offer a workshop for Louisiana university faculty to address the reform of CA. Since then, the department has presented five more statewide faculty development workshops on the reform of undergraduate mathematics education. Through these workshops and faculty presentations at regional and national meetings, the department has gained recognition for its innovation.

While some of our students begin their mathematics education in calculus or precalculus, the majority enroll in Applied College Algebra (ACA). UL Lafayette has two formats for its ACA course: a 3-h course and a 5-h course. Students with a 19 or 20 Mathematics ACT score or credit in intermediate algebra must enroll in the 5-h course, while those with scores 21–24 take the 3-h course. There are two paths students follow upon completion of ACA. One path is taken primarily by business and science-oriented majors, who are required to take more advanced mathematics courses. The second path is for non-technical majors (that is, majors other than business or science), such as those in the Colleges of Arts and Liberal Arts, whose second courses are more applied in nature. When the LBOR broadened the scope of the required first-year mathematics courses, we had the opportunity to implement an alternative course designed specifically for non-technical majors that allowed us to better serve them. Students with weak mathematical backgrounds and little interest in mathematics were learning topics in ACA such as finding zeros of a polynomial function and identifying the multiplicity of each. These same students could not compute successive discounts on an item of clothing or understand the consequences of choosing a lower loan payment regardless of the interest rate or loan term. Faculty agreed that students in non-technical fields would benefit more from a course that included quantitative reasoning than they would from a typical CA course. In addition to our personal experiences, this choice is supported by the literature.

Much has been written about the importance of QR[1] in education to produce effective workers, citizens, and consumers. Since 1983, the need to educate students to make sense of and use mathematics in their lives has been the subject of many reports issued by educational organizations in the United States: *A Nation at Risk* (National Commission on Excellence in Education 1983), *Everybody Counts* (National Research Council 1989), *Curriculum and Evaluation Standards* (National Council of Teachers of Mathematics 1989), *Principles and Standards* (National Council of Teachers of Mathematics 2000), and *College Learning for the New Global Century* (Association of American Colleges and Universities 2007). The last report defines "what contemporary college graduates need to know and be able to do" and QL is one of six essential Intellectual and Practical Skills listed (p. 12).

As editor of the seminal books, *Why Numbers Count: Quantitative Literacy for Tomorrow's America* (Steen 1997) and *Mathematics and Democracy: The Case for Quantitative Literacy* (Steen 2001), Lynn A. Steen argued that innumeracy leads to disenfranchisement: "Quantitative literacy is to mathematics what literacy is to language. In addition to the skills of reading and writing, today's society requires logical reasoning and numerical thinking." (Adults Learning Mathematics 2015). In January 2004, the Board of Governors of the Mathematical Association of America (MAA) approved the formation of a special interest group whose purpose is to advance QL at the collegiate level (http://sigmaa.maa.org/ql/about.php). Bernard L. Madison joined Steen in championing QL. They emphasized that developing adults who can reason quantitatively is a responsibility that colleges and universities cannot leave to the K-12 educators (Madison and Steen 2003, 2008). Madison (2009) stated, "Reduced interest in mathematically intensive disciplines … does not diminish the QR burden on college students in their non-mathematically intensive majors or in their everyday lives as consumers and citizens of a democracy. These circumstances do mean that QR education— absolutely necessary if we are to sustain our democratic processes—cannot be the sole province of one or two

[1] Quantitative reasoning (QR), quantitative literacy (QL), and numeracy are closely related concepts. QR is the term used to describe UL Lafayette's course. All three terms occur in the literature we cite.

disciplines. Whatever students major in, we must ensure that they learn to reason quantitatively in their contemporary world" (p. 6).

One aspect of QR is financial literacy. In summarizing eight articles devoted to financial literacy, Lusardi and Wallace (2013) concluded that financial literacy depends on quantitative literacy and is correlated with good financial practices. In order to become wise consumers, adults need to obtain some literacy concerning their personal finances while they are developing the ability to reason quantitatively. A study of QR activities at universities "suggest[s] that certain students may be at greater risk for not developing these important skills, especially women and students majoring in non-STEM disciplines" (Rocconi et al. 2013). This is exactly the audience the new QR course is designed to serve.

17.3 Development and Description of the New Course

While many universities across the nation had well-established QR courses, the concept of QR as a first-year course was new to our state. On our campus, discussions about including QR in curricula had been underway since 2006. At the request of business faculty in 2008, QL topics such as proportional reasoning and applications involving percentages had been included in the department's finite mathematics course. These changes were led by one of the authors who became interested in QR while serving on the university's steering committee for its upcoming reaccreditation. In that capacity, she had many opportunities to have conversations with faculty and administrators from numerous departments about the mathematical needs of their students.

During Fall 2012, the university's General Education Committee approved the creation of a new mathematics course called Quantitative Reasoning. Currently the new QR course is an alternative to ACA for approximately 10 % of the university's population, based on declared majors. It was important to inform the university community about the new course. Efforts included advertising through our campus radio station, an article in the alumni magazine, and meetings with advisors.

The course is essentially the creation of three mathematics faculty members: an Associate Professor with a PhD and 34 years of teaching experience, a Master Instructor with an MEd and 27 years of teaching experience, and an Instructor with an MS in Mathematics with 11 years of teaching experience. Relevant courses taught by these faculty members include not only ACA but also mathematics of finance and elementary statistics. All three faculty members have had extensive experience in curriculum design and in the implementation of several new mathematics courses.

One goal was to create a course which students find engaging and useful. Since the QR course is much less abstract than traditional algebra, it is more accessible to students. It is designed for students in non-technical fields and may be used as a prerequisite for mathematics of finance and elementary statistics, which many students take as their required second mathematics course. To provide a better foundation for these students in their sophomore mathematics courses and in life, many topics in ACA were replaced with content that is intended to strengthen the quantitative and financial literacy of the students. The course content includes typical QR material such as number sense, financial mathematics, linear and exponential growth, and graphical representation of data. Each of these topics was chosen for its importance in real-life situations or in building a foundation for future college-level mathematics courses taken by students. Depending on the students' and instructors' backgrounds, some special interest topics may be added. For example, one class investigated symmetry and shapes in architectural styles.

17.4 Implementation

After settling on the content for the course, the developers then turned their attention to pedagogy. Their goal was to create an interactive learning environment. Having taught mathematics for pre-service teachers, they knew that this could be accomplished through class activities that encouraged collaboration, classroom discussion, and the active engagement of students. Reducing the number of topics freed up class time for more in-depth discussion and in-class activities. Collaboration among students is facilitated by the design of the classroom itself. Students sit in groups at tables and attempt problems. They then discuss their solutions or questions with their group. By articulating their ideas, students learn to clarify their thinking and justify their reasoning. This is carried over to homework and exam questions. For instance, given two purchasing scenarios, they must determine mathematically the best option and support their choice through written explanation.

Class activities with manipulatives help students establish a concrete connection to concepts they are learning. In one activity, students use linking cubes to illustrate percentage change. As part of a lesson illustrating measurement conversions, an effective activity is to give students a bag containing 28 one-centimeter cubes, each having a mass of one gram. When a student holds one cube in their hand, they get a sense of what a gram feels like. Each student also compares the width of their little finger to the length of one side of the cube thus giving them a sense of what a centimeter looks like. It is then noted that a cubic centimeter can hold a milliliter of liquid. With all the cubes back in the bag, each person holds the bag to feel the weight of an ounce and learn another conversion fact.

The QR course is designed to increase a student's ability to recognize, use, and appreciate mathematics outside the classroom. Instructors have used various methods to reach this goal. One addition to the course is a presentation or two by a representative from the campus credit union. The lessons they learn about types of bank accounts, types of bank fees including ATM fees, borrowing money, applying for a credit card, and credit scores are among students' favorites. Student projects are incorporated into the course by some instructors to encourage students to explore the connection between mathematics and a topic of interest to them. Several students have given class presentations about musical rhythm, yet approached it from different mathematical vantage points. Others have explored topics of interest outside their majors. For example, one group did an interesting project looking at automotive engines, focusing on torque and horsepower—how they are calculated and what to look for based on the planned usage of the vehicle. Some instructors add bonus points in the students' homework grade for turning in a written description of an actual application of course material to their life or for researching a topic such as the interest rate and terms of a "pay-day loan." The course's topics along with its pedagogy have resulted in an engaging course that the students perceive to be relevant to their lives as citizens and to their future professions.

17.5 Addressing Faculty Challenges

As with any new undertaking, the creation and implementation of the course presented challenges. The first of these was staffing the course sections with appropriate faculty. Because the department relies heavily on adjunct instructors and teaching assistants as teachers of record for many of our first-year classes, maintaining consistency across all sections is a departmental priority. To make teaching the QR course less daunting to faculty and to promote uniformity, the developers created numerous resources.

Template lessons that included introductory information and examples for use in class were written for the entire course. Lessons for each topic were designed using software for an interactive whiteboard. Step-by-step work for each example is animated in the software so that the class can discuss what the next appropriate step would be, students can do the work for themselves, then see only that step appear in the presentation. Instructors have the option to show additional work by writing information on the lesson presentation displayed on the interactive whiteboard. The lessons contain notes to instructors reminding them of key ideas and common mistakes. A classroom was dedicated for the course and outfitted with an interactive whiteboard and a clicker response system in order to make the template lessons accessible to all sections and to allow for more interaction with online resources. Some lessons begin with an interactive quiz covering the previous lesson and assignment using the response system.

Each instructor receives a notebook which includes both digital and hard copies of these resources: a detailed instructor syllabus, instructor lessons for use with the interactive whiteboard, supplemental exercises, sample quizzes, tests from previous semesters, sample class policies, and detailed instructions on using the technology. A template QR course is also available for use on the university learning management system (LMS). It includes student versions of the lessons, supplemental exercises, quizzes to be used as review for exams, and links to additional student resources.[2] Students complete online homework through the textbook publisher's

[2]These resources will be shared with any interested party upon request; email math@louisiana.edu.

website. A model course, with all online homework assignments created, is available for faculty to copy.

In addition to these resources, a workshop was designed and conducted for potential instructors to learn about the course, its goals, pedagogy, use of manipulatives, and interactive whiteboard technology. During the workshop, sample lessons were modeled and participants worked through activities using manipulatives. Participants were also encouraged to practice using the interactive whiteboard in conjunction with prepared lessons, as well as using the response system clickers.

17.6 Addressing Challenges with Students

The creation and implementation of the course presented several intertwined challenges for students. After offering the QR course for the first semester, it became obvious that many students lacked basic study skills and motivation. While some study skills are taught in our First-Year Experience seminar, many students are unable to transfer the skills to mathematics. A number of strategies are used to help make up this deficit. In an effort to assist students in becoming organized note-takers and to allow class time to be spent doing and discussing mathematics rather than copying problems, a printable copy of examples for each lesson is available to students. Short articles and videos about study skills are assigned. Most readings and videos have a quiz or assignment attached but others are used simply for sharing helpful tips.

For many college classes, homework is not collected or graded. As a result, students often do not complete assignments. It is important for them to learn that not every assignment must have a formal assessment and that gaining benefit from any assignment, graded or not, is to their advantage. In an effort to increase student completion of assignments, the QR homework falls into two categories: traditional and online. The online homework allows students to receive immediate feedback and to request assistance in the form of guided solutions, links to text, and a link to email the instructor. Traditional homework is completed in students' notebooks, and is sometimes collected and graded for correctness. Random homework quizzes are conducted, either on paper or using the response system. The response system both engages students and provides immediate feedback. Throughout the semester, students are encouraged to seek help by referring to the student solutions manual, posting questions to the forum in the LMS, emailing the instructor, consulting with the instructor in person, asking a classmate for help, and utilizing any of the campus tutoring options.

The small number of available sections of the QR course means some students may find it difficult to fit the QR course into their schedules. Art majors, for example, have inflexible schedules due to studio courses. As a result, many eligible students are taking ACA rather than QR. As a rule our department works at meeting the student demand for a course by opening sections as needed. Because we feel the QR course would be more beneficial to these students, we are expecting to increase the number of available sections through increased publicity and education of advisors.

17.7 Outcomes

With the omission of topics, the department was concerned about rigor. The ultimate question was: "Are the students in QR progressing through their mathematics courses at the same rate as similarly prepared students who enroll in the 5-h ACA?" Two measures were used to answer this question: comparison of grade distributions and the pass rate of each population in the sophomore mathematics courses. If these two measures show no differences, then the answer to the question is "yes." Data of various types were collected and analyses performed by an Assistant Professor with a PhD in Statistics (UL Lafayette IRB SP15-68 Math).

Four semesters of grades were compared between the QR and the 5-h ACA courses. Grade distributions were analyzed for a total of 442 QR students and 1695 ACA students. In Fall 2014, Fall 2015, and Spring 2016, there were no significant differences in failures, withdrawals, or pass rates between courses. (A grade of A, B, C, or D is considered passing.) In Spring 2015, there was no significant difference in failure rates. However, that semester there were differences in the withdrawal and pass rates. A 95 % confidence interval for difference in withdrawal rates between these courses shows the withdrawal rate for the QR course is 0.7–19.9 % higher than ACA. The pass rates were 2.5–28.6 % higher for the ACA students. One possible explanation for the different outcomes in the Spring 2015 semester is that the department experimented with smaller class sizes in the 5-h ACA course staffed with more experienced instructors. This resulted in a higher pass rate than in a typical spring semester.

The performance of the students in their second mathematics courses was also examined. Data were collected on all 156 students who passed QR or the 5-h ACA in Fall 2014 and then enrolled in their sophomore mathematics course at UL Lafayette in Spring 2015. At the end of the second course, there were no significant differences between the two groups in failures, withdrawals, or pass rates. Therefore, we can conclude that for that group, QR and ACA students were making similar academic progress with respect to mathematics.

Three additional comparisons that validated the success of the QR course were performed. In Fall 2014, two sections of QR and two sections of the 3-h ACA were taught by the same instructor. This allowed us to administer a ten-question end-of-course quiz covering common concepts (e.g., computing percentage change, determining the original price if the percentage discount and the sale price are known, finding the slope of a line given two points, and writing a linear function). There was no significant difference in the scores between the groups. In Spring 2015, pre- and post-quizzes were given to all sections of QR and to a control section of ACA. For the QR students, a 95 % confidence interval showed the average improvement to be between 1.2 and 2.8 points on the 10-point quiz. However, for the ACA students, there was no significant improvement in the scores.

Pre- and post-attitudinal surveys were conducted anonymously in Spring 2015 in all sections of QR and two control sections of ACA. There were two items with significant differences from pre- to post-surveys. On the item "I can work any math problem if I have a formula," the QR students went from 40 % agreement to 61 % agreement. There was no significant difference in the ACA course. This change for

QR students, while not expected and initially disconcerting, is likely the result of student success with the financial literacy unit where several formulas were used. For "An understanding of mathematics is important to be an informed citizen," in the QR course the percentage of agreement with the statement changed significantly from 68 to 84 %. There was no significant difference for the ACA course.

17.8 Reactions on Campus

Student opinions of the course have varied over the semesters. It is worth noting a few of the comments from former students:

> As a course designed for students who did not necessarily have an aptitude for math, the course has been enjoyable.
>
> I think the course was most challenging because it was material I should've been familiar enough with to understand the lesson, but I wasn't. Wonderful class. Challenging but necessary for life skills.
>
> I loved this class … It made me wish I tried harder in math in high school.

After attending the instructor workshop, the Director of Freshman Mathematics commented,

> I consider this course to be more in-depth than a traditional 'survey of mathematics' course or a 'contemporary mathematics' course because students are required to not only perform the calculations of a traditional algebra course, but also understand and explain their reasoning.

The course was widely anticipated by faculty in the Colleges of Arts and Liberal Arts. In fact, once approval was given to create the course, the university requested the course be offered for the first time in Spring 2013 rather than Fall 2013 as initially planned. The QR course is different from what most students have seen before in mathematics courses and some students and administrators mistakenly took that difference to mean the course would be easy to pass with little effort. An Assistant Dean in the College of Liberal Arts commented, "… when [QR] was first rolled out, some of the students who took it that first semester it was offered, found it no easier than [ACA]. I have not heard that complaint for at least a year now." We believe this change occurred because each semester student expectations are becoming more in line with those for other mathematics courses. One Performing Arts faculty member stated:

> Our students are not usually big Math fans, so they seem to have accumulated a dread of Algebra. It's very helpful to be able to give them an alternative in Math that better fits their needs. Thank you and the Math faculty for being willing to offer it for students like ours.

17.9 Reflections

The faculty in our department consists of 18 research professors who teach from 3 to 6 h a semester and 21 instructors and teaching professors who teach 12–15 h a semester. As in many other departments, some faculty members are innovative and others are more traditional. We are fortunate that even through changes in administration, we have been allowed to pursue the teaching methods and course designs that we believe best serve the students regardless of their mathematical background or choice of major.

Determining the impact of the project on the authors' careers is rather difficult to do. One of the designers, a tenured teaching professor, retired in 2016. Her involvement in the development of the course was a continuation of her dedication to improving mathematics service courses at the university and a satisfying final project. The other two are mid-career instructors whose previous experiences led them naturally into the creation, implementation, and sustaining of this QR course. While the course was not a departure from their existing paths, the entire process increased their confidence and leadership skills. Since the QR course is only seven semesters old, its full impact on their careers is yet to be seen.

Acknowledgements A Louisiana Board of Regents grant was awarded to the developers to purchase technology and supplies. The grant also provided faculty summer support for work on course development. The monies awarded through this program were invaluable to the project. (Louisiana Board of Regents Support Fund Program for Enhancement of Undergraduate Education, "A Freshman Mathematics Course to Develop Quantitative Reasoning" Contract Number: LEQSF(2014-15)-ENH-TR-27).

References

Adults Learning Mathematics. (2015). Mathematics and democracy: The case for quantitative literacy. Retrieved June 14, 2016 from http://www.alm-online.net/useful-links/publications/mathematics-and-democracy-the-case-for-quantitative-literacy/.

Association of American Colleges and Universities. (2007). College learning for the new global century. Retrieved June 5, 2016 from https://www.aacu.org/publications-research/publications/college-learning-new-global-century.

Lusardi, A., & Wallace, D. (2013). Financial literacy and quantitative reasoning in the high school and college classroom. *Numeracy, 6*(2), Article 1. Retrieved March 10, 2016 from http://dx.doi.org/10.5038/1936-4660.6.2.1.

Madison, B. L. (2009). All the more reason for QR across the curriculum. *Numeracy, 2*(1), Article 1. Retrieved March 10, 2016 from http://dx.doi.org/10.5038/1936-4660.2.1.1.

Madison, B. L., & Steen, L. A. (Eds.). (2003). *Quantitative literacy: Why numeracy matters for schools and colleges*. Princeton, NJ: National Council of Education and the Disciplines.

Madison, B. L., & Steen, L. A. (Eds.). (2008). *Calculation vs. context: Quantitative literacy and its implications for teacher education*. Washington, DC: Mathematical Association of America.

National Commission on Excellence in Education. (1983). *A nation at risk: The imperative for education reform*. Washington, DC: U.S. Department of Education.

National Council of Teachers of Mathematics. (1989). *Curriculum and evaluation standards for school mathematics*. Reston, VA: National Council of Teachers of Mathematics.

National Council of Teachers of Mathematics. (2000). *Principles and standards for school mathematics*. Reston, VA: National Council of Teachers of Mathematics.

National Research Council. (1989). *Everybody counts: A report to the nation on the future of mathematics education*. Washington, DC: National Academic Press.

Rocconi, L. M., Lambert, A. D., McCormick, A. C., & Sarraf, S. A. (2013). Making college count: An examination of quantitative reasoning activities in higher education. *Numeracy, 6*(2), Article 10. Retrieved March 10, 2016 from http://dx.doi.org/10.5038/1936-4660.6.2.10.

Steen, L. A. (Ed.). (1997). *Why numbers count: Quantitative literacy for tomorrow's America.* New York: College Entrance Examination Board.

Steen, L. A. (Ed.). (2001). *Mathematics and democracy: The case for quantitative literacy.* Princeton, NY: Woodrow Wilson National Fellowship Foundation.

Chapter 18
A Story of Teaching Using Inquiry

Christine von Renesse

Abstract This chapter tells my story of learning how to teach mathematics using inquiry and becoming a facilitator of professional development (PD) workshops on inquiry-based learning (IBL) for teachers of kindergarten to graduate school. I am an associate professor at Westfield State University in Massachusetts and an integral part of the project "Discovering the Art of Mathematics" (DAoM). In this chapter, I describe a salsa rueda activity used to teach mathematics via inquiry to liberal arts students who often are not interested in or even fear mathematics. I present a vignette of a PD workshop activity designed to teach participants in an inquiry-based way how to teach using inquiry. The chapter also summarizes results of students' beliefs and attitudes surveys as evidence of the effectiveness of IBL. I close with plans for future work and a reflection on the challenges I face as I step into a leadership role.

Keywords Inquiry-based learning • IBL • Discovering the art of mathematics • Salsa rueda • Inquiry-based learning workshop

18.1 Introduction

This chapter tells my story of learning how to teach mathematics using inquiry and becoming a facilitator of professional development (PD) workshops on inquiry-based learning (IBL). According to Laursen et al. (2014),

---

MSC Codes
97U30
97U20
97B50

C. von Renesse (✉)
Department of Mathematics, Westfield State University,
577 Western Avenue, Westfield, MA 01086, USA
e-mail: cvonrenesse@westfield.edu

J. Dewar et al. (eds.), *Mathematics Education*, Association for Women in Mathematics Series 7, DOI 10.1007/978-3-319-44950-0_18

> IBL methods invite students to work out ill-structured but meaningful problems … Following a carefully designed sequence of tasks rather than a textbook, students construct, analyze, and critique mathematical arguments. Their ideas and explanations define and drive progress through the curriculum. In class, students present and discuss solutions alone at the board or via structured small-group work, while instructors guide and monitor this process (p. 407).

This description of IBL fits both my mathematics classrooms and my workshop environment.

Section 18.2 relates my educational background to how I came to teach inquiry-based mathematics classes at Westfield State University in Massachusetts. Section 18.3 describes the project "Discovering the Art of Mathematics" (DAoM) that promotes teaching mathematics for liberal arts courses using inquiry-based teaching and learning. Through DAoM, I started facilitating PD workshops for professors and teachers on IBL across the US. The mathematics for liberal arts course I teach, "Mathematical Exploration", is explained in detail in Sect. 18.4, including student learning goals, the inquiry-based activity "salsa rueda dancing" as a teaching sample, and student beliefs and attitudes evaluations. In Sect. 18.5 I reflect on collaborating with colleagues on teaching techniques. Section 18.6 describes the PD workshops and their goals, includes a workshop vignette, and outlines some future plans for workshop evaluations. The chapter concludes with a plan for future work and some personal reflections on project leadership.

18.2 Personal Background

18.2.1 Education in Germany

Besides being educated as an elementary school teacher, I minored in music at the HDK (university of arts) and received my Diplom (master) of mathematics at the Technical University Berlin. I did my practicum for the elementary licensure in a 4th grade class that was taught using methods of discovery. The children decided themselves what they would work on during the day and the teacher acted as a coach rather than a knowledge dispenser. This teaching method inspired me and set the base for my desire to teach using discovery or inquiry-based methods in my own classes. Parallel to my mathematics education I pursued music and dancing. I led an a capella choir, competed in ballroom dancing and composed my own songs. At this point in my career, the arts (music and dancing) and mathematics were parallel paths that seemed to have little connection with each other.

18.2.2 Graduate Education in the US

In 2003 I immigrated to the US to enter the PhD program in mathematics at the University of Massachusetts. I also danced several times a week and learned how to teach salsa rueda, a Cuban dance form in which couples dance synchronously in a circle. After I received my PhD in algebraic geometry, I was finally ready to move on to full-time teaching. I was still interested in teaching at the K-12 level but also really craved the academic freedom of a college professor. In the German university system there are no tenured *teaching* positions and so I applied for jobs in the US.

18.2.3 Teaching Position at Westfield State University

Westfield State University (WSU) is a public university in Western Massachusetts with about 4500 students, mostly undergraduate. Many students are first generation university students and many enter college with fairly low SAT scores. Traditionally a teaching college, more than half of our mathematics majors are pursuing their secondary education teaching license. The mathematics department consists of 12 full time professors all of whom use progressive teaching methods. The teaching position at WSU was, and is, a perfect fit for me. Most of my new colleagues were open to collaboration, sharing materials, discussing student learning and new teaching ideas: I could finally develop into the teacher I wanted to be. I could not have become the teacher and facilitator I am today without their support and feedback.

At WSU I now teach mathematics for liberal arts (MLA) classes in which I use my music and dance background to motivate mathematical ideas (see Sect. 18.4). I was particularly drawn to the teaching methods of my colleagues Julian Fleron, Phil Hotchkiss, and Volker Ecke in their MLA classes. Together we started the project "Discovering the Art of Mathematics" (DAoM, www.artofmathematics.org) with the goal of bringing authentic mathematical inquiry into MLA classes. I describe DAoM in the next section.

My first years at Westfield I avoided the calculus sequence. It seemed more natural (and easy) to teach using inquiry in MLA, classes for prospective teachers, and in upper-level courses. Thanks to the availability of many resources for teaching calculus using inquiry (e.g., www.iblcalculus.com) I now feel very comfortable (von Renesse 2014). There are always aspects to improve, and treating all the topics on the syllabus is a struggle (Yoshinobu and Jones 2012). However, compared to the way I used to teach calculus, I notice that my students understand the material more deeply now. My broad experience in using inquiry methods in many different courses and levels enables me to coach other teachers and faculty in adopting inquiry.

I also teach courses for future K-12 teachers and spend a lot of time coaching and co-teaching in K-12 classrooms. Working with K-12 teachers and the workshops I lead for college faculty have a surprising overlap. It seems to me that, at the core, teaching mathematics using inquiry relies on the same principles, regardless of whether the students are children, young adults, college students, teachers or professors. The many seemingly disconnected areas (music, dance, and mathematics) that I have studied over the years have finally come together and I enjoy deeply that I can use "all of me" in my current position.

I focus the remainder of this chapter on my work with DAoM. It has had the biggest influence on my development as a facilitator of learning mathematics and it connects all the professional ideas that I feel passionate about.

18.3 Discovering the Art of Mathematics

DAoM was founded in 2009 by Julian Fleron (PI), Phil Hotchkiss, Volker Ecke and myself with this vision:

> *Mathematics for Liberal Arts students will be actively involved in authentic mathematical experiences that*
>
> - *are both challenging and intellectually stimulating,*
> - *provide meaningful cognitive and metacognitive gains, and,*
> - *nurture healthy and informed perceptions of mathematics, mathematical ways of thinking, and the ongoing mathematics not only on STEM fields but also on the liberal arts and humanities.*

DoAM's website (www.artofmathematics.org) provides a wealth of resources to help interested instructors realize this vision in their MLA courses: a library of 11 inquiry-based learning guides, extensive teacher resources, and many professional development opportunities. Thanks to the National Science Foundation (NSF) and Mr. Harry Lucas, we can offer our materials and workshops at no cost. DAoM was the perfect way to connect my interest in music and dancing to my passion for the teaching and learning of mathematics. Over time DoAM's goals have broadened and we now support faculty in including more inquiry-based techniques into *all* their mathematics classes, not just MLA.

18.4 Mathematics for Liberal Arts Course at WSU

The MLA course "Mathematical Explorations" at WSU is "an introductory course designed to provide the liberal arts major with an opportunity to develop a broader appreciation of mathematics by exploring ways in which the artistic, aesthetic, intellectual, and humanistic aspects of mathematics are as important as its utility" (official WSU course description). The professor can choose the specific content that he or she would like to use for this purpose. There is no prerequisite for this course and students tend to be weak in algebra. The course is part of WSU's common core curriculum and we offer about 6 sections of 30 students each semester. DAoM has developed specific student learning goals for this course:

1. Students will appreciate mathematics as a human endeavor, which is one of our most fundamental intellectual pursuits.
2. Students will understand that mathematics is a vital, rapidly growing field of inquiry with a dedicated cohort of practitioners.

3. Students will understand the continued impact of mathematics in shaping history, culture, logic, philosophy, and knowledge, as well as its role as a humanistic and aesthetic discipline.
4. Students will understand the ubiquitous role of mathematics in the world around them.
5. Students will strengthen their reasoning skills and become better problem solvers.
6. Students will strengthen their skills in reading, writing, argumentation and speaking.
7. Students will become more self-monitoring, reflective learners and take greater personal responsibility for their learning.
8. Students will approach mathematics more positively and gain a balanced perspective of mathematics.
9. Students will improve their mathematical confidence.
10. Students will develop awareness of the negative impact of broadly-held societal views of mathematics.
11. Students will be capable of and interested in considering mathematics outside of the confines of the classroom, understanding the value of life-long learning in mathematics.

Students in our course are being assessed through observing their abilities and progress during group work and presentations, larger student projects (e.g. posters, papers, art work, solving the Rubik's cube), journal writing, regular conceptual homework and sometimes a final exam (Fleron et al. 2014).

18.4.1 Dancing Salsa Rueda

Using just the existing DAoM materials, an instructor could teach different topics for 11 semesters. There are so many because we keep inventing activities around topics we feel passionate about (von Renesse 2012; von Renesse and Ecke 2011, 2016; Fleron and Ecke 2011; Livingston and Fleron 2012; Fleron 2012; von Renesse & Ecke 2015).

One of my favorite activities is dancing salsa rueda with my students. Movement can be very helpful in learning (Jensen 2000). In my class, the movement motivates the mathematical questions (see example below). It also helps to engage all students, bringing lightness and laughter into the classroom and building a strong classroom community. Moving together for the first time can be frightening for the students and the instructor. Many students have shared with me that they are as afraid of dancing as they are of mathematics! But the dancing we do is really about moving, with no attachment to "doing it perfectly". Doing the exact steps at the right time is not as important as the desired position in the circle. Similarly, performing arithmetic is not as important in my class as understanding a big idea. Students are often afraid of making mistakes instead of valuing them. Using dancing

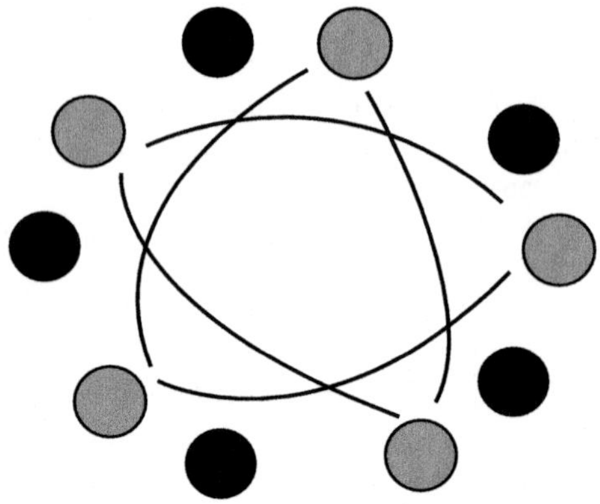

Fig. 18.1 The path of a rueda dance using only dame dos

I can model how to enjoy learning from mistakes and being persistent. In my experience this translates to valuing mistakes and being more persistent when we are learning mathematics: this is important for student learning (Kapur 2011).

In salsa rueda,[1] pairs of dancers stand in a circle and dance the same salsa moves simultaneously. There is a leader of the circle who calls out the moves and everyone is supposed to listen and react in time. Each pair has a leader and a follower, which students determine by preference and confidence, instead of by gender.

The mathematical content goals for this activity include understanding how greatest common factors and least common multiples are related, finding patterns in star polygons, and making conjectures and proving conjectures about star polygons.[2] During the dance move "dame" ("give me" in Spanish) each leader passes their follower to the next leader to the left and gets in turn a new follower from the right. This can be done skipping one or several leaders, leading to an interesting pattern. In fact, star polygons emerge from following the dancer's path. See Fig. 18.1 for a path of "dame dos" ("give me the second"–skipping one leader), with the black circles representing leaders and the gray circles followers. Notice that in each pair the follower stands on the leader's left.

In the language of the dance we are wondering:

1. Does each follower return eventually to his or her original leader? Does the answer depend on the number of dancers?
2. Does every follower get to dance with every leader in the circle? Does the answer depend on the number of dancers?
3. How many times does the follower move around the circle before returning to his or her original leader (if that happens).

[1] The video at https://artofmathematics.org/media/video-497 shows students dancing salsa rueda in my Mathematical Explorations class in Fall 2013.

[2] We create an (n,k) star polygon by taking n equidistant points on a circle and connecting each point to its kth neighbor (going to the right around the circle).

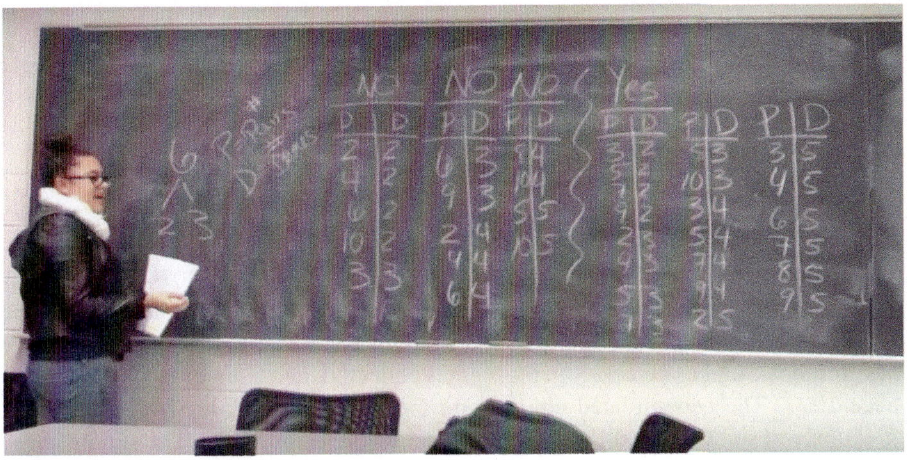

Fig. 18.2 Student presenting her work on question 2

Students spend several class periods looking for patterns, creating conjectures, and trying to figure out if and why their conjectures are true. At various points, whole class discussions allow the students to build on their classmates' ideas.[3] Figure 18.2 shows a table where a student recorded the number of partners (P) and the kind of dame (D). The "NO" table records all examples where followers don't dance with every leader, the "YES" table all others. She noticed how, for most examples, D divides P but that there are some "weird ones" like 6 and 4.

18.4.2 Creating Investigations

It is a difficult task to invent investigations that are "just right" for our students. If we guide the students too much, they don't do mathematics and lose sight of the larger picture. If we don't guide them enough, they get frustrated and give up. Mairead Greene and I (Greene and von Renesse 2016) described the processes we use to fine-tune our tasks and make sure that they align with our goals. The investigations lead students to make conjectures and eventually prove them. A proof doesn't have to be formal, but it has to establish an explanation that helps us make sense of *why* the conjecture is always true (Fleron et al. 2016). In my class, I encourage the students to alternate between thinking about the mathematics and returning to the movement to test their conjectures. Students need to consider greatest common factors and least common multiples to state their conjectures and

[3] In the video clip at https://artofmathematics.org/media/video-394, a student from my class presents her work on question (2). It shows how she explains her process and how open she is about what she doesn't know (yet).

their sense-making often includes models like the number line, factor trees and star polygons. Student work from my class is included in von Renesse (2016).

18.4.3 Student Evaluation: Does Inquiry-Based Learning Work?

Research indicates that active learning is effective (Freeman 2014; Kogan and Laursen 2014). But not many studies have been done on the MLA audience. In our project we have used pre- and post-surveys to measure changes in students' beliefs and attitudes changes since 2009. Fleron has collected data from over 1000 students since 1997. The accumulated changes on the survey all occurred in the desired direction and some of the differences were statistically significant. The survey questions we used and many of our results appeared in Ecke (2015). One of our results was that students are much more likely to perceive beauty in mathematics after the course: the percent of students agreeing that there is something in mathematics that they think is beautiful and that they could describe increased from 15 to 55 % (see Fig. 18.3). We are also interested in having but struggling to find a written test that measures students' skills (reasoning, creativity, persistence, flexibility, etc.) independently of the content treated during the semester.

In addition to data from the pre-and post-surveys such as that found in Fig. 18.3, we have collected student journals over the years. Journal entries like the following show that many students have achieved the specific learning goals DAoM developed for the course:

> From the very first moment we started the Stone Game, I was able to understand it and was actually excited to discover more ways in which a player could win. After the class ended, I remember I went back to my dorm room, tore up a piece of paper into little squares, and

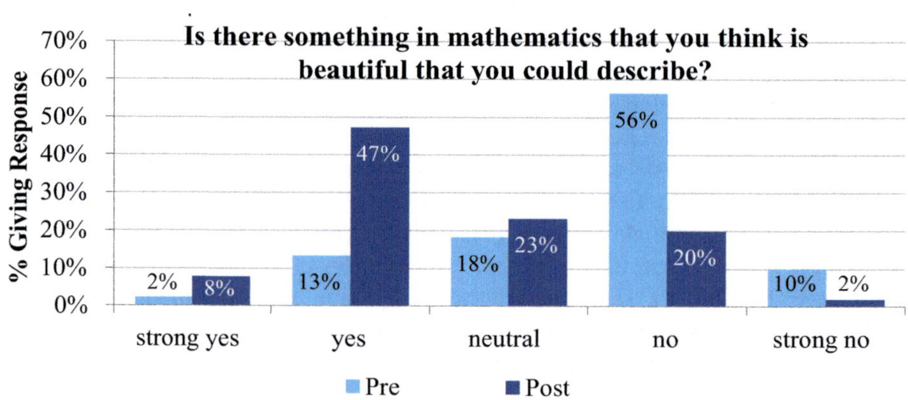

Fig. 18.3 Evaluation result

began to search for more ways that a player could win. After I was finished working through some of my ideas, I looked around at all the little scraps of paper that were covering my desk, and I was amazed with myself. That had been the first time I had ever gotten so excited about working on a math problem outside of the classroom.

DAoM Student, Fall 2015

I agree with Lockhart[4] that students should explore math for themselves and discover equations so that they understand it better than being given a formula sheet and told to memorize them. In our class we did not use any equations or formulas and I feel that I learned more in this class than any other math class before.

DAoM Student, Fall 2015

18.5 Collaborating on Teaching Techniques

Being part of the DAoM project has taught me much about teaching techniques. Visiting my colleagues' classrooms and collaborating weekly to create inquiry materials led to many questions, disagreements and discussions. This process helped me discern who *I* want to be in the classroom and why *I* prefer some teaching techniques to others. Even though all four of us clearly support inquiry, our teaching methods vary. I like to use a "large" investigation prompt to start off a new topic and then mix group work with whole class discussions to reach a conclusion (von Renesse and Ecke 2013). Fleron, on the other hand, prefers to give his students handouts with smaller prompts for the groups to work on (Fleron 2014), often without whole class discussions. The guiding principle is the same: we act as coaches and the students do the mathematics. The difference lies in how much guidance we give our students. Through many conversations we have seen the advantages and disadvantages of each approach and have found ourselves trying out and liking each other's methods of facilitating inquiry. Lately Fleron has been using my approach of the larger prompts in his classes and has really enjoyed the discoveries that his students made (Fleron 2015).

Our approaches to motivating the class also differ. I tend to pick topics that seem "easy" and engaging, (music, dance, games,…) or investigations that I know will lead students to disagree with each other, for example, wondering if $0.999\ldots=1$ or not. Then I tried out some of Fleron's materials and noticed that my students were not as excited about mathematics. I didn't know that he likes to choose topics that are rooted in pure mathematics (understanding the infinite, number theory,…) and generates motivation by regularly telling students about "cool" connections in mathematics (von Renesse 2015). Engaging a crowd by giving exciting minilectures

[4] The students read "A mathematician's lament" by Paul Lockhart (see https://www.maa.org/external_archive/devlin/LockhartsLament.pdf).

doesn't come naturally to me yet—but I am improving. The deep collaboration in our project over the last eight years is clearly visible in the peer-reviewed publications and blogs that we have written together.

18.6 Professional Development Workshops

18.6.1 History of Our Workshop Development

When our project DAoM made our learning guides freely available (www.artofmathematics.org/books) we assumed that now anyone could replicate what we do. But it is not easy for teachers and professors to change their teaching style. In addition to the curriculum materials, there are many teaching decisions that we make unconsciously or that we had not communicated in our learning guides. This realization led to the creation of our classroom web page and our traveling workshops.

While we developed these resources, I took on more and more the role of a leader in our group. As the most senior faculty member of our project, Fleron had been the natural leader of our efforts but his interests were more focused on curriculum development than leading workshops or creating pedagogical support. My background and interest in leading K-12 professional development workshops put me, the most junior faculty member and only woman of the project, suddenly in a leadership position.

18.6.2 Workshop Goals and Description

Our short traveling workshops usually include nine official contact hours. Additionally we like to plan at least one group meal to give us time to connect and network. We also offer to visit classes before the workshop to watch, model, or co-teach with the teaching participant. The classroom visits allow us to meet some of the participants and assess their needs before the workshops starts. The goals of our traveling workshops are to help faculty:

- Experience as a student what mathematical inquiry can feel like in a MLA class,
- Investigate particular content areas that might resonate with their students,
- Reflect on the interaction of teacher, student, mathematics, and inquiry materials in the classroom.

The 20–30 participants of a typical workshop are mostly professors from 2- to 4-year colleges and universities near the workshop site. We also ask the on-site workshop coordinator to invite some local K-12 teachers and graduate students since having a variety of participants promotes richer discussions and collaborations.

We believe that one learns best when using inquiry. Therefore the guiding principle for the workshop is to not lecture about how to teach using inquiry but to facilitate activities what will lead the participants to discover the teaching ideas themselves. As part of our activities we let participants experience what it feels like to be students in an inquiry-based learning environment. Participants report regularly how eye-opening this was for them and how much it motivated them to change their own teaching style. We also let participants practice their teacher moves[5] with the other workshop members, as in the workshop vignette in Sect.18.6.3.

My current research includes finding or inventing more workshop activities that help teachers and professors become successful in teaching using inquiry. The salsa rueda activity would be too long for a workshop activity, so we use shorter activities that are appropriate in difficulty level for the audience of teachers and professors. As long as the topic is approached via inquiry and the question is posed well (Weiss 2003) the specific content is not relevant for our purpose. The teacher moves that the participants engage with can be used with any mathematical topic.

In our workshops, we also use classroom videos containing other content from our MLA learning guides. For example, the video clip https://artofmathematics.org/media/video-395 shows how a week after the salsa rueda activity, I am recording students' conjectures about relating the greatest common factor and the least common multiple. This video clip is useful at workshops to demonstrate how I record ideas that are mathematically incorrect or irrelevant without losing my "poker face". The next section contains a vignette that presents a typical interaction that could have happened at any of our workshops.

18.6.3 Workshop Vignette

We just watched several classroom video clips to see how teachers use "talk moves"[6] to facilitate a good discussion. The 10 workshop participants (high school teachers and professors from 2- to 4-year colleges) described the teacher moves they observed, which include the talk moves from Susan Chapin's book Math Talk (Chapin et al. 2003):

- **Revoicing**: Let a student repeat what another student said
- **Rephrasing**: Facilitator rephrases what he/she heard the student say
- **Agree** or **Disagree**: Ask the class whether they agree or disagree with a statement
- **Adding On**: Invite students to add any observations, thoughts or questions

[5]A "teacher move" is any choice the teacher makes during class to enhance student learning. Examples are standing in a particular part of the room or clapping a rhythm to get students' attention.

[6]A "talk move" is a particular phrase the facilitator uses during a whole class discussion or a mathematical conversation with a student.

- **Wait Time**: Do nothing for at least 10 s, the hardest thing to do for most teachers
- **Sharing**: Decide who gets to share their ideas (and when)
- **Record**: Write/represent the essence of the discussion on the board

Some of the participants look a bit nervous since they know they will soon lead a discussion using talk moves in our group. I explain the goal of the role-playing: "This is your chance to practice facilitating a discussion in a space where it doesn't matter when we mess up. There are no students here; we are not trying to cover anything. Just see what it feels like." The first facilitator, Professor G, hands his problem to the groups of "students." They are supposed to decide if 0.999.,…=1 or not. While he walks between the groups I notice that he seems to "hover" over the groups with his arms crossed. I join him and suggest that he can pull up a chair to a group and tell them that he is "listening in." This little meta-conversation with Professor G doesn't distract the groups from playing with different proofs for different audiences. I observe that a few participants argue against the equality but are quickly shut down by their group members.

After 10 min of group work Professor G calls his "students" together as a whole "class". He decides to start the discussion with a group that uses the trick of multiplying $x = 0.999\ldots$ by 10 so that $10x - x = 9.999\ldots - 0.999\ldots = 9$, which makes $x = 1$. After they explain their idea he asks another participant to repeat their thinking. The next group he calls on uses a calculus approach with a geometric series. As they take over the board I wonder how the participants who don't believe in the equality are doing. Did Professor G. notice that they are there? Does he have a plan to bring them in? It seems to me that he will continue to let the groups share their thinking without connecting the ideas or bringing in disagreement. So I decide to interrupt him and address the group: "We have heard a lot of arguments in favor of equality. I wonder if there is anyone in this room who thinks 0.999… and 1 cannot be equal?" One participant immediately raises his hand to share that it doesn't feel right to him that these numbers could really be equal. "It just can never reach 1, something with the algebra must be wrong. Maybe we can't subtract the infinitely many places the way we did?" Now the energy rises in the room and a member of the first group starts to defend her algebraic thinking. Professor G. is busy recording their thinking and calls on a third group to repeat where the conflict lies.

When our allotted time for the discussion is over it is hard for me to get the participants to stop discussing the mathematics: "Take a deep breath, let go of the mathematics, and put your teacher hats back on." Pause "Thank you. Professor G., how did the discussion feel to you?" He shares that it was hard not to tell the "students" who was right and to let the conflict develop. His instinct was to call on the students who knew the correct answer and avoid the students who seemed to struggle. The question comes up when to step in and give feedback. Can we agree to disagree? Do we need to come to a conclusion before "class is over"? The discussion touches on many important aspects and challenges of an inquiry-based classroom and I try not to tell the participants how I would deal with this situation. After all, this is their moment to inquire about teaching, and not mine.

18.6.4 Workshop Evaluation

We are currently using pre- and post-workshop evaluations for our short workshops to measure the impact right after a workshop and a year later. Results are not compiled yet. Comments on our post-surveys suggest that our workshops are well received. Conversations with department chairs in the year after a traveling workshop indicate that participants use our ideas to change their class structure and work on their teaching moves. Written comments from workshop participants include:

> I believe the most effective aspect [of the workshop] was experiencing IBL as a student and unpacking what it is like as the facilitator. I think it gave me a real sense of what kinds of activities to utilize with my students.
>
> Workshop Participant, 2016

> [The most effective parts of the workshop were] explicitly looking at and practicing the teaching moves with a group of eager learners before I try it with my reluctant students.
>
> Workshop Participant, 2016

> This workshop exceeded my expectations. With a post-secondary focus, I expected a more typical 'quick fix' message. I felt the work we experienced here is much more fully and authentically aligned with current education research than any post-secondary PD I have ever experienced.
>
> Workshop Participant, 2015

I hope in the future to use a classroom observation protocol like O-TOP (Wainwright et al. 2004) or M-SCAN (Merritt et al. 2010) and videotape some classes that participants teach before and after having taken our workshop. Maybe this would allow us to measure in greater detail what they learned at our workshop.

18.7 Larger Audiences and Future Work

While the work of DAoM was intended to reach MLA instructors nationally we have seen huge interest in our materials and workshops from a much larger audience. Over the last 4 years, many K-12 schools have reached out to us for professional development. Mathematics departments from community colleges through universities are interested in our traveling workshops, international schools and universities have found our materials, and we have almost 900 followers of our monthly pedagogy blog. We are at the point where four people can't handle the work anymore, and we need more faculty leaders to help with implementing IBL workshops in mathematics.

The DAoM team just started collaborating with the Academy of Inquiry Based Learning (AIBL). The goal of the new project PRODUCT—PROfessional Development and Uptake through Collaborative Teams: Supporting Inquiry Based Learning in Undergraduate Mathematics—is to "train the trainers" so that more IBL workshops can be offered in the future. PRODUCT, under the leadership of Stan

Yoshinobu and Sandra Laursen, is supported by an NSF IUSE grant to increase the number of facilitators of IBL workshops.

DAoM is leading the short workshop development for PRODUCT, Professional Development and Uptake through Collaborative Teams. We will continue to offer short workshops at no cost until 2020 and develop materials that can support professors and teachers in offering short IBL workshops themselves. While DAoM and AIBL have been offering IBL workshops for the several years, our guiding principles differ from one another. Articulating our goals clearly, sharing our facilitation experience and deciding on joint strategies will be challenging and productive.

18.8 Personal Reflections

While I feel very comfortable teaching a class or leading a workshop, I still struggle with taking on a leadership role in a project among colleagues. How do we hold each other accountable for our respective work on the project? What do we do if we disagree on something? Where is the line between honesty and diplomacy? How does friendship intersect with work relationships? Who determines what is "fair" and how we get compensated for our respective contributions? There are more questions than answers at this point. I know that my strengths are planning ahead, working efficiently, getting my work done on time, and being an honest and clear communicator. But I am still learning how to be supportive of a colleague and give honest feedback (at the same time!), how to be a cheerleader instead of a taskmaster, and how to determine a reasonable workload instead of overcommitting.

Acknowledgements Discovering the Art of Mathematics has been supported by the NSF under awards NSF0836943, NSF1229515, NSF1525058, and by a generous gift from Mr. Harry Lucas. I would like to thank my colleagues on the Westfield State University campus for mathematical and teaching-related guidance and inspiration, especially Profs. Julian Fleron, Phil Hotchkiss and Volker Ecke. I am grateful to both my immediate and extended family for continual support and love.

References

Chapin, S. H., Anderson, N. C., & O'Connor, M. C. (2003). *Classroom discussions: Using math talk to help students learn, grades 1-6*. Sausalito, CA: Math Solutions Publications.

Ecke, V. (2015). The story of our evaluation. Retrieved June 2, 2016 from www.artofmathematics.org/blogs/vecke/the-story-of-our-evaluation.

Fleron, J. F. (2012). Radon-Kaczmarz puzzles: CAT Scans Meet Sudoku. *Math Horizons, 19*(3), 28–29.

Fleron, J. F. (2014). Assessment: Students creating solution sets. Retrieved June 2, 2016 from https://artofmathematics.org/blogs/jfleron/assessment-solution-sets.

Fleron, J. F. (2015). Tackling rascals' triangle—how inquiry challenges what we know and how we know it. Retrieved June 2, 2016 from https://artofmathematics.org/blogs/jfleron/tackling-rascals-triangle-how-inquiry-challenges-what-we-know-and-how-we-know-it.

Fleron, J. F., & Ecke, V. (2011). Navigating between the dimensions. *Mathematics Teacher, 105*(4), 286–292.

Fleron, J. F., Ecke, V., & von Renesse, C. (2016). Discovering the art of mathematics: Proof as sense making. In R. Schwell, A. Steurer & J. Franko (Eds.), *Beyond lecture: Techniques to improve student proof-writing across the curriculum*, Washington DC: Mathematical Association of America.

Fleron J., Hotchkiss P., Ecke V., & von Renesse, C. (2014). Assessment techniques. Retrieved June 2, 2016 from https://www.artofmathematics.org/classroom/assessment.

Freeman, S., Eddy, S. L., McDonough, M., Smith, M. K., Okoroafor, N., Jordt, H., & Wenderoth, M. P. (2014). Active learning increases student performance in science, engineering, and mathematics. *Proceedings of the National Academy of Sciences of the United States of America, 111*(23), 8410–8415.

Greene, M., & von Renesse, C. (2016). A path to creating inquiry activities. *Problems, Resources, and Issues in Mathematics Undergraduate Studies*. Advance online publication. http://dx.doi.org/10.1080/10511970.2016.1211203.

Jensen, E. (2000). Moving with the brain in mind. *Educational Leadership, 58*(3), 34–37.

Kapur, M. (2011). A further study of productive failure in mathematical problem solving: Unpacking the design components. *Instructional Science, 39*(4), 561–579.

Kogan, M., & Laursen, S. L. (2014). Assessing long-term effects of inquiry-based learning: A case study from college mathematics. *Innovative Higher Education, 39*(3), 183–199.

Laursen, S. L., Hassi, M.-L., Kogan, M., & Weston, T. J. (2014). Benefits for women and men of inquiry-based learning in college mathematics: A multi-institution study. *Journal for Research in Mathematics Education, 45*(4), 406–418.

Livingston, J., & Fleron, J. (2012). Exploring 3D worlds using Google SketchUp. *Mathematics Teacher, 105*(6), 469–473.

Merritt, E. G., Rimm-Kaufman, S. E., Berry, I. I. I. R. Q., Walkowiak, T. A., & McCracken, E. R. (2010). A reflection framework for teaching math. Evaluate the quality of your instruction by using the eight dimensions of M-Scan, an observation tool that links math standards with day-to-day practice. *Teaching Children Mathematics, 17*(4), 238.

von Renesse, C. (2012). Musical palindromes for liberal arts students. *Problems, Resources, and Issues in Mathematics Undergraduate Studies, 22*(7), 525–537.

von Renesse, C. (2014). Teaching calculus using IBL. Retrieved June 2, 2016 from https://www.artofmathematics.org/blogs/cvonrenesse/teaching-calculus-using-ibl.

von Renesse, C. (2015). Learning about "cool things." Retrieved June 2, 2016 from https://artofmathematics.org/blogs/cvonrenesse/learning-about-cool-things.

von Renesse, C. (2016). Salsa rueda and mathematics. Retrieved June 2, 2016 from https://www.artofmathematics.org/blogs/cvonrenesse/salsa-rueda-and-mathematics.

von Renesse, C., & Ecke, V. (2011). Mathematics and Salsa dancing. *Journal of Mathematics and the Arts, 5*(1), 17–28.

von Renesse, C., & Ecke, V. (2013). Whole class discussions. Retrieved June 2, 2016 from https://artofmathematics.org/blogs/cvonrenesse/whole-class-discussions-incl-videos.

von Renesse, C., & Ecke, V. (2015). Inquiry-based learning and the art of mathematical discourse. *Problems, Resources, and Issues in Mathematics Undergraduate Studies, 25*(3), 221–237.

von Renesse, C., & Ecke, V. (2016). Discovering the art of mathematics: Using string art to investigate calculus. *Problems, Resources, and Issues in Mathematics Undergraduate Studies, 26*(4), 283–296.

Wainwright, C., Flick, L., Morrell, P. D., & Schepige, A. (2004). Observation of reform teaching in undergraduate level mathematics and science courses. *School Science and Mathematics, 104*(7), 322–336.

Weiss, R. E. (2003). Designing problems to promote higher-order thinking. *New Directions for Teaching and Learning, 95*, 25–31. doi:10.1002/tl.109.

Yoshinobu, S., & Jones, M. G. (2012). The coverage issue. *Problems, Resources, and Issues in Mathematics Undergraduate Studies, 22*(4), 303–316.

Chapter 19
An Ethnomathematics Course and a First-Year Seminar on the Mathematics of the Pre-Columbian Americas

Ximena Catepillán

Abstract As mathematicians know—but, unfortunately, many students don't—mathematics can be both fun and culturally relevant. One way to reach more students is to teach ethnomathematics, i.e., mathematical thinking found among such non-Western peoples as the Maasai of Kenya and Tanzania, and the ancient Maya. This chapter describes, and provides a rationale for, two different courses in ethnomathematics: an undergraduate ethnomathematics course for non-STEM students, and a first-year seminar on the mathematics of pre-Columbian Americans. Also included are aspects of the development and structure of the courses, examples of ethnomathematics topics, and lists of projects. Feedback from students, alumni, and colleagues regarding the value of such courses is provided.

Keywords Ethnomathematics • Mapuche • Maya • Pre-Columbian • Warlpiri

19.1 Introduction

One of the biggest challenges for ethnomathematicians is to persuade secondary schools and higher-education institutions to teach about the cultural richness of mathematics. As our classroom populations grow more culturally diverse, we gain a precious opportunity: to teach mathematical ideas that are embedded in some students' heritages.

MSC Codes
01A07
01A12
01A13

X. Catepillán (✉)
Department of Mathematics, Millersville University of Pennsylvania,
P.O. Box 1002, Millersville, PA 17551, USA
e-mail: ximena.catepillan@millersville.edu

J. Dewar et al. (eds.), *Mathematics Education*, Association for Women in Mathematics Series 7, DOI 10.1007/978-3-319-44950-0_19

Frequently, non-STEM oriented students are intimidated by the prospect of taking mathematics courses required to fulfill their general education requirements. Fortunately, such students who enroll in ethnomathematics courses tend to lose their fears and eagerly master mathematical concepts. This is particularly true for minority students when they discover links between contemporary mathematics and the mathematics of their ancestors, which boosts their ethnic pride. Caucasian or Western students, who constitute 78 % of the student population at Millersville University, benefit greatly from a mathematics course that is not only diverse in its content but also in its students.

Ethnomathematics, a term popularized by Ubiratan D'Ambrosio at the 5th International Congress on Mathematics Education[1] in 1984, is the study of mathematical thinking found outside what, in Western tradition, is traditionally considered to be "mathematics." The term is used to express the relationship between culture and mathematics. This includes "ideas … involving number, logic, spatial configuration, and, more significant, the combination or organization of these into systems and structures" (Ascher and Ascher 1997, p. 25). The main reasons for studying ethnomathematics are: "(1) To de-mystify a form of knowledge (mathematics) as being final, permanent, absolute, unique…; and, (2) To illustrate intellectual achievement of various civilizations, cultures, peoples, professions, gender…" (D'Ambrosio 2007, p. 9).

My goals for developing an ethnomathematics course are to illuminate mathematical and scientific knowledge of the people of the ancient civilizations, formulate problems from these civilizations in symbolic mathematics language, increase awareness of the cultural diversity of mathematics (many of my students are surprised when I point out that our number system is Hindu-Arabic), relate mathematical ideas of Western and non-Western worlds, and reveal that mathematics can be found worldwide. Mathematics epitomizes both diversity and universality—numbers hold their value in any language.

19.2 The Mathematics in Non-Western Cultures Course

The development and structure of the course, a concise description of subjects treated, and an example on an ethnomathematics topic will be presented in this section. The section concludes with feedback from students, alumni, and colleagues.

19.2.1 Development of the Course

While a non-tenured faculty member at Millersville University of Pennsylvania, I was asked to develop a mathematics course suitable for the university's new African-American Studies minor. In 1993, a series of lectures at Millersville by Rutgers' Arthur Powell on the then-emerging field of Ethnomathematics inspired me to create the course Mathematics in non-Western Cultures, treating mathematical notions

[1] http://revistas.ucr.ac.cr/index.php/cifem/article/viewFile/10608/10010.

developed by Africans, Asians, and native North, Central, and South Americans, among other non-European peoples.

Mathematics in non-Western Cultures is a general education course designed for non-STEM students to fulfill the required 3-credit general education mathematics requirement. For students minoring in African-American Studies and Latino Studies, the course is strongly recommended. The course also qualifies as a diversity course at Millersville University since it (1) is intercultural and cross-cultural; (2) examines historical and environmental factors that underlie cultural differences; (3) helps students identify, critically analyze, and apply scholarship and experience related to cultural diversity; (4) provides academic structure in support of students' positive engagement with peoples of diverse histories; and (5) challenges them to evaluate their own worldview.

19.2.2 Course Structure

Students who take the course are non-STEM majors and fairly diverse. In a typical class of 20 students, 40–50 % are students of color, mostly African-Americans, Asians, and Hispanic (22 % of the students at Millersville University are students of color). Their ethnic backgrounds influence the topics I choose to teach and the list of projects I offer. More than merely providing a survey of the mathematics from various cultures, the course teaches students to think critically about the basis of their own intercultural differences. During a given semester, I treat at least two items under each of the topics listed in Sect. 19.2.3. The students work on two projects—each with a different partner—for presentation to their classmates. These presentations are followed by discussions and small-group activities.

I have taught the course in three different formats: a traditional face-to-face course; a technology-enhanced course, in which 22 % of the course is face-to-face and the rest online; and as a study-abroad course based in Mexico, with 20 % of the class meetings held at Millersville University (see Sect. 19.4).

19.2.3 Mathematical Topics in the Course

19.2.3.1 Number Theory

- Written numbers: They include the *grouping* system, as used by the Egyptians, in which the values of the symbols are added; the *alphabetic* system, as used by the ancient Indian katapayadi, whereby numbers are associated with letters of the alphabet; the *partially positional* system of the Indian Kharosti (Menninger 1970); and the *positional* system like the Babylonians'.
- Spoken numbers: The methods of counting used by most indigenous groups are closely connected with their language structure. Here, the biggest hurdle is the formation of large numbers with a reasonable number of words. The Bantu languages' number words of Africa offer excellent examples.

- Calendars: Calendrical forms are universal. Special attention is given to the Mesoamerican elaborated system of calendars.
- Magic squares: They originated in China and later transferred to Japan. The mysterious squares were thought to have special powers, and some cultures used them for astrological and divinatory purposes. Extended magic squares methods were developed in West Africa in the eighteenth century.

19.2.3.2 Topology

- Graphs: The sona sand-drawings of the Tshokwe of Angola, the designs of the Bushoong of Congo, and the rice flour kolam drawings of the Tamil Nadu in India are studied, along with their connections with unicursal Shongo networks.
- Mazes: Sand tracing figures—nitus—and maze-dances of the Malekula of Vanuatu in the South Pacific Islands are investigated together with their geometrical transformations.
- String figures: String figure making of the African Batwa Pygmies and Native Americans are analyzed using a series of simple movements called elementary operations (Vandendriessche 2015).

19.2.3.3 Logical Structures

- Kinship: The Warlpiri of Australia and the Tongans of the South Pacific Islands have unique kinship relation rules that are studied to gain insight into the political, social, or ritual organization of their members.
- Recording and counting aids: Artifacts like the Inca quipu, the Chinese rods, the Japanese soroban, and wooden and bone tally sticks are studied. Objects like these were instrumental prior to the invention of paper.
- Finger counting: More than 20 variations of finger counting are known in Africa among such peoples as the Maasai of Kenya and Tanzania. Students explore how finger counting differs according to region, ethnicity, and historical period.
- Body counting: The use of the body as a counting tool can be found throughout the world. For example, tribal people in Papua New Guinea use as many as 74 body parts in their counting system.

19.2.3.4 Group Theory

- Symmetry patterns: The seven groups of symmetries of strip patterns are studied using non-Western artifacts like ceramics, fabrics, jewelry, and basketry. In two-dimensional symmetry patterns—tessellations—most of the 17 wallpaper symmetry groups are present in non-Western cultures.

- Kinship relations and magic squares: We study the connections found among group theory, kinship relations and magic squares.

19.2.3.5 Probability

- Games of chance: We study dice games like the Nyout from Korea, Tablan from India, and Pulic from Central America. Some variations of the African Mancala, believed to be the world's oldest game, are played by the students.
- Games of strategy: African and Native American games of strategy, like those played with stones, bones, or other small objects, are explored.
- Puzzles: Puzzles from Africa like the Kpelle, river crossing and kinship puzzles are solved by drawing charts to represent the solutions.
- Games are examined using mathematical reasoning, problem-solving, and cultural approaches.

Not every topic on this lengthy list is taught each semester—curricula vary depending on students' interests and the projects they present in class. A project should include a historical introduction and explain the topic's mathematical connections. There are always new projects on the list, some suggested by students themselves. Student projects have included such topics as:

- Non-Western female mathematicians
- Fear of numbers in different cultures
- The mathematics on the Ishango Bone
- Patterns in American quilts
- Calendar conversions among Maya dates
- Chinese triangles
- Basic mathematics using Japanese origami
- Navigation techniques of the Yup'ik
- Golden ratio in non-Western cultures
- Pi and the Babylonians
- Tarot and mathematics
- Pythagoras' theorem in non-Western cultures
- The sixteen Hindu sutras
- Preparation of a video with finger counting methods

Creativity is key to successfully completing my course. For example, when students construct their own number system for a fictitious indigenous group, they are allowed to use words from their ancestors' languages for the spoken numbers, but they also are required to create their own symbolic numerals.

We also discuss the advantages and limitations of our own Hindu-Arabic number system, and we discuss such topics as how finger counting can reveal one's ethnicity. In counting to three, for example, Germans start with the thumb, Chinese with the index finger, and Filipinos with the little finger.

19.2.4 An Ethnomathematics Topic in the Course

Many ethnomathematics topics appeared in the lists of mathematical topics in Sect. 19.2.3. In 19.2.4.1 we describe one of those in more detail: the elaborate kin structure among aboriginal people of Australia's Northern Territory.

19.2.4.1 The Warlpiri of Australia

The Warlpiri (Walbiri) people occupied a large block of the arid country northwest of Alice Springs in the Northern Territory of Australia until their lands were invaded by Europeans in the first half of the twentieth century (Laughren 1982). Today most Warlpiris live in government-run settlements or ranches. The Warlpiri have a complex kinship system that encodes their social, political, and ritual organization and behavior.

Their kinship system consists of eight sections, each divided into a female and a male section. The females in a particular section all have the same name, beginning with N; this is called a skin name. Similarly, the males have a skin name, beginning with J. Therefore, following Ascher (1991), we have 16 skin subsections, labeled iN and iJ, $1 \le i \le 8$ and eight sections (see Table 19.1). Skin names determine people's roles, responsibilities, and obligations in relation to one another and to the elements of land, law, language, and ceremony (Ngurra-kurlu). A deceased person's name becomes taboo. Hence they are never referred to by name, but rather by means of their kin relation to a living relative.

Unlike personal names, skin names define sections rather than individuals. Having a skin name immediately gives a person a place in Warlpiri society because he or she has a known set of relationships. If I were a member of the Warlpiri and in Section 4, my name would be Ximena Nungarrayi Catepillán.

These eight sections obey the following rules:

(a) Each person is in only one section.
(b) Marriages take place with a person from another specified section.

Table 19.1 The female and male numbers and names for the sections

Section	Female section	Male section	Female skin name	Male skin name
1	1N	1J	Nakamarra	Jakamarra
2	2N	2J	Nampijinpa	Jampijinpa
3	3N	3J	Napanangka	Japanangka
4	4N	4J	Nungarrayi	Jungarrayi
5	5N	5J	Napaljarri	Japaljarri
6	6N	6J	Napangardi	Japangardi
7	7N	7J	Napurrurla	Jupurrurla
8	8N	8J	Nangala	Jangala

(c) Their children are in another section, which depends on the section of the mother.

In Fig. 19.1, which summarizes the sections' rules, an equal sign = indicates marriage partners. For example, a woman (man) in section 6 can marry a man (woman) in section 2. The symbol <> points from the father's section to his child's section and vice-versa. For example, if a boy is in section 1, his father is in section 7, and vice versa. The arrows point from the mother's section to her child's section. Hence, if I were in section 8, then both my children would be in section 5 (Laughren et al. 1996).

If a mother is in section 5, her daughter is in section 7, her granddaughter in section 6, her great-granddaughter is in section 8, and her great-great-granddaughter is again in section 5. If a boy is in section 1, his father is in section 7, and his grandfather is again in section 1, and so on.

Consequently the female sections form two 4-cycles (1, 4, 2, 3) and (5, 7, 6, 8). The females whose sections are in a single 4-cycle form a matrimoiety. Similarly the male sections form four 2-cycles (1, 7), (2, 8), (3, 6), and (4, 5). The four male sections in the first two 2-cycles form one patrimoiety and the four sections in the second two 2-cycles form another (Ascher 1991).

The following questions help students understand the rules of the Warlpiri kin system:

- Assume that you are a Warlpiri member and in section 1, in which section(s) are:
  - Your siblings? Uncles on your mother's side? Aunts on your father's side?
  - Nephews on your sister's side? Cousins on your father's side? Your great grandmothers? Sisters and brothers in law? The father of the brother of your mother in law? Spouse?
- List three members of your family that are in your section.
- Assume you are in section 1. Use the Warlpiri rules to draw a diagram with three generations of your own family tree.

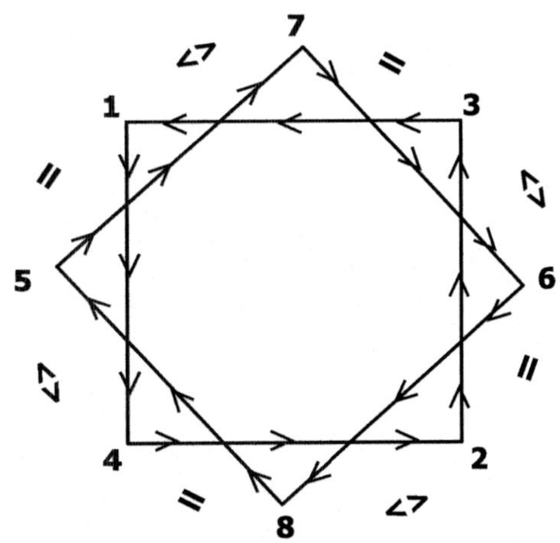

Fig. 19.1 Diagram of the eight sections' rules

Table 19.2 The number sections and their m, f representations

1	2	3	4	5	6	7	8
I	m^2	m	m^3	mf	m^3f	f	m^2f

- Is it conceivable for each of us to draw the diagram of our family tree? What are the hurdles?
- List advantages and disadvantages regarding the Warlpiri kinship rules.

Let us designate section 1 as I, and look at the other sections from this standpoint—that is, through their relationship to I. From the 4-cycle (1, 4, 2, 3) it follows that if a female is in section 1, then her mother (m) is in section 3, the mother of her mother (m^2) is in section 2, the mother of the mother of her mother (m^3) is in section 4, and the mother of the mother of the mother of her mother (m^4) is in section 1, therefore we can write $m^4=I$. It follows from cycle (1, 7) that if a male is in section 1, then his father (f) is in section 7, and since the mother of someone in 7 is in section 5, we can identify section 5 with mf. In addition, if we identify section 5 with mf, since the mother of someone in section 5 is in section 8, we can associate section 8 with m^2f. Similarly section 6 is identified with m^3f.

It is also important to clarify that not everyone in section 3, for example, is a mother. In fact, there are all types of people in every section: mothers, fathers, children, grandparents, etc. Table 19.2 identifies every section with its (m, f) combination (Ascher 1991).

This is a good opportunity to introduce the mathematical concept of *group* and its properties. I can consider the *commutativity* property by asking if mf equals fm. Students can use Fig. 19.1 to rewrite any combination of f's and m's as one and only one of the eight representations in Table 19.2. In 1978 the Australian linguist David Nash brought to Laughren's attention that the resulting structure was the dihedral group of order 8, the symmetry group of a square (Laughren 1982).

19.2.5 *Reactions from Colleagues, Students and Alumni*

The following are several comments I have received from students and alumni who have traveled to Mexico with me:

- Your [ethnomathematics] class literally altered my life trajectory, and helped me to reframe the entire lens through which I see the world today. It was no small thing!
- I truly can't overstate the impact you, the course, and our trip to Mexico had on my life both personally and professionally. It was the singular most important course I ever took at Millersville.
- What it meant to be a "history person" as I understood it, was deeply challenged by what I learned in that class.

- The course and trip to Mexico opened my eyes to other worlds and other ways of doing things that, given my exclusively Eurocentric education, I took for granted. I continue to be astonished by what I learned, years later, and still love to share the knowledge that I acquired with my students and others!
- ... [The course] expanded my knowledge on the celebrating of diversity. It caused me to think not only about math in other places, but also how that math affected the daily lives of the people who used it.

I am indebted to my colleagues from the Millersville University departments that the course serves. It is because of their advising that my course reaches full enrollment soon after registration begins. Here are some of my colleagues' comments:

- Without this course, the minor [African American Studies] would not be able to fulfill its mission.
- Students report that this course has given them a new perspective on mathematics and has changed their vantage point from being terrified to being excited and open minded about the possibilities.
- Their [Art majors] right-brained interpretation of the world seems to readily make the leap to viewing the world through the lens of the course. I have seen direct influences from [the course] in the work my Advanced Sculpture students are doing which to me is the most desirable outcome possible and broadens my students' view.
- It's exciting for our anthropology majors and minors to explore mathematics in non-Western cultures. [The course] provides an added dimension to our program's cross-cultural study. We are extremely fortunate to have this mathematics course available for our students, and it is wildly popular among them!
- The course provides speech communication majors an ideal choice to immerse themselves in cultures, including often their own native one. The course maintains the rigor of equivalent mathematics offerings while reducing the intimidation level many of our students possess.

19.3 The Freshman Seminar on Culture, Science, and Mathematics in the pre-Columbian Americas

My native South American ancestry played an important role in the development of my interest in pre-Columbian subjects. I grew up in a large family in which being a member of the Catepillán clan meant being strong-willed and proud of our indigenous blood. More objectively, I felt compelled to develop this seminar based on comments like the following from one of my former ethnomathematics students:

> American history that is pre-1776 like the Ancient Maya, is typically (and frustratingly) compacted into a single unit in American public schools, whereby teachers and textbooks fly through the entirety of pre-Columbian America as if it's something of a boring prequel to "the real story."

In this section, the development, structure, and a list of topics for the seminar will be presented. The section ends with two examples of topics for the seminar.

19.3.1 Development of the Seminar

In 2008, Millersville University encouraged faculty members to create seminars on subjects that we felt passionate about. For me, this offered a great opportunity to create a course that combined two of my intellectual passions: mathematics and pre-Columbian studies. I proposed a seminar on the mathematics and science of the civilizations of the pre-Columbian Americas. The seminar is an introduction to the study of the pre-Columbian North, Central, and South Americas, part of the broad interdisciplinary field of Native American Studies. The emphasis is on the role that science and mathematics played in the cultures of these indigenous groups. The seminar explores the pre-Columbian world through the eyes of the ancestors, as well as those of contemporary students. Special attention is given to archaeoastronomy and mathematics, on which all of the great cultures of antiquity have left a mark.

The history and culture of the pre-Columbian people motivate students' interest in the mathematical and astronomical achievements of these peoples. Every topic treated in class and each project developed by students must begin with a historical overview. The objectives guiding the development of the seminar are to demonstrate the mathematical and scientific knowledge of the people of the pre-Columbian civilizations, formulate problems from these civilizations in symbolic mathematics language, increase awareness of the cultural diversity of mathematics, and relate mathematical and scientific ideas among the people of the pre-Columbian civilizations. In comparison with the course Mathematics in non-Western Cultures, the emphasis in this course is on the indigenous peoples of the pre-Columbian Americas.

19.3.2 Seminar Structure

Students who take the 3-credit seminar are first-year students who have not declared a major. These students share a designated residence hall and have special programming to aid them during their adjustment to college life and in their search for a major.

During the semester, students work on two projects that they present to their classmates. For the first project (most semesters) they are asked to choose a Tribal College of the American Indian Higher Education Consortium (http://www.aihec.org/), study the history of the tribe and a tribal related mathematics or science course offered by the college. There are 37 tribal colleges and universities in the United States. These institutions are working to strengthen the tribal nations and make a lasting difference in the lives of American Indians and Alaska Natives. Students compare a selected tribal college with their own university and discuss the advantages and disadvantages of each.

Marlene Lang, who has Native American roots, is a regular speaker at my seminar. She is completing a doctoral dissertation that places the voices of young indigenous writers alongside theological models of the church, in search of a deeper cultural reconciliation. She described her participation and the students' reaction as follows:

> I have participated in the seminar presenting the story of my own Wisconsin Native ancestry. My father's father lived on the portion of the Menominee Reservation given to the Stockbridge-Munsee Band of Mohicans. The students were saddened to hear how my father's birth certificate was left blank where it asked for name of "father." They learned of the "age of invisibility" stretching into the mid-twentieth century, during which it was considered good to hide one's Native identity and blend in with white society. My grandfather's name was likely erased for this purpose. Having met him in person on the reservation was a singular event in my life and the students seemed to sense its importance. While my hair is black and my eyes are brown, my mother's Irish heritage left me "not looking like an Indian," to the Millersville freshmen. They were surprised to learn that I did not appreciate Columbus Day and that Native Americans have a very different view of the event we call Thanksgiving. There were gasps at this information (Personal communication, November 18, 2015).

For the second project, a list of topics is available for students' selection or they can suggest their own. The list, as with my Mathematics in non-Western Cultures course, varies each semester. Projects have focused on topics such as these:

- The mathematics of Native American pictography
- Preparation of a video on finger counting methods of Native Americans
- Zero in Mesoamerica
- Patterns in Native American quilts
- Calendar conversions among Maya dates
- The concept of infinity and the indigenous of America
- Number systems of the Patagonia becoming extinct
- Navigation techniques of the Yup'ik
- Golden ratio in the Maya culture
- Maya dates over billions of years ago
- Reckonings on the Inca yupana
- Eclipses in the indigenous cultures

Students also attend Native American Studies and Latino cultural events held at the university, and later discuss them in class.

19.3.3 Mathematical Topics in the Seminar

19.3.3.1 Astronomy in Ancient American Cultures

- Ancient pre-Columbian observatories and astronomical sites: American Southwest, Mesoamerica and Inca territory among others.

- Building alignments: Structures, temples, tombs and pyramids that were aligned with the precession of planets and the solar equinoxes.
- Calendars: Calendrical and religious observances based on phases of the moon and solar cycles.
- Planets: The planets that the ancients could see with the naked eye. Lunar and solar eclipses, and how celestial events empowered the leaders.
- Navigation: Indigenous navigation techniques. Some indigenous groups have navigated for centuries using a combination of geometry, the stars and constellations.

19.3.3.2 Pre-Columbian Number Systems

- Written numbers: They include the *grouping* system like the Aztecs, *partially positional* like the Aymará of Bolivia, and *positional* like the Maya (see Sect. 19.2.3.1). Many different bases are used by the pre-Columbian groups.
- Spoken numbers: The method of counting in most indigenous groups is closely connected with their language structure (see Sect. 19.2.3.1).

19.3.3.3 Counting Boards

- Boards and recording devices: The Inca developed counting boards, *yupanas*, to make computations before the results were recorded on a *quipu*, a recording device consisting of colored cords containing information in the form of knots. The *chimpu* of Bolivia and Peru, a quipu variation, is another recording device. Patolli, one of the oldest of games played in Mesoamerica, used a board. The Maya used grids for reckonings and games (De Landa 2005).
- Finger counting: Several variations of finger counting like the Yuki of California, the Bakairi and Bororo of Brasil, Maidu of the Northwest (Closs 1996).
- Body counting: We find this type of counting in a variety of groups throughout the Americas.

19.3.3.4 Calendrical Systems

- Celestial calendar: The Sun Dagger site and a celestial calendar of the Pueblo of the American Southwest.
- The Maya system of calendars: The Tzolkin, Haab and Round Calendars, and the Long Count. Conversions of dates among the calendars, correlations among the calendars and the Julian and Gregorian calendars.
- Other calendric forms—albeit not as complex as the Maya—are ubiquitous in the Americas.

19.3.3.5 Numerical Representations

- Hieroglyphic inscriptions: Hieroglyphic engravings and codices have enabled us to learn astronomical and calendrical data recorded by the Maya. Ancient monuments like the *stelae* show carved hieroglyphic texts displaying solar and lunar calculations, in addition to historical data. There is an enormous amount of historical information, from many pre-Columbian groups, inscribed on buildings, stairs, lintels, tombs, pottery, fabrics, etc.
- Pictographs and petroglyphs: Pictograph and petroglyph sites of the people of the Southwest, the Great Basin, the Checta of Perú and the Cave of the Hands in Argentina, among others.

19.3.3.6 Strip and Planar Symmetries

- Strip symmetries: There are seven strip symmetry groups, and every indigenous strip pattern used in arts and crafts can be identified with one and only one of the seven groups.
- Planar symmetries: A similar study can be done with planar symmetry groups and the 2-dimensional patterns developed by the pre-Columbian societies. Cultures have their own pattern styles that identify them; the meaning of these patterns is also studied.

19.3.4 Topics in Ethnomathematics of the Pre-Columbian Americas

Two examples of topics for the seminar are presented in this section. The first describes the number system of an indigenous group from South America, and the second describes strip patterns of the pre-Columbians. Both of these topics can be included in the ethnomathematics course Mathematics in non-Western Cultures.

19.3.4.1 The Mapuche of South America

The Mapuche (translation: *people of the earth*) were the first inhabitants of half of the area today known as Chile and Argentina, and are the largest ethnic group of Chile. Before the Spanish conquest, the Mapuche occupied a vast territory in the South American cone, with the population numbering about two million. In the pre-Hispanic times, the Inca conquered as far south as Cerro Chena, 20 km south of Santiago, the capital of Chile (Stehber 2015). I am of Chilean origin, and my last name, Catepillán, is Huilliche (translation: *people of the south*), a subsection of the Mapuche. Mapudungún (*language of the earth*) is the Mapuche language, and their writing number system varies slightly from region to region. The most common number words in their native language can be seen in Table 19.3 (De Augusta 1903).

It is interesting to note that *Quechua*, the language of the Inca, was used for the word for 1,000. There was contact, albeit not friendly, between these indigenous groups. The Mapudungún language adopted the Spanish word for 1,000,000. There are many examples in which the languages have been combined, a smooth merger since their number systems have the same base.

For numbers combining two or more numbers—compound numbers—the basic rules are:

In addition the higher number precedes the lower number whereas in multiplication the lower number precedes the higher number.

Some examples are:

- kayu mari $= 6 \times 10 = 60$
- mari kayu $= 10 + 6 = 16$
- meli pataka pura $= 4 \times 100 + 8 = 400 + 8 = 408$
- ailla pataka ailla mari ailla $= 9 \times 100 + 9 \times 10 + 9 = 900 + 90 + 9 = 999$
- waraŋka = mari pataka $= 10 \times 100$
- küla kechu meli $= 3 \times 5 + 4$

The following questions explore some of the mathematics involved in this number system. Many other number systems from indigenous South American cultures provide options for investigation.

- What is the base in this number system?
- Write the following numbers as Hindu-Arabic numbers:
- epu pataka reqle mari meli

Table 19.3 A list of Mapudungún number words

Mapudungún number words	Hindu-Arabic number
Kiñe	1
Epu	2
Küla	3
Meli	4
Kechu	5
Kayu	6
Reqle	7
Pura	8
Ailla	9
Mari	10
Pataka	100
Waraŋka	1,000
Pataka waraŋka	100,000
Millón	1,000,000

- pura waranka meli pataka kayu mari pura
- epu mari meli waraŋka kayu pataka epu mari kiñe
- Write the Mapudungún words for 21, 30, 44, 68, 123, and 765. In some cases there are several possible answers.
- What does $5 \times 8 + 4$ mean in this system?
- Is it possible to write a number as reqle?
- Which is the largest number that the Mapuche can write?
- Compare this system with your own and list advantages and disadvantages.

Written numbers are used by many indigenous groups. Sadly, a number of their languages and dialects are either extinct or rapidly disappearing, together with their written number systems, e.g., Chumashan of Southern California, Selk'nam of Tierra del Fuego, and Nahuatl of Mexico. A popular project among students is to choose a language or dialect not discussed in class that is extinct or threatened with extinction, and to study its number system and culture, along with possible reasons for the language's demise.

19.3.4.2 Pre-Columbian Strip Patterns

Patterns repeating along a strip extending indefinitely in both directions can be classified into seven groups according to their symmetry types (Ascher 1991). A sample pattern for each of the seven symmetry types can be found in Fig. 19.2. Strip patterns are present in every pre-Columbian indigenous community because their

	Symmetry Type	Sample Pattern
1.	Translation only	q q q q q q q q q q q
2.	Horizontal reflection	d d d d d d d d d d q q q q q q q q q q
3.	Vertical reflection	db db db db db db db
4.	Horizontal/vertical reflections	db db db db db db qp qp qp qp qp qp
5.	Glide reflection	d d d d d d d q q q q q q q
6.	Rotation 180°	q q q q q q q b b b b b b b
7.	Rotation/vertical reflection	db db db db qp qp qp qp

Fig. 19.2 The seven symmetry types

Fig. 19.3 Andean cross strip pattern (*left*) and Andean cross (*right*)

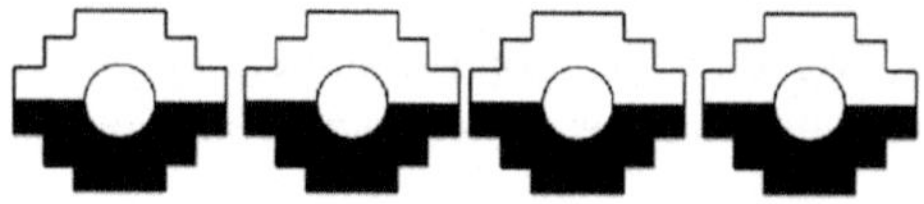

Fig. 19.4 Inca cross strip pattern

cultural beliefs are represented in their art; thus, art expression has been a form of communication for these groups. These people have an enormous respect for nature and art for art's sake is not part of their life.

Figure 19.3 depicts a strip pattern made with the Andean cross, one of the most common symbols in the Andean cultures, which represents the eternity of those cultures and is found in pottery, textiles and ponchos worn by leaders–*lonkos*–of indigenous Mapuche groups. The strip pattern has symmetry type (4).

The strip pattern in Fig. 19.4 is based upon the *chakana* Inca cross. The cross is created in such a way that only the upper half protrudes from the ground; the cross is completed by the shadow cast by the upper half. The shadow part represents the non-material world. The strip pattern has symmetry type (3).

Strip patterns abound in all pre-Columbian indigenous groups and are a great subject for student projects, as students study the embedded cultural messages, the history of the peoples who created them, and the geometrical classification of the seven symmetry types. Students enjoy finding all seven symmetry types in pre-Columbian strip designs.

19.4 Study Abroad Sessions

Each summer from 2006 to 2010, with support from the university, I conducted a special version of the Mathematics in non-Western Cultures course that included a weeklong trip to Mexico's Yucatán Peninsula to study Maya civilization. I resumed teaching this version of the course in spring 2016. Archaeologists from the Maya Exploration Center (http://www.mayaexploration.com/index.php) have helped to teach the course's study abroad component, by lecturing, conversing with students, and guiding them during visits to such sites as:

- Chichén Itzá, from the Terminal Classic to Post-Classic periods (c. 750–1200AD);
- Cobá, a settlement from the Middle and Terminal Classic periods (500–900AD);
- Ek-Balam, from the Middle Pre-Classic to Post-Classic periods (500BC–1200AD); and
- Yaxunah, from the Middle Pre-Classic through the Post-classic periods (500BC-1200AD).

One of the highlights of the Mexico trip is Stela 1 in Cobá. A stela is an inscribed stone slab used for commemorative purposes. The stela depicts an unusual Maya date, one that goes back more than 28 octillion and 679 septillion years (Stuart 2011).

During these study-abroad sessions, students immerse themselves in Maya culture. They sample Maya cuisine, meet descendants of the Maya, who speak Maya languages or Spanish, and do most of the work on their major project for the course. After we return to campus, students give presentations based on their field experiences—presentations that are always rewarding and enjoyable, because of the students' high levels of excitement and motivation.

19.5 Conclusions

Studying ethnomathematics encourages math-averse students by demonstrating the universality of mathematics and showing students how mathematics played a role in the lives of the ancestors of some of their non-European classmates. Students of European ancestry who take my ethnomathematics course truly enjoy being immersed in the non-Western cultures of their classmates.

In 2014, I co-developed a graduate version of my Ethnomathematics course for our Master of Education in Mathematics program at Millersville University. This version of the course provides the opportunity for graduate students interested in K-12 teaching to learn mathematics and mathematical activities rooted in non-Western cultures. When they become mathematics teachers themselves, graduates of the course can incorporate these cross-cultural mathematics' principles and activities in their lesson plans.

Waclaw Symanski, my co-author for the textbook *Mathematics in a Sample of Cultures* (Catepillán and Szymanski 2016) and I were writing additional chapters for a second edition when he died unexpectedly in August 2016. Work will continue on the new edition.

One of my career dreams is to teach my undergraduate and graduate ethnomathematics courses in Spanish, perhaps in an online, bi-lingual format.

Acknowledgements Millersville University has supported this project in various ways. The university granted me a sabbatical to coauthor Mathematics in a Sample of Cultures. Millersville also funded my travel to Mexico's Yucatán Peninsula to teach a study-abroad version of the course. And the continuing support of my students and colleagues made possible the writing of this chapter.

References

Ascher, M. (1991). *Ethnomathematics: A multicultural view of mathematical ideas*. Pacific Grove, CA: Brooks/Cole.

Ascher, M., & Ascher, R. (1997). Ethnomathematics. In A. B. Powell & M. Frankenstein (Eds.), *Ethnomathematics: Challenging eurocentrism in mathematics education* (pp. 25–50). Albany, NY: SUNY Press.

Catepillán, X., & Szymanski, W. (2016). *Mathematics in a sample of cultures*. Dubuque, IA: Kendall-Hunt Publishing Co.

Closs, M. P. (1996). Native American number systems. In M. P. Closs (Ed.), *Native American mathematics* (pp. 3–43). Austin, TX: University of Texas Press.

D'Ambrosio, U. (2007). Peace, social justice and ethnomathematics. In B. Sriraman (Ed.), *International perspectives on social justice and mathematics education: The Montana mathematics enthusiast, monograph 1* (pp. 25–34). Missoula, MT: The University of Montana Press.

De Augusta, F. J. (1903). *Gramática Araucana (Araucanian grammar)*. Valdivia, Chile: Imprenta Central.

De Landa, D. (2005). *Relación de las Cosas de Yucatán*. Mexico: Editorial San Fernando.

Laughren, M. (1982). Warlpiri kinship structure. In J. Heath, F. Merlan, & A. Rumsey (Eds.), *The languages of kinship in Aboriginal Australia, Oceania linguistic monograph* (Vol. 24, pp. 72–85). Australia: University of Sydney.

Laughren, M., Hoogenraad, R., Hale, K., & Japanangka Granites, R. (1996). *A learner's guide to Warlpiri*. Alice Springs, Australia: IAD Press.

Menninger, K. (1970). *Number words and number symbols, a cultural history of numbers*. Cambridge, MA: The M.I.T. Press.

Stehber, R. (2015). Confirman poder del estado Inca al sur de Santiago de Chile (The power of the Inca state south of Santiago, Chile is confirmed). Resource document. La Gran Epoca. Retrieved January 23, 2015 from http://www.lagranepoca.com/archivo/34835-confirman-poder-del-estado-inca-al-sur-santiago-chile.html.

Stuart, D. (2011). *The order of days: The Maya World and the truth about 2012*. New York, NY: Harmony Books.

Vandendriessche, E. (2015). Thomas Storer's heart-sequence: A formal approach to string figure-making. *Journal of Mathematics & Culture, 9*(1), 119–159.

Chapter 20
First-Year Seminar Writing for Quantitative Literacy

Maria G. Fung

Abstract This chapter describes a series of writing assignments from two distinct first-year seminars at Worcester State University. Both of these seminars focus on building quantitative literacy skills. Students research and write about a variety of topics related to globalization, population growth, human rights and climate change. The impact of these seminars for the students, institution, department, and instructor is discussed.

Keywords Writing • Quantitative literacy • Assignment • Examples • First-year seminar

20.1 Introduction

According to the Association of American Colleges & Universities (AAC&U 2014a), quantitative literacy encompasses all the capabilities of an individual to work with numerical data in a variety of contexts and situations. Students should be able to collect, organize, and analyze numerical or statistical evidence, as well as to communicate their findings in a number of representations, from tables, charts, and graphs to mathematical equations. After considering many different paradigms, both historical and semantic, Karaali et al. (2016) proposed that quantitative literacy be conceived as "competence in interacting with myriad mathematical and statistical representations of the real world, in the contexts of daily life, work situations,

MSC Codes
97-01
97B20

M.G. Fung (✉)
Department of Mathematics, Worcester State University,
486 Chandler Street, Worcester, MA 01602, USA
e-mail: mfung@worcester.edu

J. Dewar et al. (eds.), *Mathematics Education*, Association for Women in Mathematics Series 7, DOI 10.1007/978-3-319-44950-0_20

and the civic life". This framework forces quantitative literacy outside of the traditional mathematics college courses (such as finite mathematics, college algebra, and precalculus) into an across-the-curriculum effort, especially suited for interdisciplinary endeavors. A first-year seminar, which is at the centerpiece of the general education curriculum at many institutions, is a logical venue for focusing on quantitative literacy.

In today's data-driven world, quantitative literacy is an essential habit of mind. According to the 2003 National Assessment of Adult Literacy Survey (NAAL), 22% of American adults scored in the "below basic" category on quantitative literacy, with another 33% in the "basic" category. This constitutes an alarming percentage of Americans who are unable to perform simple arithmetic operations (such as calculating a gratuity on a bill or a figuring out a total price for postage) or interpret data (such as reading and interpreting simple charts or graphs). The burden falls on the K-16 education system to remedy this situation. University mathematics departments need to become leaders in the efforts to ensure that college graduates are numerically literate citizens. According to the recommendations of Committee on Undergraduate Programs in Mathematics (CUPM) of the Mathematical Association of America (MAA), mathematics courses for general education should "increase quantitative and logical reasoning abilities needed for informed citizenship and in the workplace" and "improve every student's ability to communicate quantitative ideas orally and in writing" (MAA CUPM 2004, p. 28).

In *Mathematics and Democracy*, Steen (2001) goes further to distinguish college mathematics from quantitative literacy knowledge. He calls for quantitative literacy instruction, neither as a replacement nor an alternative to mathematics, but as "an equal and supporting partner in helping students learn to cope with the quantitative demands of modern society" (p. 135). Mathematics faculty profit both by rethinking the general education mathematics tracks and by collaborating with colleagues from other departments on infusing general education courses with a quantitative literacy focus (Miller 2012).

20.2 First-Year Seminars at Worcester State University (WSU)

Steen (2004) urges mathematics and statistics faculty to relate quantitative literacy to other "reform" programs such as first-year seminars and teacher preparation. As a part of its general education program, the first-year program at WSU requires every incoming student with fewer than 12 college credits to complete a first-year seminar during their first semester of study. These seminars are interdisciplinary, discussion-based courses that cover a current or controversial topic within the humanities, social sciences, or Science Technology Engineering and Mathematics (STEM) disciplines. Each of the more than three dozen seminars offered every fall shares the three-part goal of increasing students' written and oral communication skills, critical thinking, and information literacy (the ability to collect, evaluate, and use appropriately a variety of supporting sources). Every seminar requires a final capstone writing assignment for students to demonstrate their development towards achieving its three-part goal. This assignment is referred to as a *signature* assignment, since it serves as an assessment tool for the institution.

First-year seminars are seen by the university community as opportunities to delve deeper into topics of joint interest to instructor and student, without being confined by a particular discipline. First-year seminars also serve a significant role in helping students to transition to college life by building up their communication and analysis skills. These seminars foster a very close relationship between students and faculty members that appears to persist past the end of the first year. Due to their interactive close-knit community character, these seminars are a significant retention tool for the institution.

As a result of their interdisciplinary and discussion-based nature, first-year seminars offer a natural venue for exploring topics in quantitative literacy. Since instructors have the freedom to weave themes from different subject perspectives, they can find ways to engage students in thinking critically about a specific topic across disciplines. Quantitative literacy is a logical extension of all of the components of the three-part learning goal for first-year seminars, since it involves: looking critically at numerical or statistical data; compiling, evaluating, and analyzing data similarly to any other information; and finally communicating one's findings both orally and in writing.

Two first-year seminars with a quantitative literacy emphasis were taught in the fall semesters of 2011 and 2014 at WSU. The first seminar was called "Disturbing Times in Worcester and the World" and it was a part of the university-wide effort of exploring Worcester's place in the world through a combination of courses, events, and speakers. The second was "The Nature of Climate Change: A Quantitative Approach." Both involved explorations of data sources, bias and reliability of data collection, and manipulation of results. A variety of written assignments and projects comprised a fundamental part of the curriculum in both courses.

20.2.1 Disturbing Times in Worcester and the World

This course had several themes running through it—globalization, population growth, human rights, international security, and environmental issues. All of these were examined both locally (in Worcester) and globally (in the world). The required texts for the course included *Disturbing Times: the State of the Planet and Its Possible Future* by Scott T. Firsing, *What the Numbers Say* by David Boyum and Derrick Niederman, *Now or Never: Why We Must Act Now to End Climate Change and Create a Sustainable Future* by Tim Flannery, and *Confessions of an Economic Hitman* by John Perkins. The last book was a required reading for all first-year students, and at the end of the semester the author gave a talk on campus.

A typical class started with discussions of assigned readings, both in small groups and as a class. Then students worked on short projects and assignments, wrote about their findings and presented their ideas. Every class concluded with a short debriefing of the ideas from the day's explorations. Occasionally, the instructor gave mini-lectures on topics from descriptive statistics, measurement and probability. Writing permeated virtually every activity in the course. It was done both in and outside of class, with feedback from peers and instructor.

The quantitative literacy focus in this course included collecting, organizing and summarizing data, reliability and bias of data sources, percentages and proportions, elementary probability theory including conditional probability (Niederman and Boyum 2003).

20.2.2 The Nature of Climate Change

This course formed a learning community together with an honors English composition course. The students were enrolled in both courses in consecutive time slots, and both classes focused on the same themes of climate change. The first-year seminar had two required readings: *The End of the Long Summer: Why We Must Remake Our Civilization to Survive on a Volatile Earth* by Dianne Dumanoski and *Climate Myths: The Campaign Against Climate Science* by John J. Berger. The English composition course used *Moral Ground: Ethical Action for a Planet in Peril*, a collection of essays edited by Kathleen Dean Moore and Michael P. Nelson. Several times during the semester, students attended lectures on environmental science and watched scientific documentaries related to climate change during the combined time blocks for the learning community. A joint final essay served as the *signature* assignment for the first-year seminar and the capstone project for the writing course.

The class structure was very similar to the "Disturbing Times in Worcester and the World" course. Each chapter of the readings was used as a starting point for learning and exploring different facets of climate change. At the end of the semester, students had the opportunity to meet Diane Dumanoski. Speaking with the author of a book that the students had carefully read and analyzed was an exceptional experience. The question-and-answer session both challenged and delighted the author, who was sincerely impressed by the depth of the students' understanding.

Similar to the other course, the quantitative literacy component focused on collecting, organizing and summarizing data, reliability and bias of data sources, percentages and proportions. It also included some modeling and prediction.

20.3 Writing-to-Learn

Writing has long been an essential pedagogical method for learning new material, for making connections and for meta-cognition (Walvoord 2014). Writing is especially critical when students work in a classroom environment where active engagement with the material, explorations and discussions are at the center of learning (Bean 1996). In college classrooms where the bulk of learning is accomplished through student activities, the writing process allows students to comprehend the problems they are working on at a deeper level, to synthesize new ideas and connections, and to arrive at conclusions, or to generate the next problem to be considered (Bean 1996).

Quantitative literacy tasks always involve communication of findings, thus making writing a natural way of learning and expanding knowledge. Meier and Rishel's (1998) three categories of mathematical writing—personal, reflective and expressive—apply to the realm of quantitative literacy in a way that corresponds directly to the level of complexity and demands on the student writer. Examples of personal or informal writing include personal reflections on readings or in-class activities, and journals. Expository writing ideas include short project papers, essays on specific topics from the course, or beginner's research papers. Expressive writing examples include the midterm research projects and the *signature* writing assignments from each of the seminars. More details on the quantitative literacy writing assignments in each of the three categories follow.

20.3.1 Informal Writing

The informal writing in both first-year seminars took the form of daily reflections, extensions of the assigned readings, journals, check-in or exit cards (short written responses to questions posed at the start or end of class), and free writing. Most of the time, students had to come to class with a short paper that included a summary of the reading for the day and their reaction to it. Sometimes students had to write a letter to a parent, peer, or political figure able to make policy decisions.

Here is an example of a typical "Disturbing Times in Worcester and the World" assignment:

- Read Chap. 4 in "*Disturbing Times…*" on protecting human rights. Write a two-paragraph summary of this chapter. Focus on one of the topics in the chapter and research it further. Prepare a five-minute presentation on this topic of your choice to share with the class. Focus on using reliable data sources to support your claims. Your audience will be your classmates.
- Research the organization mentioned in the reading, *Transparency International*. Spend some time researching their work. Find three quantitative things of interest to you and write a small paragraph for each. Did any of these findings shock you? Discuss.

Early in the semester the students in "The Nature of Climate Change" were asked to find two current online articles that dealt with climate change and to summarize each article in several paragraphs. They also had to determine if the articles referred to data sources or other quantitative information and to provide reference for the sources. Students read and commented on each other's papers, a process that helped them assess their own work. Then there was a whole class discussion about the reliability and bias of sources, and therefore in the presentation of information to the public.

In their bi-weekly journal assignments, students in the learning community were asked to write freely about the most significant ideas they have encountered and how these ideas were related to both of the linked courses. One requirement of each journal entry was to include quantitative information, and to think critically about where this information comes from and how this information influences the way we perceive and think about the issues at hand.

Exit and check-in cards are a way to get immediate student feedback on their understanding and struggles. Short prompts such as "One numerical idea I am still struggling with is....", "The most exciting quantitative concept discussed is....", or "Statistics allows us to...." are an effective way to start a conversation or to revisit a difficult concept.

Free writing was typically done in class as a way to focus students' attention on a topic. For instance, students write for five minutes all their ideas about human rights' violations both locally in Worcester and globally in the world. Alternatively, students write for ten minutes about how technology might be able to help us combat climate change. The goal of these assignments is to initiate students' thinking about a topic and to make them aware of their own background knowledge and limitations. It is important for free writing ideas to be shared with a group, so that a productive discussion can ensue.

Informal writing assignments get students engaged in the material and invested in their learning. These assignments represent simple and immediate ways in which quantitative literacy topics could be brought to the attention of students to prime them for exploring these topics further.

20.3.2 Expository Writing

Expository writing for both first-year seminars comprised longer writing assignments that might take from several days to several weeks to complete. The two assignments in "Disturbing Times in Worcester and the World" included a Census Data project and a population growth project. In the former, the students were tasked to compare Worcester, MA Census Data from 2000 to 2010. The first part of this assignment was to explicitly discuss any changes that could be observed with appropriate numerical measures such as percentage growth. Students had to use appropriate graphs and charts to show their comparison. For the second part of the assignment, students had to find data about the racial and ethnic distribution of students in the Worcester public schools from 2010. At the heart of this part of the project was the fact that Hispanics, a large group in Worcester, MA, can report as any race on the Census, thus skewing the numbers of Whites. So, students were asked to respond to these questions: Was there a discrepancy between the Census and the Worcester Public School population data? How can the discrepancy be explained or resolved? Could students support their claims with some further research?

In the population growth project, students looked at the populations of Worcester, of the state of Massachusetts, and of the world from 1900 to 2010 in 10-year increments. They collected and organized this data in a way that was easy to understand, employing a variety of graphs. Using two different functions of best fit in *Excel*, they were able to predict how the populations would change with time. Which models are more realistic? Why or why not? Which models have larger errors for intermediate points? Students wrote about the limitations of simple models (linear, exponential and logistic) to capture a dynamic process that involves migration and a number of external factors that can directly alter a population, such as food and water shortages, urbanization, and disease spread.

In "The Nature of Climate Change" course, students worked on climate data from the Worcester, MA project. The instructions for this project were:

- Go to the website: weather-warehouse.com
- Find out information about the average monthly temperatures in Worcester (collected at the regional airport) from the past 50 years. Use this information to obtain yearly average temperatures for Worcester from the past 50 years. Enter all your data in an Excel spreadsheet, where you record the years in the first column and the yearly average temperatures in the second column.
- Using Excel, construct a plot of yearly average temperatures on the y-axis. This is called a time-series (the years will appear on the x-axis). What trends do you see?
- Using Excel, construct a "scatter with straight lines and markers plot" of the average yearly temperatures—it will consist of separate dots for each data point and every adjacent pair of points is connected with a straight-line segment.
- Calculate the line of best fit for the scatter plot. What do you observe about its slope? What does this tell you about the trend in temperature for the past 50 years?
- Now split the data up into two groups of 25 years, meaning, for instance, that you start with the 1963–1988 first and study it independently. Then do the same, with the 1988–2013 group. Repeat the steps of finding the best fit line. What do you notice? Is there a difference for the first 25 years versus the last 25 years? How can we explain this difference?
- Now calculate the standard deviations in all three cases. How do they compare? What could you conclude from your calculations?
- Write a two-to-three-page presentation of your observations that could be shared with a climate change doubter.

Compared to personal writing assignments, expository writing assignments provide more complex and sophisticated ways of encouraging students to reflect on what they are learning, and to synthesize new ideas. They are an important tool in any educator's repertoire and can easily be adapted to a variety of quantitative situations.

20.3.3 Expressive Writing

Expressive writing assignments often serve the role of capstone projects, whether at midterm or at the end of the semester. They are among the most complex independent written work students produce in a course. These assignments allow students choices for the ideas they want to explore and much more creativity in structuring and presenting their findings.

In "The Disturbing Times in Worcester and the World" seminar students worked on exploring and comparing five different educational challenges of Worcester, MA, and of Pakistan, supporting their claims with ample quantitative evidence. Then they presented solutions for addressing these challenges and for bringing about improvement. The students had to quantitatively estimate the effect of their proposed solutions.

Another assignment from the same course focused on creating a population reduction program for Sudan, a country with high population growth. Students had read about the importance of education, government policies, and availability of birth control as factors for curbing population growth. They considered the hypothetical impact that improved universal educational programs, better governmental policies, and ample donations of means of birth control from the United States would bring to Sudan within a decade, 50 years, and a century. All essays focused on different measures, but shared good predictive numerical calculations about change in population growth. Errors of estimation were also discussed.

In "The Nature of Climate Change" seminar, students completed two creative writing assignments. After they had become familiar with the main factors causing climate change, students wrote an essay about the impact of one of Mexico, Brazil, China, India, Indonesia, or Russia on climate change. These countries were selected based on their large populations and growing economic impact. The essays included the following sections:

- Introduction: summarize some information about the country's history, geography, and economics.
- Climate change impact through:
  - Use of fossil fuels and other materials that promote greenhouse gases
  - Economic development that requires the use of increased resources and specifically leads to deforestation, etc.
  - Population change and water/food resources
  - For the climate impact, provide data that you find and deem reliable; this means using at least 3–5 sources (online sources are acceptable but make sure the data are reliable).
- Projections of how these factors will impact climate change if things are left unchanged: please show your reasoning, support your claims with quantitative evidence, and describe the immediate and long-term impact of keeping the status quo.

- Ideas for change, e.g., policies (both local, regional and global), international programs, local initiatives, etc. and how these ideas will impact the climate. Include short (up to 5 years) and long-term (10–50 years) projections of the effects of your proposed interventions. Support all your claims with numerical calculations and estimations.
- Conclusion: discuss what have you learned from this project.
- Bibliography—you can use MLA, APA, or Chicago but be consistent throughout.
- Graphs and tables: include them as an appendix rather than in the body of your paper.

The final shared assignment of the learning community also fits the category of a creative writing assignment. Assignment details are provided in the Appendix. The topics students chose varied from the production of tidal power, to the systematic reduction of red meat consumption, to universal composting, to the creation of a national K-12 climate change curriculum.

20.3.4 Brief Comments on Assessing Writing

The personal writing assignments are most easily evaluated by a simple check-off for work completion and immediate comments in the margins of the paper. The expository and creative assignments could be assessed on a simple 3-point system, where 1 is minimal, 2 is emerging and 3 is thoroughly developed. Several components of the assignment such as organization, completeness, and clarity could be assessed with this 3-point system. Instructors could certainly create more complicated rubrics to fit any particular assignment. It is critical to provide detailed information about the expectations for each assignment, so that students understand how to be successful.

20.4 Outcomes of Quantitative Literacy Writing

20.4.1 Impacts on Students

Students from both groups found the numeracy focus useful (as noted in their comments to the professor) and some shared their enthusiasm with the instructor several years after the date of the course's completion.

In terms of grades, students from both quantitative literacy seminars received grades with an average not different in a statistically significant way from that of the entire first-year seminar cohort. However, the 2014 students' scores on critical thinking using the AAC&U rubric (AAC&U 2014b) in a university assessment effort were consistently above the benchmark for first year students. The English professor in the learning community on climate change was impressed with the increased detail to argument and greater clarity in the students' writing, which could be partially attributed to their work with quantitative sources.

In the fall of 2015, the ten students from the 2011 seminar were surveyed by email to answer the following questions: What was the most memorable part of the course? What were the most important things you learned in the course? A similar survey is planned for the students in the 2014 seminar.

Four students responded to this survey. A qualitative analysis showed the emergence of three different themes: critical look at numbers and data, community-building, and awareness of the need for social engagement. The topics of climate change and global warming, sustainability, the effects of globalization on world culture, and critical consideration of numbers and data were mentioned explicitly in the responses.

One student noted she was most influenced by " the inconsistency of numbers (data, surveys, stats) in contrast to my prior understanding of its absolutism- and as you can imagine that lesson transferred to the rest of my studies and understanding of the world." Another student wrote about how "it is important to question numbers since they can be easily miscalculated or presented in misleading ways."

Collaboration was also important to students since the course was "a warm class and it allowed us to build a mini-community." A student wrote about how "an individual contribution can make all the difference" in resolving the most pressing issues of today.

20.4.2 Department and Institutional Support

The mathematics department at WSU has been very supportive of this work. Several members of the department led a university-wide initiative to have quantitative literacy across the curriculum as a larger part of the general education curriculum, as opposed to the more narrowly defined "quantitative reasoning" category, which required a high mathematics placement test score and impeded many departments in the humanities and social sciences from participating in these efforts. There were several presentations by mathematics faculty at university-wide conferences about the importance of quantitative literacy and the significance of every discipline's involvement in the efforts to create numerically competent graduates. The mathematics department revised the general education mathematics course (MA 105 Survey of Math) for students in the humanities to include a unit on statistics, by replacing the section on different numerical systems and bases.

There is increased institutional support for quantitative literacy initiatives, involving grant money to develop quantitative literacy modules in a variety of general education courses. WSU is a Liberal Education and America's Promise institution (LEAP 2014), and the mathematics department will continue to lead and oversee efforts to incorporate more quantitative literacy emphasis into the general education curriculum.

20.4.3 Impact on the Author

As author of this article, I have learned much from my students since the beginning of my quantitative literacy work. My passion for social engagement and my interests in writing in mathematics courses and in collaborative learning naturally led me to designing and teaching first-year seminars focused on quantitative literacy. Teaching a first-year seminar based on discussion is a challenging yet exciting experience that impacts my pedagogical decisions in teaching any course. I am incorporating more discussion and writing assignments into all of my mathematics courses. I have infused a quantitative literacy emphasis into the sequence of three courses for elementary school teachers that I often teach. This has taken the form of focus on mental calculation, estimation, probability and statistics projects, and applied proportional reasoning and percentage problems.

There is another first-year seminar planned for the fall semester of 2016. It is called "What the Numbers Say" and it will focus throughout on quantitative literacy. Once again, it will be linked with an English composition course with common thematic material and a shared *signature* assignment. Students will investigate quantitative scenarios from the news, politics (amidst the Presidential elections), medicine, and sports.

The two main texts for the course will be *What the Numbers Say: A Field Guide to Mastering Our Numerical World* by David Boyum and Derrick Niederman and *The Numbers Game: The Commonsense Guide to Understanding Numbers in the News, in Politics, and in Life* by Michael Blastland. Students will be given a different problem every few weeks that would require their use of quantitative methods for modeling and generating a proposed solution. They will work collaboratively to produce a feasible result. They will discuss, present, and write about their solutions.

I have also collaborated with members of the history, music, and art departments on developing materials for their quantitative literacy modules. I am currently involved with a Science Technology Engineering Art and Mathematics (STEAM) K-6 school in Groveland, MA, helping the mathematics teachers there to incorporate more quantitative literacy and writing into their work.

20.5 Conclusion

As a result of the interdisciplinary focus of the two first-year seminars, students received plentiful opportunities to explore the significance of quantitative information—data compilations, percentage statistics, trend graphs, and numerical modeling, among others—in the context of a diverse set of political, environmental and humanitarian issues of imminent global concern. The three thematic writing assignments discussed above sharpened the students' abilities to analyze, synthesize, predict, and conclude on the basis of collected quantitative data. Similar assignments could be used successfully in a variety of mathematical and general education contexts.

Appendix: Assignment for an Integrated Project

Essay Project 3: Proposing a Solution

Part 1

Project 3 is an integrated project between LC193H: The Nature of Climate Change and EN250: Creative Thinking and Critical Writing. This project asks you to offer a solution to an identified problem associated with the environment, climate, or sustainability.

Your solution has to be ***feasible***. That is to say, your solution should be focused, supported by evidence, and have the possibility of being implemented at some level. To this end, you may want to focus on a smaller implementation of a solution, such as at the state, local, or even campus level. For example, you could propose (argue) that Worcester State University would benefit from a compost pile. In your essay, you would demonstrate through argumentation and the appropriate use of evidence that a compost pile is beneficial to the campus community. Another example would be to propose (argue) that the colored garbage bag program found in cities such as Worcester and towns such as Shrewsbury and Northborough be expanded to the entirety of the State of Massachusetts. Once again, you will have to establish that there is an exigence that needs to be addressed, and after establishing the exigence you then will present your solution. ***One way to think about this project is as a problem/solution essay***.

Particulars

Five to seven pages, double-spaced.

You should use at least two scholarly sources and appropriate **quantitative** and **statistical** evidence to support your claims.

You will be assessed on the strength of your writing (complexity of main claim and supporting claims, topic sentences, paragraph construction, fluidity and transitions, clarity, and persuasive appeals).

And on the quality and strength of your sources and evidence, particularly your quantitative evidence. Indirectly, you will also be assessed on your ability to find quality sources for use as evidence, as you will have to demonstrate the strength of authority of your sources in your essay.

Part 2

For LC193H, you will need to make a ***poster*** based on the first part of the project. Your poster will be displayed at the First-Year Experience Showcase on December 9th from 2:00 to 4:00 in the May Street Building. More information on the poster session will be provided in class.

For EN250, you will need to re-purpose your information and arguments from your essay into a ***letter*** directed to a specific political leader or business leader. You will have to research which leader you will write to and plan on explaining how you are adapting your rhetorical strategy to a different audience and a different genre.

WRITE WELL!

References

AAC&U. (2014a). Quantitative literacy VALUE rubric. Retrieved June 25, 2016 from https://www.aacu.org/value/rubrics/quantitative-literacy.

AAC&U. (2014b). Critical thinking VALUE rubric. Retrieved June 25, 2016 from https://www.aacu.org/value/rubrics/critical-thinking.

Bean, J. C. (1996). *Engaging ideas: The professor's guide to integrating writing, critical thinking, and active learning in the classroom*. San Francisco, CA: Jossey-Bass.

Karaali, G., Villafane-Hernandez, E., & Taylor, J. (2016). What's in a name? A critical review of definitions of quantitative literacy, numeracy, and quantitative reasoning. *Numeracy, 9*(1). doi: 10.5038/1936-4660.9.1.2.

LEAP States Initiative. (2014). Retrieved May 26, 2016 from https://www.aacu.org/leap/states.

MAA CUPM. (2004). *Undergraduate programs and courses in the mathematical sciences: CUPM curriculum guide 2004*. Retrieved May 26, 2016 from http://www.maa.org/sites/default/files/pdf/CUPM/cupm2004.pdf.

Meier, J., & Rishel, T. (1998). *Writing in the teaching and learning of mathematics*. Washington, DC: Mathematical Association of America.

Miller, J. A. (2012). *Serving two masters: A study of quantitative literacy at small colleges and universities*.(Unpublished doctoral dissertation). Montclair State University, NJ.

National Assessment of Adult Literacy (NAAL). Demographics—overall. Retrieved May 26, 2016 from https://nces.ed.gov/naal/kf_demographics.asp.

Niederman, D., & Boyum, D. (2003). *What the numbers say: A field guide to mastering our numerical world*. New York, NY: Broadway Books.

Steen, L. A. (2001). *Mathematics and democracy: The case for quantitative literacy*. Princeton, NJ: National Council on Education and the Disciplines.

Steen, L. A. (2004). *Achieving quantitative literacy: An urgent challenge for higher education*. Washington, DC: Mathematical Association of America.

Walvoord, B. E. F. (2014). *Assessing and improving student writing in college*. San Francisco, CA: Jossey-Bass.

Chapter 21
Tactile Mathematics

Carolyn Yackel

Abstract Tactile mathematics, defined as recognizing deep mathematical concepts through engagement with physical objects, can be used to help students discover mathematics for themselves. This paper discusses the design of tactile learning activities, the insertion of such activities into existing courses, and special considerations for courses to be taught almost entirely with tactile activities. We explain a specific example activity for a group theory course. A collection of mathematics faculty members experienced in tactile learning contribute their thoughts on the implementation of largely tactile mathematics courses. We end with the role of tactile mathematics in the author's career.

Keywords Tactile mathematics • Active learning • Mathematical art

21.1 Introduction

Imagine being in a classroom listening to a traditional graph theory lecture about Hamiltonian cycles—paths in a graph that visit each vertex exactly once, starting and ending at the same place. Now imagine attending the same class, but sitting down to a piece of stryrofoam with pushpins for the vertices of the graph. You are instructed to attempt to form Hamiltonian cycles with yarn. The professor has carefully crafted a set of questions through which the class together begins to discover the basic theorems about Hamiltonian cycles.

MSC Codes
97D40
00-xx
97B40
97H40

C. Yackel (✉)
Department of Mathematics, Mercer University,
1501 Mercer University Drive, Macon, GA 31207, USA
e-mail: yackel_ca@mercer.edu

J. Dewar et al. (eds.), *Mathematics Education*, Association for Women in Mathematics Series 7, DOI 10.1007/978-3-319-44950-0_21

The second classroom scenario gives an example of tactile mathematics: the process of coming to understand deep mathematical ideas through engagement with physical objects. Instances of tactile mathematics can include physical models made specifically for the teaching of various mathematical concepts called manipulatives. Other examples of tactile mathematics involve having students engage in a craft, such as origami, and noticing specific aspects of the results of their work. Key features of the process are twofold: the hands-on (tactile) facet of the activity and the thoughtful mathematical reflection upon the result of the process.

The purpose of this chapter is to introduce the notion of tactile mathematics. Section 21.2 is intended to help readers understand how to design lessons involving tactile activities. We demonstrate this lesson design method with a concrete example of a tactile learning activity for a traditional abstract algebra classroom in Sect. 21.3.1. Section 21.3.2 contains a list of resources for teachers interested in inserting a tactile mathematics activity into the classroom. Section 21.4 discusses special considerations for instructors preparing to implement an entire course predicated on the use of tactile activities. Section 21.5 recounts the role of tactile mathematics in the author's career.

21.2 Lesson Design

We take a multi-layered approach to class planning involving identifying the mathematical skills and the mathematical concepts to be learned, as well as the mathematical maturity goals for our students. We develop our course or lesson design to meet these learning goals. The teaching method for each lesson is selected to maximize student learning of the concepts for that lesson. Some material might be best taught with the lecture method, whereas other material might be most easily absorbed by students if they are actively engaged. The process of working from goals to lesson plans has been labeled backward-design.

The pedagogical use of tactile mathematics is one form of what is sometimes called inquiry-based learning. Inquiry-based learning is a method characterized by presenting students with problems for which the solutions require students to first assemble new results from information and ideas presented in class and in readings (Prince and Felder 2007). The results-building phase often involves an experimentation process in which examples play the role of experiments. Some students find working with examples comprised of numbers and functions to be too abstract. Examples grounded in physical objects or actions can be more helpful to these learners. Making the mathematical model of the physical scenario explicit is the most important and frequently the most difficult part of the process. If tactile mathematics activities are to be effective, instructors must help students navigate appropriate use of mathematical ideas, terms and notation (Alfieri et al. 2011).

As with most inquiry-based learning, tactile activities can take up a significant amount of class time. The tactile aspects of the activities need to be simple, allowing the majority of class time to be spent on mathematics. Careful planning ensures that activities are targeted at specific learning objectives or concepts. The instructor needs to plan by envisioning several ways the students might develop the ideas as they work through the material (Steffe and D'Ambrosio 1995). Each sequence of idea development is called a learning trajectory.

21.3 Incorporating the Tactile into Existing Courses

We begin this section with an example and follow up with a list of resources for the reader intrigued by the pedagogy and looking for source material.

21.3.1 Example of a Tactile Activity to Determine the Symmetries of the Equilateral Triangle

In this section we describe a textbook-independent tactile activity for the first day of an abstract algebra course with an enrollment of about 15 students. This activity has also proven effective half-way through an introduction to proofs course of the same size, as we study number systems. We also utilize this activity for mathematics enrichment for interested high school students in groups of up to 20. As described in this section, the activity will exhibit high levels of student autonomy commensurate with the mathematical maturity expected from juniors and seniors enrolled in abstract algebra. For use with students at an earlier mathematical stage, we provide more support in terms of initial definitions, notation use, and structuring discussion.

The explicit objective of the activity is for students to determine the symmetry group of the equilateral triangle. This group is known both as the dihedral group of order six, D_3, and the symmetric group on three objects. Abstract algebra students need to have a good understanding of all groups of small order. Having a physical interaction with the symmetries of the equilateral triangle helps students internalize the surprising fact that the group is not commutative, which runs counter to the majority of their previous experiences in mathematics. The activity's counterintuitive result is also a nice hook for an enrichment program. The symmetry group is an important example for students first learning to think carefully about definitions, such as in an introduction to proofs course.

For the remainder of this section, we focus on the abstract algebra scenario, trusting the reader to make the necessary adaptions to other situations. The division between skill objectives and concept objectives can be a fine line. Below we include a list of each followed by a list of student and mathematical maturity goals. The lists are provided in keeping with the lesson design method outlined in Sect. 21.2.

Skill objectives for the activity: students will

1. Recognize and discover symmetries of objects,
2. Determine symmetry groups of equilateral triangles,

3. Make a group multiplication table,
4. Learn to use multiple notations for the same idea,
5. Learn 2-line notation (explained below), and
6. Learn group generator notation (explained below).

Concept objectives for the activity: students will understand

1. The idea of equivalence of two symmetries,
2. Composition as a binary operation we call multiplication,
3. The notion of closure of a set under an operation,
4. The role of the identity element in a group, and
5. That not all groups are commutative.

While investigating the symmetry groups of equilateral triangles through the activity, I expect my students to be working on a number of student and mathematical maturity goals as well:

1. Learning how to hold a mathematical discussion with peers,
2. Learning to give mathematical evidence, such as through examples,
3. Learning how to ask questions of one another and of me,
4. Learning how to grapple with uncertainty, such as when one hasn't been told exactly what to do,
5. Learning how to keep records of mathematical experiments, such as keep track of the results of the symmetries,
6. Gaining mathematical authority, such as by making conjectures and supporting these with evidence,
7. Gaining mathematical autonomy, such as by working individually or disagreeing with the group,
8. Taking ownership of mathematics, such as by figuring out how mathematics is working for oneself, and
9. Understanding the need for mathematical notation.

To help my students investigate the symmetry group of the equilateral triangle, I hand out paper triangles with labeled vertices from Fig. 21.1 (right), along with a labeled outline triangle copied onto an otherwise blank piece of paper, as in Fig. 21.1 (left), that helps students to keep track of the triangle's initial position. Teachers must have a flexible understanding of many possible learning trajectories so that they are able to quickly and accurately interpret student comments and questions in class, thereby giving helpful guidance as described by Liping Ma (Ball and Bass 2000). I will describe a typical classroom scenario that includes a learning trajectory through this activity.

First, students must think through what constitutes a symmetry. How do they determine when a symmetry occurs? What does the result of a symmetry look like? Typically after about a minute of bafflement, a class conversation emerges. Perhaps one student will already know at least one example of a symmetry, which the student will present to the others. Alternatively, I will present the idea of rotation by 120° counter-clockwise, showing this motion of the moving triangle on top of the

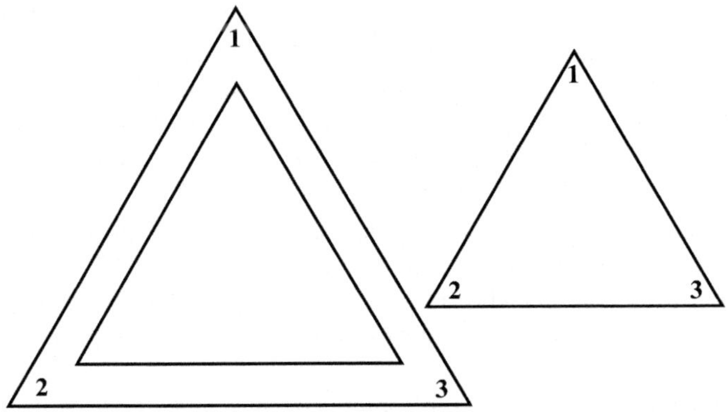

Fig. 21.1 Fixed triangle (*left*) and triangle to move (*right*) for investigating D_3

fixed background. At this point, more discussion ensues over what other motions would constitute symmetries. If necessary, I interject that we can also flip over the triangle.

Next, I ask how many symmetries exist and how they can be obtained. After an initial discussion, the individuals settle down to make a list. At some point fairly early in the process, we have a discussion about how some of the symmetries have the same result as others. For example, rotation through 480° counter-clockwise has the same result as rotation through 120° counter-clockwise, so we call it the same symmetry. Our exchange allows a winnowing of the lists. A further discussion of the identity ensues. Because this activity typically marks my students' first exposure to formal groups, they are often surprised that we will include the identity element in our set. A nice discussion of the importance and usefulness of closure under composition of symmetries results.

Now we are ready to discuss a representation for our symmetries, if this has not already happened. Usually two-line notation is the easiest for students to understand first. Recall that with the triangle as pictured in Fig. 21.1, rotating the interior triangle from the starting position of matching vertex labels by 120° in the counter-clockwise direction gives the two-line notation $\begin{pmatrix} 1 & 2 & 3 \\ 2 & 3 & 1 \end{pmatrix}$, because after rotation, vertex 3 is now in the location previously occupied by vertex 1, vertex 1 is now in the location previously occupied by vertex 2, and vertex 2 is now in the location previously occupied by vertex 3.

Now I lead a classroom discussion on the number of symmetries of the equilateral triangle. After a bit of debate, students explain to one another the reason that there are six. I then point out that if we perform one symmetry and then perform another symmetry, we should still have a symmetry of the triangle, thus introducing the idea of composition as multiplication. Next I ask if this might give us more symmetries than we had previously considered, again raising the issue of closure. This question allows me to bring in the idea of starting a multiplication table, known as a Cayley table.

The class members can now begin to fill out their Cayley tables individually, not knowing if more than the initial six elements will need to be added. Oftentimes one or more students will realize the helpfulness of conceiving of the symmetries as rotations and reflections. If so, I have individual or small group consultations in which I encourage adoption of a different notation, such as the typical group generators and relations $D_3 = < r, f \mid r^3 = f^2 = e, rf = fr^2 >$, where e denotes the identity, r denotes rotation by 120° counter-clockwise, f denotes a specific reflection, and these two symmetries satisfy the relation $rf = fr^2$. Notice the differentiation of instruction between students who are ready for the concept of the group as described by generators and relations and students who need more time to play with groups in terms of two-line notation.

After 5–8 min, we can discuss the contents of the Cayley table. This is an important step, as some of the students will have made mistakes based on the assumption of commutativity of multiplication, on tracking errors as they manipulated the triangle, and on notational errors as they attempted to record the result of the symmetries. The discussion is always rich with insights as students realize that: this set is closed under composition; multiplication by the identity does nothing; multiplication is not commutative in this group meaning that multiplication does not need to commute(!); every element has an inverse in the set; and every row and every column of the Cayley table contains each element.

Finally, we discuss various representations we might have wanted to use instead of two-line notation. Depending on the class members, we may have written the Cayley table in terms of a different representation. In all cases, we are sure to use generator-relation notation and cycle notation. Remember that if we rotate the interior triangle from the starting position of matching vertex labels by 120° in the counter-clockwise direction then in cycle notation, we write (123) because vertex 1 is sent to the location of vertex 2, vertex 2 is sent to the location of vertex 3, and vertex 3 is sent to the location of vertex 1. If we reflect across the vertical altitude through vertex 1 so that vertex 1 is fixed, the symmetry would be (23) because vertex 2 is sent to vertex 3 and vice-versa. Vertex 1 is not mentioned because it is fixed.

The triangle activity can easily take an entire class period, which could be considered slow coverage of material. However, the results are excellent, as evidenced by subsequent one-on-one conversations, student classroom comments, in-class work, homework and exams. All students

1. are highly participatory,
2. gain a thorough grounding in what constitutes a symmetry,
3. know at least one way to represent a symmetry through notation,
4. understand the difficulties with choosing a notation, which is a fundamental problem of mathematics and certainly an issue for people trying to use mathematics in their lives,
5. understand the physical representation of the dihedral group,
6. understand and can use the notion of a Cayley table,
7. understand and can use the notion of an identity element,

8. have a reference multiplication table for D_3, which we name at the end of the class period, and
9. comprehend that I think they can explore not just this aspect of group theory, but mathematics in general, make conjectures, and figure out ideas on their own. This point addresses student autonomy and ownership of the subject matter.

Most students can

1. subsequently operate with all dihedral groups in their heads, but those who cannot make their own physical manifestations to manipulate,
2. remember that the group is not commutative and realize the implication that not all groups are commutative,
3. subsequently use the group generator notation and two-line notation, and
4. understand composition as a form of multiplication.

In the next class period we are quickly able to generalize to state the definitions of all the finite dihedral groups. Perhaps more importantly, students pick up on the foreshadowing of the ideas and theorems mentioned above:

- Every group is closed under multiplication.
- Every group has an identity element.
- The definition of an identity element.
- Every element of a group has an inverse.
- Every row and every column of a Cayley table contains each element exactly once.

I like to call activities that set the stage for a number of further such results *touchstone activities*, because students remember them and we can easily refer back to the activity and the corresponding mathematical realizations. For example, if I say, "Remember the triangle activity?" The students recall not just manipulating the triangles, but the notation, Cayley tables, non-commutativity of the group, and so forth. Tactile activities make excellent touchstone activities because it is simple to make them unique by varying the objects or even the color of paper used. The very fact that students are using their hands to manipulate objects is novel in a university mathematics class, rendering the activity memorable.

21.3.2 Resource List

Readers interested in using tactile mathematical activities in the classroom may reference numerous books connecting mathematics and other subjects to develop interesting activities. Below we include a non-comprehensive list of books with ideas ready for the teacher. Individuals will still need to flesh out specifics for classroom use. Full bibliographical citations for these books can be obtained from the chapter end bibliography.

Books

- *Tactile Learning Activities in Undergraduate Mathematics: A Recipe Book for the Classroom* (Barnes and Libertini 2016).
- *Project Origami: Activities for Exploring Mathematics* (Hull 2006).
  This detailed book, soon to have a new edition, is written for the classroom instructor using inquiry-based learning. Each activity is clearly described, a variety of handout choices is included, and these are followed by a thorough solution and discussion of pedagogy.
- *Viewpoints: mathematical perspective and fractal geometry in art* (Frantz and Crannell 2011).
- *Experiencing Geometry: Euclidean and non-Euclidean with History* (Henderson and Taimiņa 2005).
- *Crocheting Adventures with Hyperbolic Planes* (Taimiņa 2009).
  Chapter two concentrates on hyperbolic geometry and provides fairly clear instructions for how to develop an activity.
- *Making Mathematics with Needlework: Ten Papers and Ten Projects* (belcastro and Yackel 2008).
- *Crafting by Concepts* (belcastro and Yackel 2011).
  Each chapter of *Making Mathematics with Needlework* and *Crafting by Concepts* addresses one or more mathematical concepts through the medium of fiber arts. Section three of each chapter contains ideas for how to teach the material, often with suggested activities. The first book includes many specific exercises for students, whereas the second includes possible open-ended projects.
- *Throughout Geometry from Africa: Mathematical and Educational Explorations* (Gerdes 1999).

21.4 Courses Designed Around the Tactile

This section discusses special considerations for planning a course designed specifically around the use of tactile activities, such as a mathematics and origami course. I interviewed three mathematics professors experienced with this pedagogy and will offer our collective wisdom. Eve Torrence, of Randolph Macon College, has taught several mathematics and art courses, including the mathematics of design, mathematical origami, and a course employing mathematical fiber arts. This last course utilized the volumes by belcastro and Yackel, and by Taimiņa mentioned above. Torrence is a mathematical sculptor with talent in multiple craft domains. She has a wealth of experience translating between the realms of mathematics and the tactile. Tom Hull, author of *Project Origami*, listed above, teaches at Western New England University and has many years of experience teaching gifted high school students at the summer programs Hampshire College Summer Studies in Mathematics and Mathily. His expertise is mathematical origami, which he has taught as course insertions and as full courses at many levels. Ruth Favro, of Lawrence Technical University, developed a course in Geometry in Art. The author teaches at Mercer University, co-edited the volumes by belcastro and Yackel, and created a special section of the college's liberal arts mathematics course that presented essentially the material through fiber arts projects. The summary of individual conversations with the members of this group of experienced practitioners follows.

Each teacher firmly believes in using tactile mathematical activities to motivate and illustrate mathematical principles. Favro designed her Geometry in Art class for a client population of architecture majors and visual thinkers. This made her course a perfect fit for the students. Contrastingly, Torrence and Hull developed classes around the mathematical material in the art. They mentioned student excitement about the method of delivery of the mathematical content. Students in future sections of the courses signed up on the recommendation of past students. Special sections of standard liberal arts courses, such as the one taught by the chapter author, can use student advising to increase the likelihood that students register for tactile mathematics sections because of an interest in a craft. By allowing students to self-select into courses dedicated to tactile mathematics, concerns about students who are averse to tactile learning methods are mitigated.

A fear common to instructors considering using tactile activities is that some students will lack the requisite manual dexterity to create the physical models. In response, we note that origami and needlework, such as knitting, crocheting and embroidery, require only basic manual dexterity. These crafts have been taught to grade-school aged children for centuries. To become a skilled artisan naturally requires a great deal of practice and training; therefore, an expert's level should not be expected of the student. Yet a person who has mastered the skill of writing with a pencil should be able to perform each of these skills adequately, and the average adult should be able to reach basic proficiency with only a few hours of practice.

We recommend steering to different courses students with disabilities that will prevent their completion of the tactile component because the point of these courses is to integrate the tactile with the mathematics. The Americans with Disabilities Act requires that accommodations in the form of alternative assignments or assessments be given to students with disabilities as long as the component is not what is termed an *essential element* of the course. In the case of a tactile mathematics course, the tactile component is arguably essential, so the student who cannot participate in this component could take a different course to satisfy the mathematics requirement. If no alternative is available, the instructor will be compelled to offer alternatives to each tactile component, thus designing a parallel course. The case of students with disabilities electing to take a course that clearly does not fit their abilities is rare, and is addressed here only for the sake of completeness.

The first big hurdle to learning a physical skill is believing in one's capacity to acquire the skill so that one will engage in sustained practice essential for picking up muscle memory. The second is paying attention to detail while performing the actions involved so that one can observe helpful nuance and learn from mistakes. The idea that attention to detail enables learning is one aspect of the concept of metacognition, which is sometimes described as thinking about one's thinking (Schoenfeld 1987). Interestingly, engaging in sustained practice and developing metacognitive skills are both important for learning mathematics as well as for learning to craft. As a result, introduction of a tactile dimension into a mathematics course poses the same hurdle to accessing mathematical concepts instead of providing an additional entryway to those concepts. In fact, we often see that the metacognitive abilities and the self-discipline required for practice are prerequisite for both the standard mathematics learning and the craft approaches. Thus, students lacking in metacognition and self-discipline often have a difficult time in tactile mathematics courses, although such courses tend to work on building students' skills in metacognition and in their self-awareness of the need for self-disciplined practice.

Teaching an active learning section is typically thought to require more work on the instructor's part in the classroom than teaching in the lecture style, because working with spontaneous student comments and questions to help students develop the mathematical concepts for the day requires a great deal of ingenuity. More concretely, the teacher must have a deep grasp of the subject matter from multiple perspectives to allow her to correctly interpret the students' words. Liping Ma calls this teacher knowledge *profound understanding of fundamental mathematics* (Ball and Bass 2000). The teacher must also have a flexible notion of how students might put the mathematical ideas together to reach the goal concepts. This involves having a hypothetical learning trajectory for individual students and being able to change that trajectory in real time (Steffe and D'Ambrosio 1995).

Experienced instructors, Torrence, Hull, Favro, and I, agree that teaching a tactile mathematics course is even more challenging for the instructor than teaching a typical active learning style course. In addition to the usual challenges of active learning, the instructor must contend with classroom management issues of distributing materials and helping each student to understand the directions for the physical activity. In lecture classrooms and some active learning classrooms, students who get lost often simply quit paying attention. In a tactile activity classroom such disengagement is unlikely. Instead, confused students frequently demand help from the instructor or fellow students. Theoretically involvement of all students is fantastic, but diversion of the instructor to focus solely on one student's struggles for a long period of time can quickly derail a lesson. Establishing productive classroom norms is imperative. Norms should address the need for paying attention when directions are given, the speed with which students follow directions, and the amount of persistence required before asking for help. Aspects of classroom management relevant for teaching with tactile activities can be learned. Experienced instructors new to tactile mathematics will likely require several semesters to gain this skill set.

Teaching tactile mathematics often requires one to teach a particular skill, such as origami or needlework. As origamist Hull points out, "folding takes time" (Hull 2006). In our conversation for this chapter, he elaborated that during each class period time needed to be allotted for teaching students the folding they would need that day and for students to complete the day's folding. A similar statement is true regardless of the tactile activity. Lesson plans for tactile mathematics courses need to allocate time for skill teaching, activity completion, mathematical reflection, and class discussion. Learning to predict an appropriate pace for each component is another aspect of gaining expertise in facilitating tactile mathematical activities.

A common approach to address all three issues—students lacking manual dexterity, students who are hopelessly confused, and the time factor—is to have the students work in groups. If the physical objects are such that different people can construct different parts, assembling the pieces at the end, then often the more skillful students can help those who are less skillful gain skill, whether that skill lies in the manual realm or the conceptual realm. Groups serve the further function of requiring group members to explain and justify the mathematics that they have individually developed through interaction with the physical.

Active learning using tactile activities to motivate applications can be natural for teachers who are already familiar with the underlying crafting technique, such as origami, knitting, etc. Moreover, such teachers are usually passionate about their craft, and by bringing their craft into the classroom they are sharing an additional facet of themselves with their students. As with any discipline, teachers need to know the subject at a level substantially deeper than the level at which they plan to teach. We include this comment because some readers may inadvertently underestimate the time they will need to devote to mastering the crafting techniques so that they will be able to teach them to others. Hull notes that the first instance of teaching an origami mathematics course takes a great deal of preparation time for the instructor, especially if that person is not an origami expert. To mitigate the time factor, it is best to keep the in-class projects very simple.

Students will ask wonderful questions about the relationship between the craft and mathematics. While the teacher does not need to know every answer, a depth of experience considering the craft through a mathematical lens will be invaluable to answering such queries. Having experience performing the techniques will allow the teacher to predict the most likely questions.

21.5 Personal Experience with Tactile Mathematics

I have been a zealous fiber artist since I was a child. After my grandmother taught me two crochet stitches, I taught myself to crochet, knit, tat, and make Japanese temari balls from reading books. Throughout this time, I was equally fascinated by mathematics, which I was learning in parallel to needlework. During graduate school, sarah-marie belcastro and I began discussing the similar cognitive skills that are required of both successful needleworkers and mathematicians.

At the Joint Mathematics Meetings in 2001, belcastro and I hosted the first annual Knitting Circle to bring together mathematical fiber artists of all craft persuasions (knitting, crochet, embroidery, quilting, beading, etc.) to practice their crafts and converse about the mathematics they observed in their crafting. The initial group of fewer than a dozen has grown to a yearly gathering of about 50 mathematicians. These include a few just learning a craft, some explaining the mathematics they perceive in their work, and many showing off their craft work. While participants craft together, they mentor one another in crafting, teaching, writing, and mathematics. The Knitting Circle has encouraged crafters who think deeply about the relationship of mathematics to their craft to share those thoughts with other mathematicians. Other crafters have begun to consider how mathematics relates to the construction of their art and how to make mathematics explicit in their finished projects. The activity spurred by interaction with colleagues prompted belcastro and me to organize American Mathematics Society (AMS) Special Sessions on Mathematics and Mathematics Education in Fiber Arts in 2005, 2009, and 2014. Our edited volumes mentioned earlier together with a forthcoming volume expand greatly on the work presented in those sessions.

Connecting my passions for fiber art and mathematics has been intellectually stimulating, a source of significant community engagement, and an avenue for mathematical writing and editing. Presentations of my own mathematical art work have included venues beyond AMS Special Sessions on Mathematics and Mathematics Education in Fiber Arts. The Mathematical Association of America's Special Interest Group for Mathematics and the Arts (http://sigmaa.maa.org/arts/) provides excellent opportunities, as does the Bridges Organization (http://www.bridgesmathart.org). The relatively small mathematical art community is enormously supportive. I am grateful for the camaraderie and profusion of opportunities afforded to me by the network of organizations and individuals comprising this community.

Acknowledgements The author wishes to thank Mary Sandoval and the anonymous reviewers for helpful suggestions.

References

Alfieri, L., Brooks, P. J., Aldrich, N. J., & Tenenbaum, H. R. (2011). Does discovery-based instruction enhance learning? *Journal of Educational Psychology, 103*(1), 1–18.

Ball, D. H., & Bass, H. (2000). Interweaving content and pedagogy in teaching and learning to teach: Knowing and Using Mathematics. In J. Boaler (Ed.), *Multiple perspectives on the teaching and learning of mathematics* (pp. 83–104). Westport, CT: Ablex.

Barnes, J., & Libertini, J. (2016). *Tactile learning activities in undergraduate mathematics: A recipe book for the classroom*. Washington, DC: Mathematical Association of America.

belcastro, s.-m., & Yackel, C. (Eds.). (2008). *Making mathematics with needlework: Ten papers and ten projects*. Wellesley, MA: AK Peters, Ltd.

belcastro, s.-m., & Yackel, C. (Eds.). (2011). *Crafting by concepts*. Natick, MA: AK Peters, Ltd.

Frantz, M., & Crannell, A. (2011). *Viewpoints: Mathematical perspective and fractal geometry in art*. Princeton, NJ: Princeton University Press.

Gerdes, P. (1999). *Geometry from Africa: Mathematical and educational explorations*. Washington, DC: Mathematical Association of America.

Henderson, D., & Taimiņa, D. (2005). *Experiencing geometry: Euclidean and non-Euclidean with history* (3rd ed.). Upper Saddle River, NJ: Pearson Prentice Hall.

Hull, T. (2006). *Project origami: Activities for exploring mathematics*. Wellesley, MA: AK Peters, Ltd.

Prince, M., & Felder, R. (2007). The many faces of inductive teaching and learning. *Journal of College Science Teaching, 36*(5), 14–20.
Schoenfeld, A. (1987). What's all the fuss about metacognition? In A. Schoenfeld (Ed.), *Cognitive science and mathematics education* (pp. 189–215). Hillsdale, NJ: Lawrence Erlbaum.
Steffe, L., & D'Ambrosio, B. (1995). Toward a working model of constructivist teaching: A reaction to Simon. *Journal for Research in Mathematics Education, 26*(2), 146–159.
Taimiņa, D. (2009). *Crocheting adventures with hyperbolic planes*. Wellesley, MA: AK Peters, Ltd.

Chapter 22
Incorporating Writing into Statistics

Katherine G. Johnson

Abstract This chapter describes the work of a statistician who, after a career outside the academy, became a teacher of statistics in an urban university. To determine if the inclusion of write-to-learn activities improved her working adult students' ability to communicate the results of hypothesis tests in context, the author reviewed the literature on using writing-to-learn and conducted a research study on incorporating such activities in an introductory statistics course. Students' perceptions of the write-to-learn activities were also obtained. The study showed that the write-to-learn activities helped students become better at communicating statistical results in context, and students had positive impressions of the writing activities, claiming that the writing activities were helpful in their learning. Suggestions are offered on how to best incorporate write-to-learn activities in an introductory statistics class.

Keywords Write-to-learn • Statistics education • Introductory statistics

22.1 Introduction

After a career as an applied statistician, I have chosen to teach statistics because I want others to have a greater appreciation for my field of study. Data and statistics have become prevalent in our society, so my goal as a statistics educator is for students to develop a better understanding of statistics and be able to successfully interpret and discuss the uses of statistics they encounter in their lives. In particular, I want to enable them to be critical thinkers, and to evaluate and question statistics being reported. I am continually researching ways to best accomplish this goal.

MSC Codes
62-01
62-02

K.G. Johnson (✉)
Department of Mathematics, Metropolitan State University,
700 E. Seventh Street, St. Paul, MN 55106, USA
e-mail: katherine.johnson@metrostate.edu

J. Dewar et al. (eds.), *Mathematics Education*, Association for Women in Mathematics Series 7, DOI 10.1007/978-3-319-44950-0_22

In this chapter I describe my background and the environment in which I teach. I explain what prompted me to review the literature on using writing in the teaching of introductory statistics and decide to include writing-to-learn in my classes. I then describe a study I undertook of the effects of using writing-to-learn in two of my classes.

22.2 Context for This Work

22.2.1 Author's Background

After completing a Master of Science in Statistics, I worked for over 20 years as an applied statistician. I designed studies and analyzed data for cancer research trials, created a store location model and analyzed marketing research data for a grocery wholesaler, and designed clinical studies and analyzed data as a statistical consultant in the medical device industry. In these positions I did as much writing and communicating about my analyses as I did carrying them out. Effective communication is a necessary skill in statistics.

I decided to change careers and obtained a secondary mathematics teaching license while getting a Master of Arts in Education. I taught high school mathematics for one year and then was hired to teach statistics in a tenure-track position in the department of mathematics at Metropolitan State University in St. Paul, MN.

22.2.2 Institutional Setting

I teach at a public, urban university known for serving adult students, offering small classes and a flexible class schedule. In 2013 there were approximately 11,000 students, 90 % undergraduate. The university has a high enrollment of first generation students, minorities (38 % students of color), and older working adults (average age is 32 years), with many attending part-time (64 % of students).

The mathematics department has 11 faculty members, including three in statistics. We offer bachelor's degrees in applied mathematics and mathematics education, and minors in applied mathematics and applied statistics. The statistics minor provides students with knowledge and skills needed for a future career involving data evaluation and analysis. It offers a program of study in core areas of statistics with an emphasis on applications including statistics programming, regression analysis, analysis of variance, biostatistics, categorical data analysis, nonparametric methods, environmental statistics, and probability. The department also supports other programs at the university by offering foundational courses, including introductory statistics. We currently offer 11 sections of Statistics 1 each semester, with 24 students in each class. The students in Statistics 1 major in a variety of programs, including nursing, accounting, management, marketing, economics, social work, law enforcement, and psychology.

22.3 Teaching Statistics 1

22.3.1 Description of Statistics 1

A student who successfully completes Statistics 1 will know the principles and methods of statistics used in the collection and analysis of data, including design of experiments, sampling methods, descriptive statistics, normal distributions, regression and correlation, probability, confidence intervals, and significance tests.

The course meets once a week for 3 h and 20 min. Classes may involve direct instruction and small-group problem solving, along with homework assignments and weekly quizzes as formative assessments. The assignments require students to perform calculations and graph data using online statistics software and also to summarize the meaning of their results in the context of the data being analyzed. Summative assessments include a midterm, a final examination, and a final project that requires students to analyze a set of data and write a report summarizing the statistical methods used, the results of the analysis, and the interpretation of the results in context.

Many of the adult students are anxious about being in the statistics class, and they struggle to understand the more difficult concepts. I use techniques from my education courses, for example scaffolding of all material used during instruction and detailed examples, which help students understand the new topics and master the calculations required in the course. Despite having done this, I found that the students still had difficulty communicating about statistics; they could perform a hypothesis test, but had difficulty writing what the answer obtained means in relation to the question being asked. These experiences prompted me to investigate how I could help my students gain a deeper understanding of inference and be more effective in writing about statistics in their final projects.

22.3.2 Review of Literature

In determining how to facilitate students' deeper understanding of statistics, I consulted the Guidelines for Assessment and Instruction in Statistics Education (GAISE), prepared by the American Statistical Association (ASA 2005). The ASA guidelines[1] recommend that educators use classroom activities that promote student inquiry, problem-solving, and decision-making; focus on conceptual understanding; and integrate real world data. They advise instructors to foster active learning, engage students in their own learning, and encourage students to explore and ask questions. The explorations can be through verbal and written communication with instructors and peers. Effective communication is a large part of the practice of statistics because the practitioner must describe, explain, clarify, and interpret results.

[1] The ASA has updated the guidelines and published a new GAISE report (ASA 2016). The recommendations involving communication of statistics are the same as in 2005.

In my effort to help my students become better communicators I also investigated how writing can be used to help students develop a deeper understanding of statistics. This type of writing is called write-to-learn. In researching writing in statistics I found several publications (Beins 1993; Delcham and Sezer 2010; Kågesten and Engelbrecht 2006; Radke-Sharpe 1991; Rothstein and Rothstein 2007; and Taylor and McDonald 2007) that outline the benefits of write-to-learn activities for improving students' statistical thinking and learning. Writing encourages students to think about the concepts they are learning, connect them to what they know, and then communicate them to others. The research showed that if students think about what they will write, get their thoughts on paper, review what they write, reflect on how the reader will perceive it, and make corrections as needed, they gain a deeper understanding of statistics and become better at communicating statistical results.

According to Daniels et al. (2007, p. 22), write-to-learn activities are short, spontaneous, exploratory, informal, and non-graded. The idea is that for students to learn and understand new concepts, they need to grapple with the ideas, transform them, and put them into their own words. Delcham and Sezer (2010, p. 611) said that exploratory writing encourages meaningful reflection. Students are able to think about the statistical concepts they are learning and develop a deeper understanding of them. In addition, these writing activities give students opportunities to practice writing and help students develop confidence for more formal writing.

Another exploratory write-to-learn activity is journal writing. It engages students in critical reflection on their learning. Langer (2002, p. 339) said journals help students gain a better understanding of abstract ideas. The writing allows students to reflect on their own progress in understanding topics, to document their successes and failures, to summarize what they have learned, and to raise questions they were hesitant to ask or had trouble formulating in class. In addition, journal writing gives students opportunities to strengthen communication skills as well as write about anxieties or frustrations with what they are learning (Hammett 1993, p. 6).

However, journal writing may not be easy to incorporate. Langer (2002) claimed journal writing can be difficult for non-traditional students because they may not understand the term "reflection." In his study he found adult students were less likely to produce journals that are "qualitatively reflective and collaborative" (p. 349). He contended they have not had much exposure to the journal process. Russek (1998) shared examples of prompts to help students reflect on their learning. For example, having students finish the statements "I learned that I," "I was surprised that I," or "I discovered that I" can help students become reflective. Bean (2001, p. 101) describes this type of writing as "guided journals." In addition, he contended that outlining the reason for the writing assignments and including the expectations for students are important to enable them to learn new material. Bossé and Faulconer (2008, p. 12) agreed, claiming the purpose of the assignment must be clear for students to gain a better understanding of any new concepts they are learning.

Another way to include exploratory writing is for students to write during class, for example by asking them to respond to a question during the last few minutes of class. Exit slips, as this type of writing is often called, get students used to writing about what they are learning (Daniels et al, p. 35). Holmes (2012) claimed that using "concept checks" deepens students' conceptual understanding of statistics. This assignment differs from exit slips in that after writing, students discuss their responses in class. Having students listen to other students' responses is helpful for learning how others are conceptualizing the material. This writing activity can be modified to ask students to compare and contrast statistical concepts, like type I and type II errors. These activities enable students to review material, but they also provide instructors information on students' conceptual understanding of statistics. The writing activities can be used as an alternative assessment of students' learning.

The research also indicated why students have difficulty with formal writing like project papers. Writing about statistics is not similar to other types of collegiate writing. It requires students to translate statistics into conversational language, turning data and information into a narrative (Rothstein and Rothstein 2007, p. 22). Students need to communicate conclusions in "natural language" (Forster et al. 2005, p. 1), taking a complex idea and explaining it clearly to a general audience. Also, students are not used to presenting arguments supported by numerical reasoning (Stromberg and Ramanathan 1996, p. 160). In their observational study, Lipson and Kokonis (2005, p. 8) found report writing was difficult for students. They claimed students think numbers can speak for themselves and do not need to be explained. Students have difficulty translating numbers into words because the writing requires them to have a deeper understanding of the statistics they are interpreting. Kågesten and Engelbrecht (2006, p. 708) said that to be successful technical writers, students need to organize their material and be confident in their understanding of difficult concepts.

22.3.3 Using Writing in Statistics 1

In the Statistics 1 classes I studied (see Sect. 22.4), all class activities provided opportunities to discuss and write about statistics. After the midterm, activities were designed to help students describe results in context. For example, when introduced to confidence intervals, students were given a two-part assignment: first to define the parameter they were estimating and calculate the interval; then to provide a written interpretation of the interval in context. Working in small groups, students discussed their work, and then each student provided a written response. The first part was difficult for some students. They thought the sample proportion provided in the problem was the correct response. Other students were able to provide a written description of the parameter but were not able to write about the interval in context. They did not make the connection between the two parts of the assignment and did not interpret their answer to the first part in context.

In addition, when learning about hypothesis tests students were given scenarios and asked to describe what the researchers were trying to determine, provide a written description of the parameter, and then write the hypotheses using correct notation. The separate questions encouraged students to think about what the problem was asking before they tried to develop a hypothesis. The assignment also asked students to describe their results in context. Asking them first to describe the parameter helped them think about the language to use in answering this final question.

Students had many opportunities to write in class and also wrote journal entries outside of class. Prompts for writing assignments appear in Tables A.1 and A.2 in the Appendix.

22.4 Classroom Research Study

The prior research on incorporating writing into statistics showed it helped traditional students' learning and understanding of statistics. I decided to investigate if write-to-learn activities would help the first generation, minority, and adult students in my introductory statistics classes understand and use statistical inference and become better at communicating the results of their analyses. I hoped the activities would enable the students to advance from level one to level five of Garfield's "Model of Statistical Reasoning" (Garfield 2002). In level one, students use statistical terms without understanding; in level two they can describe a concept but cannot apply it correctly. Level three students correctly identify a dimension of a statistical process, but cannot fully integrate the dimensions of the process. The fourth level students can identify the parts of the statistical process without being able to integrate them correctly. Finally level five students have a complete understanding of statistical inference.

The primary objectives of my study were to determine if the activities improved students' ability to:

1. Set up accurate hypotheses,
2. Carry out a hypothesis test,
3. Explain the results of a hypothesis test in context.

22.4.1 Participants and Setting

The research study was carried out in two sections of introductory statistics during the 2013 fall semester at the university. Students self-selected into the course; the protocol was approved by the Human Subject Review Board and students consented to participate in the study. Students had the option of excluding their data from the study.

Forty-six students were included in the study; 21 (46 %) students were enrolled in one section and 25 (54 %) were in the other. The participating students were from a variety of majors and most were older than traditional college students; nearly half of the students (45 %) were over 30 years of age, 15 % of them were older than 40. Almost two-thirds (65 %) were enrolled full-time. Overall, 33 % had taken a previous statistics course.

22.4.2 Materials

Both classes were taught by the same instructor and had the same curriculum, including the textbook used at that time, *The Practice of Statistics*, fourth edition, by Starnes et al. (2010). All statistical calculations were performed on a programmable calculator. Weekly assignments and quizzes, two exams, and a final project were used to evaluate student performance. The study was designed to capture data from two quizzes and five questions from the final exam. The quiz and exam questions are included in Tables A.3 and A.4 in the Appendix.

The write-to-learn activities were added to the existing curriculum. All students were asked to complete the write-to-learn activities and received points for participating. Journal writing was done outside of class and was graded according to a rubric designed to encourage thoughtful responses to the writing prompts. The writing activities were included as 10 % of the final grade for the course.

In addition, students completed a survey that asked for gender, age, primary language, and enrollment status, prior knowledge of statistics, and level of anxiety in taking statistics. A survey at the end of the course asked for students' level of anxiety and their impressions of including writing in the course.

22.4.3 Procedures

During the first week of the 15-week semester, the study was explained, informed consent was obtained, and students completed the initial survey. The first quiz, given during the second week, was used as a measure of students' initial statistical knowledge before completing write-to-learn activities.

During the first half of the course, students completed three journal-writing activities that were intended to get students used to writing about statistics. They also completed write-to-learn activities during class, usually at the end of the class. These were used to evaluate students' content and procedural knowledge. Students learned about graphing, summary statistics, confidence intervals, and were introduced to hypothesis tests involving proportions.

Week ten's quiz was the baseline for evaluating the objectives of the study. During the remaining weeks students completed write-to-learn activities on one-sample and two-sample tests involving means and proportions, correlation, linear regression, and Chi-square tests. The second survey was completed in week 14, and the final examination was given in week 15. Students' scores on the baseline quiz were compared to corresponding questions on the final examination. Data from students who had not consented were not included in the analyses.

22.4.4 Statistical Methods

An analysis was performed to determine if students' initial knowledge was a possible confounder in evaluating the objectives of the study. Using an independent *t*-test, the mean percent correct on the five final exam questions was compared for students who scored at least 80 % and students who scored less than 80 % on the first quiz. In addition, a Chi-square test was used to determine if there was an association between students' scores on the first quiz and the baseline quiz at week 10. If prior knowledge were a confounder then the remaining analyses would incorporate students' scores on quiz one.

To evaluate each objective, analyses were done to determine if students who did not correctly answer questions on the baseline quiz answered corresponding questions correctly on the final. Since the data were paired binomial, McNemar's Test was used to test no association between students' responses on the baseline and final. Exact *p* values were reported.

22.4.5 Study Results

The *t*-test showed there was no significant difference in the mean percentage correct on the baseline quiz between students who scored at least 80 % and students who scored less than 80 % on the first quiz ($t(42) = -0.16$, $p = 0.87$). Similarly, there was no significant difference in the mean percent correct on the five final exam questions between the groups ($t(42) = -1.64$, $p = 0.11$). Thus, students' prior knowledge was not a confounder.

Forty students took both the baseline quiz and the final examination. Evaluation of objective (1) showed there is no statistical evidence that write-to-learn activities helped students set up accurate hypotheses. On the baseline quiz 68 % and on the final 81 % of students defined the hypotheses correctly. There was no statistically significant difference between these proportions ($p = 0.30$). The power to detect the 13 % difference was only 27 %.

Objective (2) was evaluated by determining if more students calculated the correct *z*-statistic and the correct *p* value for the final exam. On the baseline quiz 72 % and on the final 93 % of students calculated the *z*-statistic correctly. There was a significant difference in these proportions ($p = 0.02$). However, 70 % of the students on the baseline quiz and 83 % on the final correctly calculated the *p* value; these proportions were not statistically different ($p = 0.18$). Examining why the *p* value was not calculated correctly revealed that students calculated the statistic correctly but determined a p value for a two-sided alternative hypothesis. Most students were successful in performing the calculations, and the write-to-learn activities were beneficial in helping students perform a hypothesis test.

Objective (3) was evaluated by determining if more students were able to correctly interpret the p value and write the conclusion in context on the final exam. On the baseline quiz and final, 70 % and 83 % of students, respectively, correctly interpreted the p value; these proportions were not statistically different ($p=0.27$). However, on the baseline quiz 18 % and on the final 55 % of the students explained the result of the test in context; these proportions are significantly different ($p=0.01$). The results indicated the write-to-learn activities made a significant difference in helping students learn to explain the results of a hypothesis test in context.

22.4.6 Discussion

All class assignments had students discuss answers in their small groups. This enabled them to clarify their understanding of hypothesis tests and was intended to prepare them to write their own hypotheses. Student feedback indicated that they liked working in small groups and felt discussing the problems with their peers helped them understand what they were learning. Based on the results from the study, the use of writing-to-learn was successful. Most students recognized and correctly defined the parameter on the final exam, and defined the null hypothesis correctly. The common mistake made by students was not recognizing a one-sided alternative was a more appropriate test to answer the researcher's question.

Students also wrote about inference outside of class. For example, they described a situation where they would use a sample to infer something about a population parameter. Students had gotten used to writing about statistics and most were able to provide clear explanations. Students' responses indicated they had developed an understanding of how hypothesis testing can be used.

The last journal assignment asked students to explain in their own words the meaning of a reported p value. Some responses indicated students did not have a clear understanding of p values; they parrotted technical language from the textbook rather than using their own words. Students' difficulty writing about the p value was consistent with their responses on the final exam. Although several students wrote about the results in context, not all had mastered this skill. Students understood a small p value meant evidence against the null hypothesis, but the inability of most to write in context was an indication they did not have a complete understanding of statistical inference.

Students successfully completed their projects writing about the methods and the results used in their analyses. At the beginning of the semester, students were not engaged in the topic of statistics. For example, in critiquing a published study, students used incomplete sentences and, despite providing their impressions of the study, they did not give a reason for their responses. As the class progressed, the journal entries became more detailed, students used terminology correctly, and they seemed more engaged in learning. For example, a student who was unable to write about the standard Normal distribution in week 2, was able to describe a medical-related scenario where a hypothesis test could be used in week 9; she correctly described the population, sample, the parameter, and the statistic in the scenario. Most impressive was that she connected what she learned to her employment experience in the medical industry.

22.4.7 Conclusion

This study supports previous findings that writing helps students become better at communicating statistical results (Beins 1993; Kågesten and Engelbrecht 2006). There was an increase in the number of students who correctly performed the calculations for the hypothesis test and wrote about the results in the context of the data. Students were actively engaged in the write-to-learn activities and were successful in communicating the methods and results in their final project. Including write-to-learn activities is a valuable tool for engaging students in learning, improving students' communication skills, and assessing students' learning.

Limitations of the study include not being generalizable to other statistics classes. The two classes cannot be considered equivalent because students self-selected into them. There was not a concurrent control group, and there was not enough statistical power to detect the small increases for some objectives. Future research would benefit from use of a no-writing control comparison and a larger sample size.

22.5 Suggestions

Students studying statistics should be expected to discuss and write about statistics. Write-to-learn activities enable them to learn about statistics and learn to communicate effectively about statistics. Based on my experience incorporating write-to-learn activities into my statistics course I offer the following suggestions:

- Find opportunities for students to discuss and write about what they are learning, encouraging them to reflect and write on their understanding.
- Assign more writing activities in the first half of the semester so students become familiar with writing about statistics and become more confident writers.
- Demonstrate how to write about confidence intervals and hypothesis tests in context; provide an outline of a written summary of an analysis.
- Have students analyze real world data and write a paper explaining their methods and results.

Appendix: Assessment Questions and Writing Prompts

Table A.1 Journal writing prompts

Week	Prompt
1	Write continuously for ten minutes answering the following question: "What is statistics and how is it relevant?"
3	The story titled "Researchers find women's risk of cancer increases with height" which is shown below was published in the *Pioneer Press*, July 26, 2013. Read the story and then write for ten minutes in your journal reflecting on the following questions • Was the study an experiment, sample, or an observational study? • Identify the key variables that they discussed in the story. • What confounding variable(s) might be present? • What potential bias exists in the study (something about the process that might make the conclusion unreliable or incorrect)? • What is your over-all impression of these results? If you were tall should you be concerned about your risk of getting cancer?
5	Writing continuously for 10 min, in your own words describe what a random variable and a probability distribution are. Also explain the difference between a discrete random variable and a continuous random variable. Consider including your own example to help you understand/explain these concepts.
8	The journal writing exercise for week 8 is about confidence intervals for an unknown parameter, the proportion of successes in a population. The following is an example of a possible scenario in which you might need to use inference to estimate an unknown proportion. Read the following paragraph, which ends with a few questions that can guide you in your reflection, and then write continually for ten minutes about estimating a population proportion using a sample proportion while also addressing the questions.
9	The underlying principle of all statistical inference techniques is that one uses sample statistics to learn something (i.e. infer something) about population parameters. Convince me that you understand this statement by writing a paragraph describing a situation in which you might use a sample to infer something about a population parameter. Clearly identify the sample, population, statistic, and parameter in your example. Be as specific as possible.
10	There are several symbols used in hypothesis testing. This week I am asking you to write about the symbols we have used most often. First, fill in the following table for each of the symbols, then choose three symbols and write how they are used together in hypothesis testing.
11	Does eating more fiber reduce the blood cholesterol level of patients with diabetes? A randomized clinical trial compared normal and high fiber diets. Here is part of the researcher's conclusion: "The high fiber diet reduced plasma total cholesterol concentrations by 6.7 % ($p=0.02$), triglyceride concentrations by 10.2 % ($p=0.02$), and very low density lipoprotein cholesterol concentrations by 12.5 % ($p=0.01$)." A doctor who knows no statistics says that a drop of 6.7 % in cholesterol isn't a lot—maybe it's just an accident due to the chance assignment of patients to the two diets. Explain in language understandable to someone who knows no statistics how "$p=0.02$" answers this objection.

Table A.2 Class writing prompts

Week	Prompt
2	In your own words describe the Standard Normal Distribution and how it is used.
4	Describe what the Law of Large Numbers mean, in your own words.
6	In your own words define "sampling distribution of the mean." Describe in your own words how the standard error of the mean (i.e., the standard deviation of the sampling distribution of the mean) is calculated?
8	A recent Gallop Poll conducted telephone interviews with a random sample of adults. Data were obtained for 1000 people. Of these, 37 % said that football is their favorite sport to watch on television. Define the parameter p in this setting. Explain why we can't say that 37 % of all adults would say that football is their favorite sport to watch on television. Construct a 95 % confidence interval for p. Interpret the confidence interval in context.
9	A bank is testing a new method getting delinquent customers to pay their credit card bills. The standard way was to send a letter (costing about $0.46) asking the customer to pay. That worked 30 % of the time. They want to test a new method that involves sending a DVD to customers encouraging them to contact the bank and set up a payment plan. Developing and sending the video costs about $10.00 per customer. In your own words describe what the bank hopes to show is true. How can they do this? What is the appropriate symbol to use to describe the value 30 %? Or in other words, what is the parameter of interest in this scenario? Describe what the parameter represents (define the parameter). Describe in your own words the alternative hypothesis that should be used to test their new method. Now, write the null and alternative hypotheses using symbols.
10	Trying to encourage people to stop driving to campus, the university claims that on average it takes people 30 min to find a parking space near campus. You don't think it takes that long to find a parking spot. What are you hoping to show? Define the null and alternative hypotheses to test your claim. Hemoglobin is a protein in red blood cells that carries oxygen from the lungs to body tissues. People with less than 12 g/dl of hemoglobin are anemic. A public health official in Jordan suspects that Jordanian children are at risk of anemia. He measures a random sample of 50 children. What is the Jordanian official hoping to show? Define the null and alternative hypotheses
11	(a) The drug AZT was the first drug that seemed effective in delaying the onset of AIDS. Evidence for AZT's effectiveness came from a large randomized comparative experiment. The subjects were 870 volunteers who were infected with HIV, but did not yet have AIDS. The study assigned 435 of the subjects at random to take 500 mg of AZT each day and 435 to take a placebo. At the end of the study, 38 of the placebo subjects and 17 of the AZT subjects had developed AIDS. Researches want to test the claim that taking AZT lowers the proportion of infected people who will develop AIDS in a given period of time. Define the hypotheses to test this claim. Carry out a test of this claim at a significance level of $\alpha=0.05$. Interpret your result in context. Construct and interpret a 99 % confidence interval for the difference in the proportion of infected people who will develop AIDS for those who took AZT and those who took the placebo. (b) Millions of dollars are spent each year on diet foods. Trends such as low-fat diet or the low-carb diet have led to a host of new products. A study was conducted that compared the weight loss between obese patients on a low-fat diet and obese patients on a low-carb diet. Let μ_1 represent the mean number of pounds obese patients on a low-fat diet lose in 6 months, and μ_2 represent the mean number of pounds obese patients on a low-carb diet lose in 6 months. State the null and alternative hypotheses if you want to test whether or not the mean weight loss between the two diets are equal. Suppose that a sample of 100 obese patients on a low-fat diet lose a mean of 7.6 pounds in 6 months with a standard deviation of 3.2 pounds, while a sample of 100 obese patients on a low-carb diet lost a mean of 6.7 pounds with a standard deviation of 3.9 pounds. Is there evidence of a difference in the mean weight loss of obese patients between the low-fat and low-carb diets? Use a 0.05 level of significance. Construct and interpret a 95 % confidence interval estimate for the difference in treatment means.

(continued)

Table A.2 (continued)

<table>
<tr><th>Week</th><th>Prompt</th></tr>
<tr><td>12</td><td>The marketing manager of a large supermarket chain would like to use shelf space to predict the sales of pet food. A random sample of 12 equal sized stores' pet food is selected, with results shown in the following table:
<table>
<tr><td>Store</td><td>Shelf space (feet)</td><td>Weekly sales ($100 s)</td></tr>
<tr><td>1</td><td>5</td><td>1.6</td></tr>
<tr><td>...</td><td>...</td><td>...</td></tr>
<tr><td>12</td><td>20</td><td>3.1</td></tr>
</table>
Construct a scatterplot of the data and describe the relationship between shelf space and weekly sales. Calculate the regression equation for the data. Describe the meaning of the coefficients. Graph the residuals against the independent variable. Are there any concerns about the regression model? Explain</td></tr>
<tr><td>13</td><td>For a recent year, the following are the number of homicides that occurred each month in New York City: 38, 30, 46, 40, 46, 49, 47, 50, 50, 42, 37, and 37
Use a 0.05 significance level to test the claim that homicides in New York City are equally likely for each of the 12 months. Is there evidence to support the police commissioner's claim that homicides occur more often in the summer when the weather is better? Explain</td></tr>
</table>

Table A.3 Quiz questions

<table>
<tr><th>Week</th><th>Questions</th></tr>
<tr><td>2</td><td>If data for a research study follow a Normal distribution $N(30, 2)$ and the researcher transferred all his data to Z scores, what distribution do those Z scores follow? Circle the correct response
(a) $N(32, 2)$
(b) $N(30, 2)$
(c) $N(0, 1)$
(d) $N(2, 30)$
Eleanor scores 680 on the SAT Mathematics test. The distribution of SAT scores follows a Normal distribution with mean 500 and standard deviation 100.
What is Eleanor's standardized score? What is the proportion of students who got SAT Mathematics scores lower than Eleanor's score?</td></tr>
<tr><td>10</td><td>A gallop Poll report on a national survey of 1028 teenagers revealed that 72 % of teens said they seldom or never argue with their friends. Yvonne wonders whether this national result would be true at her college. She surveys a random sample of 150 students at her school.
(a) Write the appropriate null and alternative hypotheses to test Yvonne's question. Make sure to define the parameter you are using.
(b) For Yvonne's survey, 96 of the 150 students in the sample said they rarely or never argue with their friends. Carry out a significance test using $\alpha=0.05$. Include the values and the calculator function you used for the test. Provide the test statistic and p value from the test.
(c) Do the data provide convincing evidence against the null hypothesis? Interpret this result in context of the problem.</td></tr>
</table>

Table A.4 Final examination questions

Number	Question
1	A researcher performed a one-sided hypothesis test to see if the newly developed drug is more effective than the standard drug. At the end, he reached the conclusion of failing to reject the null hypothesis based on the significance level of 0.05. However, his colleague found that a two-sided test will be more appropriate to test the claim. Based on the information, what decision will the researcher's colleague reach if performing a two-sided test? (a) Reject the null hypothesis based on the significance level of 0.05 (b) Fail to reject the null hypothesis based on the significance level of 0.05 (c) Reject the null hypothesis based on the significance level of 0.01 (d) The researcher's colleague cannot make any decision
2	Why do very small p values indicate that the evidence against the null hypothesis is strong? (a) Because the p value is the probability that the null hypothesis is true (b) Because the small p value indicates that the data lie within the confidence interval (c) Because the small p value indicates that data like ours would be very uncommon if the null hypothesis were true (d) Because the small p value indicates that data like ours would be very common of the alternative hypothesis were true
3	The weight of newborn babies in United States follow a Normal distribution with mean 6.5 pounds and standard deviation 1.5 pounds. Emma is a newborn baby and weighs 7 pounds. The standardized score for Emma's weight is Z. (a) What type of distribution should Z follow? (b) What are the mean and the standard deviation of this distribution?
4	Resting pulse rate is an important measure of the fitness of a person's cardiovascular system, with a lower rate indicating greater fitness. The mean pulse rate for all adult males is approximately 72 beats per minute. A random sample of 50 male students currently enrolled in the Business School at a major university was selected and the mean resting pulse rate was found to be 76 beats per minute with a standard deviation of 12 beats per minute. The experimenter wishes to test if the students are less fit, on average, than the general population. Following the given steps in (a) through (e), carry out an appropriate test to examine whether the students are less fit than the average male. a) Formulate the null and alternative hypotheses, defining the parameter used. b) What type of test would you choose? c) Find the test statistic. d) Find the p value. e) Make a decision in context to the problem using a significance level of $\alpha = 0.05$
5	Economists often track employment trends by measuring the proportion of people who are "underemployed," meaning they are either unemployed or would like to work full time but are only working part-time. In the summer of 2010, 18.5 % of Americans were "underemployed." The mayor of Our Town wants to show the voters that the situation in his town is not as bad as it is in the rest of the country. His staff takes a simple random sample of 300 residents and finds that 50 of them are underemployed. Following the given steps in (a) through (e), carry out an appropriate test to examine the mayor's claim. a) Formulate the null and alternative hypotheses. b) What type of test would you choose? c) Find the test statistic. d) Find the p value of your test. e) Make the appropriate conclusion in context of the problem; use a significance level of 0.05

References

ASA. (2005). Guidelines for assessment and instruction in statistics education: College report. Alexandria, VA. Retrieved June 28, 2016 from http://www.amstat.org/education/gaise/GaiseCollege_full.pdf.

ASA. (2016). Guidelines for assessment and instruction in statistics education: College report. DRAFT February 2016. Alexandria, VA. Retrieved February 28, 2016 from http://www.amstat.org/education/gaise/collegeupdate/GAISE2016_DRAFT.pdf.

Bean, J. C. (2001). *Engaging ideas: The professor's guide to integrating writing, critical thinking, and active learning in the classroom*. San Francisco: Jossey-Bass.

Beins, B. C. (1993). Writing assignments in statistics classes encourage students to learn interpretation. *Teaching of Psychology, 20*(3), 161. doi:10.1207/s15328023top2003_6.

Bossé, M. J., & Faulconer, J. (2008). Learning and assessing mathematics through reading and writing. *School Science & Mathematics, 108*(1), 8–19. doi:10.1111/j.1949-8594.2008.tb17935.x.

Daniels, H., Zemelman, S., & Steineke, N. (2007). *Content-area writing: Every teacher's guide*. Portsmouth: Heinemann.

Delcham, H., & Sezer, R. (2010). Write-skewed: Writing in an introductory statistics course. *Education, 130*(4), 603–615.

Forster, M., Smith, D. P., & Wild, C. J. (2005). Teaching students to write about statistics. *Proceedings of the IASI Satellite Conference: Statistics education and the Communication of Statistics*. Sydney, Australia. Retrieved October 16, 2012 from http://iase-web.org/documents/papers/sat2005/forster.pdf.

Garfield, J. (2002). The challenge of developing statistical reasoning. *Journal of Statistics Education, 10*(3). Retrieved June 28, 2016 from http://www.amstat.org/publications/jse/v10n3/garfield.html.

Hammett, J. E. (1993). Writing to learn statistics: Maintaining learning journals in order to identify and address undergraduate students' misconceptions. *The Proceedings of the Third International Seminar on Misconceptions and Educational Strategies in Science and Mathematics*. Ithaca, NY: Misconceptions Trust. Retrieved June 28, 2016 from http://www.mlrg.org/proc3pdfs/Hammett_Statistics.pdf.

Holmes, K. Y. (2012). Tips for incorporating writing into an introductory statistics course. *Observer, 25*(1). Washington, DC: Association for Psychological Science. Retrieved June 28, 2016 from http://www.psychologicalscience.org/index.php/publications/observer/2012/january-12/tips-for-incorporating-writing-into-an-introductory-statistics-course.html.

Kågesten, O., & Engelbrecht, J. (2006). Supplementary explanations in undergraduate mathematics assessment: A forced formative writing activity. *European Journal of Engineering Education, 31*(6), 705–715. doi:10.1080/03043790600911803.

Langer, A. M. (2002). Reflecting on practice: Using learning journals in higher and continuing education. *Teaching in Higher Education, 7*(3), 337–351. doi:10.1080/13562510220144824.

Lipson, K., & Kokonis, S. (2005). The implications of introducing report writing into an introductory statistics subject. *Proceedings of the IASI Satellite Conference: Statistics Education and the Communication of Statistics*. Sydney, Australia. Retrieved June 28, 2016 from http://iase-web.org/documents/papers/sat2005/lipson.pdf.

Radke-Sharpe, N. (1991). Writing as a component of statistics education. *The American Statistician, 45*(4), 292–293. doi:10.2307/2684457.

Rothstein, A., & Rothstein, E. (2007). Writing and mathematics: An exponential combination. *Principal Leadership: High School Edition, 7*(5), 21–25.

Russek, B. (1998). Writing to learn mathematics. *Writing Across the Curriculum, 9*, 36–45.

Starnes, D., Yates, D., & Moore, D. (2010). *The practice of statistics* (4th ed.). New York: W.H. Freeman and Company.

Stromberg, A. J., & Ramanathan, S. (1996). Easy implementation of writing in introductory statistics courses. *The American Statistician, 50*(2), 159–163. doi:10.2307/2684429.

Taylor, J. A., & McDonald, C. (2007). Writing in groups as a tool for non-routine problem solving in first year university mathematics. *International Journal of Mathematical Education in Science & Technology, 38*(5), 639–655. doi:10.1080/00207390701359396.

Chapter 23
An Infusion of Social Justice into Teaching and Learning

Priscilla Bremser

Abstract We present a narrative account of the effects of adopting a social justice perspective on one mathematician's career path. We offer geographic and institutional context, explore interpretations of "social justice" and its intersections with mathematics, and describe implications for teaching and professional learning. We illustrate our explorations of this perspective in teaching first-year seminars, number theory for in-service teachers, mathematics for pre-service teachers, as well as some standard mathematics courses such as abstract algebra and linear algebra. The chapter ends with some reflections on the author's professional development.

Keywords Social justice • Education • Active learning • Inquiry-based learning

23.1 Introduction

There is a story about mathematics in which the discipline exists apart from the messiness of human society. Answers are right or wrong, debates are not necessary, and judgments are impartial. This story was a comfort to me as a teenager; it suggested a calm, if clinical, fairness. Mathematics was a refuge from conflict and disagreement, and positive feedback from my teachers and standardized tests afforded a sense of safety.

The pure objectivity story is, of course, incomplete, and fairness in the practice of mathematics is far from a necessary outcome. Mathematics happens in the context of human societies. In this chapter, I explore intersections between mathematics

MSC Codes
97A40
97B40
97B50

P. Bremser (✉)
Department of Mathematics, Middlebury College,
14 Old Chapel Road, Middlebury, VT 05753, USA
e-mail: bremser@middlebury.edu

J. Dewar et al. (eds.), *Mathematics Education*, Association for Women in Mathematics Series 7, DOI 10.1007/978-3-319-44950-0_23

and the search for equity in communities, and describe the profound effects such explorations have had on my professional life.

The dictionary definition of "justice"—the quality of being fair and reasonable—is often applied to individual people or specific incidents, evaluated according to societal norms. The term "social justice" goes further. I take it to mean fairness in social systems. It suggests that we develop norms by which we evaluate institutions in their treatment of individuals and groups. As an example of this distinction, consider lending practices. Laws prohibiting usury convey a standard, held by many societies over many centuries, that interest rates above a certain level are unjust. Those laws, however, did not prevent the Federal Housing Authority and the mortgage industry from redlining, effectively putting home ownership beyond the reach of non-white residents in cities across the United States between 1934 and 1968. This is an affront to the ideal of social justice, with lasting consequences (Madrigal 2014).

The evolution of my understanding of justice has been shaped by my own experience. When my high school calculus teacher made it clear that we four girls weren't welcome in his class, I must have been confident enough to assume that the problem was his, not mine, and naïve enough to regard him as a dinosaur nearing extinction, not an agent of persistent attitudes about women in mathematics. When I moved from a supportive mathematics department at a women's college to a graduate school department with an all-male faculty, the loss of role models affected me deeply. By the time my graduate school advisor told me that because I was getting married I wouldn't need a job, I understood that women in mathematics didn't just have a role model problem; in fact we could never assume that we would be taken seriously.

At the time, "social justice" was not part of my vocabulary, but working for fairness in society was part of my life. During the summer of 1981, when I was in graduate school, I volunteered at the National Organization for Women headquarters in Washington, D.C., as the ratification deadline for the Equal Rights Amendment loomed. I have always voted for candidates whose platforms affirm goals of gender equity and human rights, and have supported organizations working to hold our institutions to standards of fairness.

These activities were, in my mind, distinctly different from my work as scholar and teacher of mathematics, though the boundaries blurred at times. During the drive to a number theory conference with colleagues, a man asked me why there were so few women in mathematics. Resenting the expectation that I should have a ready response—I did not—I probably said that the mathematics profession was not immune to the biases visible throughout our society. At least I hope that's what I said.

Perhaps it was that conversation that caused me to pay more attention to questions of gender in mathematics education. Once my sons started school, I paid more attention to K-12 mathematics education in general. Meanwhile, I was finding scant satisfaction in pure mathematics research, and began to direct more intellectual energy toward educational issues, with the explicit goal of finding constructive ways to get involved.

By this time, I had gotten to know my adopted state. Vermont ranks 49th in population (about 626,000 at present), and 95 % of residents identify as non-Hispanic and white, including descendants of French-Canadian, Irish, and Italian immigrants. Some relatively urban areas have welcomed more recent refugees from Bosnia and Herzegovina, Rwanda, and Somalia. The 2010 the US Census determined that 61.1 % of Vermont residents lived in rural areas (United States Census Bureau 2012). Poverty shows up in pockets. One local elementary school has 10 % of students receiving free or reduced-price lunch; another, feeding into the same high school, has 59 % (Vermont Agency of Education 2015). While agriculture does not dominate Vermont's economy as it once did, farming is a significant part of the economy and landscape of Addison County, home of Middlebury College. Local dairy farms and orchards employ hundreds of migrant farm workers, some undocumented.

Middlebury College, founded in 1800, is a highly selective liberal arts institution with 2450 undergraduates and a 9:1 student to faculty ratio. Admission is need-blind for domestic students, and Middlebury is committed to meeting each student's full financial need (with grants and loans) for all 4 years of study. To earn tenure, a faculty member must meet high expectations in teaching, scholarship, and service. The review procedures call for evaluation of scholarly productivity by external experts alongside class visits by colleagues in other disciplines. I joined the faculty in 1984, and in 1990 became the second woman awarded tenure in the natural science division.

Almost immediately, I was named chair of the mathematics department at Middlebury. Enjoying the culture of mutual respect and congeniality for which the department is known, I began to shift away from the lecture format in one course, a process I discuss in the final section of this chapter. Three years on the college's Committee on Reappointment had me visiting dozens of classes across many disciplines each semester, which informed my own teaching significantly. An upper-level Chinese language class is a fascinating place to observe and reflect on body language, eye contact, and who is talking when.

During my 2000–2001 sabbatical, immediately after my service on that committee, I took part in an Algorithmic Number Theory program at the Mathematical Sciences Research Institute in Berkeley, California, where the liberal-arts college is a largely unfamiliar concept. While I relished the intellectual recharge of that year, I missed the interactions with students. Upon my return to Middlebury I found myself on the Educational Affairs Committee, overseeing curricular changes and evaluating requests for new faculty positions. Once again, election to a major committee meant a heavy workload and difficult decisions, but also new insights into educational systems and how they might support or impede student learning.

In the spring of 2006, I saw an announcement for a course development workshop called "Mathematics and Social Justice" at Lafayette College in Pennsylvania, organized by Rob Root. Intrigued by the pairing of those terms, I signed up. During the opening session, I felt a wall start to crumble in my brain. I heard about service learning projects in which undergraduates in one class helped low-income neighbors with their tax forms, and in another designed a traffic flow plan for an environmental museum. Participants presented descriptions of writing assignments connecting quantitative literacy with participation in democratic society.

In 2007 I co-organized and hosted a sequel workshop, where Charles Hadlock, editor of *Mathematics in Service to the Community* (Hadlock 2005), spoke. The 22 participants formed small working groups to outline course modules on such topics as "lending and access to money" and "criminal justice."

One outcome of the workshops was the publication of "Mathematics of, for, and as Social Justice," a chapter (Bremser et al. 2009) in a volume focused on post-secondary social justice education (Skubikowski et al. 2009). Co-writing the chapter helped frame my conception of the intersections between mathematics and social justice. We describe the mathematics *of* social justice as a set of quantitative analytical tools necessary to engage social questions. The redlining of real estate regions presents questions in need of such tools. People use mathematics *for* social justice to effect change by, for example, helping consumers assess the risks of various loan options that target specific groups. We regard math *as* a social justice issue because of the disparities in access to quality mathematics education. That chapter goes on to offer case studies of college courses residing in those intersections.

Participants in the two workshops have gone on to develop and refine courses with social justice considerations. Sheila Weaver, at the University of Vermont, developed and taught an entry-level Mathematics and Social Justice course with a service-learning component. Students worked with the Vermont Campaign to End Childhood Hunger to collect and analyze data on food stamp recipients, many of whom lack transportation to the office distributing the stamps (Bremser et al. 2009). Weaver also offers a special section of an entry-level probability with statistics course focusing on social justice issues, including income distribution and poverty, political representation, and job discrimination. This course satisfies a diversity distribution requirement at the university (S. Weaver, personal communication, February 23, 2016).

Andrew Miller of Belmont University in Tennessee, also a participant in both workshops, has presented his work at a Mathematical Association of America MathFest and elsewhere. In his courses dealing with quantitative literacy and consumer finance issues, Miller assigns group projects with significant writing components. For example, he asks students to compare the costs of two credit cards, including finance charges and late fees, and construct a spreadsheet and guide for consumers. For another assignment, the team gathers data about student debt, using it to write a report for high school seniors on paying for college. A third calls for students to serve as consultants to a fictional company by constructing an automatic enrollment retirement plan, including a strategy for advertising the plan to employees with detailed evidence for the advantages of participation (A. Miller, personal communication, May 10, 2016).

A significant contribution to the genre comes from Thomas Pfaff of Ithaca College, a speaker at the second workshop who maintains a website called "Sustainability Math" as a repository of curriculum materials (Pfaff n.d.). These include units on "The Gini Coefficient" and "CO_2 Levels" for first-semester calculus, and on "Sustainable Fisheries" for differential equations. In making his case that sustainability education is imperative, Pfaff states that "(h)umans have reached a state where we are negatively impacting the ability of future generations to meet their needs and aspirations."

In the remainder of this chapter, I describe particular ways in which engaging the intersections of mathematics and social justice has affected my professional life. In the next three sections I describe creating and teaching a First-Year Seminar (Sect. 23.2), teaching number theory to in-service teachers (Sect. 23.3), and developing and teaching a mathematics course for pre-service teachers (Sect. 23.4). In Sect. 23.5 I describe the broader impact of this work on my teaching of mathematics, and I conclude with reflections on my professional development.

23.2 The First-Year Seminar

Middlebury College requires every entering student to enroll in a First-Year Seminar (FYS), and all departments contribute courses to the FYS program. Each seminar is limited to 15 students, and must have a significant writing component. The 2006 workshop gave me some resources and the courage to develop a new first-year seminar, incorporating my interest in mathematics and in equity, and to include a service-learning project. Here is the course description.

Mathematics for All What kinds of mathematical knowledge are necessary for full participation in contemporary democratic society? How well, and how fairly, do our schools educate students in quantitative skills and reasoning? By what measures might we judge success? We will learn about different approaches to mathematics education in light of these questions. Readings will include selections from *Mathematics for Democracy: The Case for Quantitative Literacy* (L.A. Steen, Editor [Steen 2001]), as well as recent articles by education researchers. To connect theory and actual practice, students in this class will conduct a service-learning project in a local school. All are welcome, regardless of mathematical background.

In 2007, when I first offered the course, Vermont was using the New England Common Assessment Program (NECAP) to measure schools' effectiveness in mathematics and English. The seminar project that year was to learn as much as we could about NECAP, visit a local elementary school to interview students and teachers, and then to prepare a brochure for parents. We learned that tests designed to assess schools may have limited value for assessing individual students, and that there was a correlation between NECAP scores and household income (measured by eligibility for free and reduced-price lunch).

The project in 2010 took place at a local high school, where we observed classes and interviewed teachers and students before designing a website of mathematics resources for them. As it happened, the teachers' union was in tense negotiations with their school board at the time, so we got an unanticipated dose of reality in our conversations about whether an equitable education system can be built on local funding mechanisms.

In the 2013 version of Mathematics for All, we read and wrote about mathematics learning for pre-kindergarten children, and had an introduction to Head Start, a federal school readiness program for students from low-income families. We volunteered at the Head Start classroom in town. After each of my students had visited at least once, they designed some math games and then led the children in playing them. Among my goals for my students were to gain some familiarity with the mathematical thinking of 3- to 5-year-olds, an understanding of the importance of play for that age group, and appreciation of the complexities of learning and teaching, all of which were supported by the visits. One unexpected lesson from that project was that many children in the program were from households led by women, and the young men in my group were greeted with enthusiasm every time they entered the classroom. Another was that some migrant workers on dairy farms in Vermont have small children, who may speak only Spanish at home.

While I haven't done a systematic study of the long-term impacts of the seminar on my students, I have heard back from a few former students. One, who is now a secondary school teacher, reports that the seminar "was the perfect first class for a prospective math teacher as it exposed me to the various issues and debates in the math and education worlds, and provided me with valid arguments for the importance and relevance of math in education and life." Another refers to his research paper on math anxiety, and says, "I think a big part of studying math, perhaps even bigger than in other fields, is the critical role that being able to teach it plays in studying it. So in a lot of that teaching to friends, family and classmates that has come along with studying math at an advanced level, I have noticed a lot of math anxiety and seen first hand how crippling it can be. So I guess doing all that research I did on math anxiety helped me really target the anxiety and make the math I taught as accessible as I could, while trying not to use too many big and fancy but intimidating words and methods." This student, now a mathematics major, conceived and organized a tutoring program at the local high school.

According to a third student from my seminar, "the biggest impact Mathematics for All had on me came through the connection to the Head Start classroom. I really enjoyed working with the kids in that classroom, and I continued volunteering that year and the following on my own time. This may sound silly, but working with that classroom made my opinion of children change from a relatively negative one to a positive one."

23.3 Teaching Number Theory to In-Service Teachers

In considering mathematics *as* social justice, and contemplating what I could offer to K-12 mathematics education, I saw two possible arenas: the preparation of future teachers, and professional development of those already in classrooms. Because I decided to begin my contribution with the latter, I knew that I needed to meet Ken

Gross, founder of the Vermont Mathematics Initiative (VMI) (University of Vermont n.d.; Vermont State Mathematics Coalition n.d.). VMI is a 3-year master's degree program for practicing teachers, who devote 6 weekends during the school year and two full weeks in the summer to content courses. In addition, each teacher designs and implements an action research project, and works with a mentor to become a mathematics leader in her school. The content courses are taught by teams that include both mathematicians and experienced classroom teachers. While my original intent was to learn about the program, I ended up signing on as an instructor in 2007.

Schoolteachers have up-close views of inequities in our society (Bremser 2014b). Their stories of the disadvantages some children face strengthen my resolve to do what I can to support good mathematics instruction across my home state. In May 2016, VMI held its first conference, including presentations from some graduates of the program. There I learned that VMI has now graduated sixteen cohorts in Vermont, reaching schools in every county, along with two cohorts from an offshoot program in Cincinnati, Ohio. I also heard directly from some graduates that they had used what they'd learned from my number theory course at VMI in their classrooms.

The educational equity question that is most present for me at VMI concerns the teachers themselves. Because VMI was originally designed for K-8 teachers, most of the participants are women, and many of those women are uncomfortable with mathematics. Their own stories include being told "you'll never be good at math" as children, and being directed into less challenging mathematics courses, or away from mathematics altogether. One teacher I met in Cincinnati had never taken any algebra in high school; when she was in her teens, African-Americans were expected to perform only menial work after graduation.

As the lead instructor of the number theory course, I introduced a daily written reflection, asking participants to write 100–200 words on how their mathematical thinking had been affected by the day's class work, discussions, and homework problems. This idea grew from what I'd discovered in preparing for the seminar about the role of metacognition from Bransford et al. (2000): when we monitor our own thinking and learning processes, we become better at learning. The teachers benefit, I expect, from taking some time each evening to articulate their own intellectual growth. The assignments also serve as a formative assessment that complements the class discussions and exit questions (questions or prompts that students respond to in writing at the end of class).

23.4 Mathematics for Pre-Service Teachers

Continuing my exploration of mathematics *as* social justice, and informed by my VMI experience, I developed the first mathematics content course specifically for future teachers at Middlebury College. The course is co-listed with our Department of Education Studies, and I have taught four versions. Each time I've had a largely female group; again their mathematical autobiographies at the start of the term often feature frustration and discouragement.

Here, too, a periodic reflection assignment has proven useful. I'm often touched by the thoughtfulness and honesty that my students show when given the chance. To read that some students admit that they never really understood place value until this course is shocking, at first, but makes me appreciate their determination to get it right this time. Students who started learning mathematics in Spanish or Vietnamese have interesting stories to tell about the transition to English.

Unlike my other mathematics courses, Mathematics for Teachers enrolls mostly students from groups underrepresented in mathematics. Now that I've started another term as department chair, I am engaging my colleagues in the questions that have emerged as I worked with these students. What is our obligation to such students at Middlebury? If the practices of mathematical thinking are beneficial, what opportunities can we offer those who have had negative mathematics experiences before college?

23.5 Reconsiderations of Standard Mathematics Courses

In the early 1990's, when I first started organizing my abstract algebra course in what we now call an Inquiry-Based Learning (IBL) model, I did so because it made intuitive sense. I thought about the difference between sitting passively in a lecture and working actively on a problem set. I remembered an evening as an undergraduate when I borrowed a friend's plastic cube, a Polaroid photo on each face, to work out the symmetry group. I couldn't remember anything that professor had said in class. Teaching at a college with small classes, fewer than 20 in most upper-level mathematics courses, it seemed natural to replace lectures with guided group activities.

My search for readings for the first-year seminar led me to the Common Core State Standards for Mathematics (National Governors Association 2010), an introduction to cognitive science (Bransford et al. 2000), Liping Ma's influential comparison between elementary teachers in the US and China (Ma 2010), and the work of mathematics educators like Deborah Ball and Heather Hill (Ball and Hill 2009). In absorbing those detailed analyses of the work elementary school teachers do, I realized that I'd never inspected my own teaching work as closely.

Once I added my work at VMI to the mix, my nagging discomfort with my own lecture model for calculus and linear algebra grew to a level that I could not ignore. Sitting next to elementary school teachers as they struggled with problems that I had thought were straightforward made me face the fact that I had too little information about how my calculus and linear algebra students were thinking. Reading about different kinds of knowledge confirmed that even students who were adept at various procedures might be missing the concepts underlying and connecting those procedures. Visiting fourth-grade classrooms reminded me that just because students are quiet does not mean they are engaged.

For years I had grumbled that most of my students didn't actually read the textbook. But why should they? Most of the tasks I was asking them to do—exercises at the back of each section—could be accomplished by watching me perform similar ones (after carefully explaining why they worked) and then flipping through the section for worked examples. I had regarded the class sessions as highly interactive, because students could ask questions at any time, and I threw out questions as well, but in retrospect it is clear that I was doing most of the heavy idea lifting.

I now realize that by limiting participation to students who raised their hands, I was not providing an equitable learning environment. Cold-calling on quiet students in front of the whole class was never an option for me, but listing "participation" on the syllabus as a component of the grade fell far short of facilitating quality engagement with the material for all students.

Now I'm committed to providing an IBL environment in all of my classes. In linear algebra, for example, I abandoned the textbook and its formulaic exercises in favor of homework assignments with more discovery-oriented and open-ended questions. Students discuss their work in groups of three or four for the first half of the class, and then we come together as a whole to share solutions and tie the new concepts to previous ones. As an antidote to some of my students' notions about the nature of mathematics, I stress the idea that there can be several acceptable solutions to many problems, as I've learned to do at VMI. I also listen carefully to my students' conversations with one another, and I have a pretty good idea of where everyone, even the quietest, is without having to wait for the next exam.

The case for active learning is growing. In fact, a meta-analysis published in 2014 concluded with the statement that "(t)he analysis supports theory claiming that calls to increase the number of students receiving STEM degrees could be answered, at least in part, by abandoning traditional lecturing in favor of active learning" (Freeman et al. 2014). Also of interest is the evidence in Kogen and Laursen (2013) that women may be more likely to persist in mathematics after taking an IBL course. What started as an effort to improve my teaching in general turns out to have social justice implications.

23.6 Reflections on My Professional Development

The strands of my professional life are interwoven rather than parallel. Their interactions are mutually beneficial, and remind me of something a senior colleague in history once told me about the role of her scholarship: "I can't be a good teacher unless I'm learning at the same time."

Just as my students' learning improves if they are actively engaged in appropriately challenging cognitive tasks, my own learning about education improves if, among other things, I write and speak about it. Some of my writing over the past 10 years has grown out of agitated responses to items in the news (Bremser 2013a, b). My "Listening to Teachers" essay in the AMS *Notices* (Bremser 2014b) grew out of impatience with blanket complaints about mathematics instruction that I'd heard from some fellow mathematicians, and my sense that we as a professional community could take more responsibility.

My writing generated invitations to speak about my VMI work and my Math for Teachers course at meetings of the American Mathematical Society (AMS). At one of those sessions, I heard about large universities where there was little communication between mathematicians and mathematics educators, despite the need for future teachers to take classes in both areas. As a result, I made a special effort to coordinate with my own colleague who teaches mathematics and science teaching methods in our Education Studies program, and rearranged my teaching schedule so that students could take my course and then hers.

My talks, in turn, generated an invitation to be one of the editors of "On Teaching and Learning Mathematics," an AMS Blog (AMS n.d.). Writing and editing posts for that blog has deepened my understanding of mathematics education at all levels, and supports my own teaching. For example, our series on active learning has encouraged me to evaluate my in-class and homework assignments more carefully in terms of their cognitive demand. Am I asking students to recite, apply, explain, evaluate, or create mathematics? What balance of tasks have I included?

Looking back at my first shift toward an active learning model in abstract algebra, I have to wonder whether I would have made such a fundamental change before tenure. Probably not, to be honest with myself. Though my department grants faculty members a great deal of autonomy, teaching evaluations play a large role in the college's review process, and I didn't know how students would respond to working in small groups each day. I needn't have worried. The evaluations for my first version in the new format were almost uniformly positive, the one exception being a student who wrote, "You are the expert. You should show us."

I've since learned to be more explicit, early and often, about why I structure my courses the way I do. My syllabi now contain this quote:

> Trying to come up with an answer rather than having it presented to you, or trying to solve a problem before being shown the solution, leads to better learning and longer retention of the correct answer or solution, even when your attempted response is wrong, so long as corrective feedback is provided (Brown et al. 2014, p. 101).

My professional activities related to social justice and K-12 education developed later, when the personal satisfaction they offered outweighed whatever I imagined might result from colleagues' disapproval. In this case, it is probably better that I concentrated my energy on number theory research early on. Beyond Middlebury, I have been engaged in many conversations about the concerns of untenured mathematics faculty who are—or want to be—involved in education-related work. Advice from senior colleagues already on board ranges from "Don't touch it until after tenure" to "We need you to contribute now" (I'm paraphrasing; for a more nuanced conversation, see Bremser (2014a), including the comments).

My own advice would be this: start with a careful assessment of your own department and institution, as well as your tolerance for risk. If you decide to jump in, be prepared to argue for the quality of your new work; your pure research colleagues may know nothing of the education landscape, or have misconceptions about it. If you're feeling cautious, look for small ways to lay the groundwork for a more secure time. For example, through a speakers' bureau, I gave over a dozen talks in schools over the years. I made contacts with dynamic teachers, and saw fascinating variations among classrooms, both of which informed my later work.

Trained as a pure mathematician, and retaining my appreciation for the beauty of abstract algebra and number theory, I have found a new cohesion among the components of my professional life by asking human questions. How do we measure fairness? What can I do to address inequities in my State, and in my own classroom? I've found more intellectual satisfaction from exploring these questions than I could have imagined when I started.

References

AMS. (n.d.). On teaching and learning mathematics, an AMS blog. Retrieved March 10, 2016 from http://blogs.ams.org/matheducation/.

Ball, D. L., & Hill, H. (2009). The curious—and crucial—case of mathematical knowledge for teaching. *Phi Delta Kappan, 91*(2), 68–71.

Bransford, J. D., Brown, A. L., & Cocking, R. R. (Eds.). (2000). *How people learn*. Washington, DC: The National Academies Press.

Bremser, P., Kimber, C., Root, R., & Weaver, S. (2009). Mathematics of, for, and as social justice. In K. Skubikowski, C. Wright, & R. Graf (Eds.), *Social justice education: Inviting faculty to transform their institutions* (pp. 131–147). Sterling, VA: Stylus.

Bremser, P. (2013a). Practice standard #3: Construct viable arguments and … math sugar off. personal blog. Retrieved March 10, 2016 from https://mathsugaroff.wordpress.com/2013/06/12/practice-standard-3-construct-viable-arguments-and/.

Bremser, P. (2013a). Aftermath: Limits to growth. *Math Horizons, 21*(2), 35.

Bremser, P. (2014a). Do mathematicians need new journals about education? American Mathematical Society, on teaching and learning mathematics blog. Retrieved March 10, 2016 from http://blogs.ams.org/matheducation/2014/07/10/do-mathematicians-need-new-journals--about-education/#sthash.2FCtX5ZC.RF3oHN13.dpbs.

Bremser, P. (2014a). Listening to teachers. *Notices of the American Mathematical Society, 61*(5), 512–514.

Brown, P. C., Roediger, H. L., & McDaniel, M. A. (2014). *Make it stick: The science of successful learning*. Cambridge, MA: Harvard University Press.

Freeman, S., Eddy, S. L., McDonough, M., Smith, M. K., Okoroafor, N., Jordt, H., & Wenderoth, M. P. (2014). Active learning increases student performance in science, engineering, and mathematics. *Proceedings of the National Academy of Sciences, 111*(23), 8410–8415.

Hadlock, C. R. (Ed.). (2005). *Mathematics in service to the community*. Washington, DC: Mathematical Association of America.

Kogen, M., & Laursen, S. (2013). Assessing long-term effects of inquiry-based learning: A case study from college mathematics. *Innovative Higher Education, 39*(3), 183–199.

Ma, L. (2010). *Knowing and teaching elementary mathematics*. New York: Routledge.

Madrigal, A. C. (2014, May 22). The racist housing policy that made your neighborhood. The Atlantic. Retrieved May 18, 2016 from http://www.theatlantic.com/business/archive/2014/05/the-racist-housing-policy-that-made-your-neighborhood/371439/.

National Governors Association. (2010). Common Core State Standards for Mathematics. Retrieved June 26, 2016 from http://www.corestandards.org/wp-content/uploads/Math_Standards1.pdf.

Pfaff, T. (n.d.). Sustainability math. Retrieved March 10, 2016 from http://www.sustainability-math.org/Materials.html.

Skubikowski, K., Wright, C., & Graf, R. (Eds.). (2009). *Social justice education*. Sterling, VA: Stylus.

Steen, L. (Ed.). (2001). Mathematics and democracy: The case for quantitative literacy. National Council on Education and the Disciplines.

United States Census Bureau. (2012). *2010 census of population and housing, population and housing unit counts, Vermont*. Washington, DC: U.S. Government Printing Office. Retrieved May 10, 2016 from https://www.census.gov/prod/cen2010/cph-2-47.pdf.

University of Vermont. (n.d.). Vermont mathematics Initiative. Retrieved March 10, 2016 from http://www.uvm.edu/~cems/mathstat/?Page=VMI/default.php&SM=VMI/_vmimenu.html.

Vermont Agency of Education. (2015). Free or reduced lunch percentages by school. Retrieved May 16, 2016 from http://education.vermont.gov/data.

Vermont State Mathematics Coalition. (n.d.). Vermont Mathematics Initiative. Retrieved June 1, 2016 from http://www.vtmathcoalition.org/math-initiative.

Part V
Benefitting the Public and the Larger Mathematical Community

Chapter 24
Popular Culture in Teaching, Scholarship, and Outreach: *The Simpsons* and *Futurama*

Sarah J. Greenwald

Abstract Subject to thoughtful analysis of the benefits and challenges, popular culture can be an ideal source of fun ways to connect students and the general public to mathematics. My colleague Andrew Nestler and I created, class-tested, and widely shared activities related to the Twentieth Century Fox television show *The Simpsons*. The scholarship of teaching and learning (SoTL) provides us with an analytic framework to develop, improve, and share our activities. We designed the activities to introduce or review important mathematical concepts and engage students. Later I expanded my interest into *Futurama*, another Twentieth Century Fox television show. I will describe informal outreach activities connected to both programs, including our educational website Simpsonsmath.com and my interactive lecture that audiences have accessed worldwide from a *Futurama* DVD. I will summarize the reception of my work by departmental colleagues, the institution, and the mathematical community. I will reflect on how this work has affected students and general audiences. I will also consider the direct and indirect impacts on my career and the unique challenges and rewards of working with popular culture in teaching, scholarship, and outreach.

Keywords Popular culture • Outreach • SoTL • *The Simpsons* • *Futurama*

MSC Codes
00A09
00A66
97A40
97A80
97D40

S.J. Greenwald (✉)
Department of Mathematics, Department of Cultural, Gender, and Global Studies, Appalachian State University, 121 Bodenheimer Drive, 326 Walker Hall, Boone, NC 28608, USA
e-mail: greenwaldsj@appstate.edu

J. Dewar et al. (eds.), *Mathematics Education*, Association for Women in Mathematics Series 7, DOI 10.1007/978-3-319-44950-0_24

24.1 Introduction

Educators have a variety of methods available to them to deliver mathematical content and facilitate student engagement and learning. Mathematical cartoons can be a fun way to introduce or review concepts and reduce student anxiety (Schacht and Stewart 1990). The audio and video components of animated cartoons can also help make material more memorable.

The Simpsons (Groening and Brooks 1989–present) and *Futurama* (Cohen and Groening 1999–2009) are animated sitcoms that include hundreds of mathematical references. One reason is that a number of the writers have mathematical backgrounds, including college and graduate degrees (Greenwald 2007). For example, Ken Keeler, who has a PhD in applied mathematics and originally worked at Bell Labs, wrote for both television shows.

For over 15 years, my co-author Andrew Nestler and I have been not only engaging students but also sharing with teachers mathematical humor from the television show *The Simpsons*. The Scholarship of Teaching and Learning (SoTL) gives us a framework in which we create valuable activities, analyze their benefits and challenges, refine them, and share them (McKinney 2006). A SoTL project may take many forms, and in our case we mainly use reflection and analysis based on observational research and student evaluations. We have found that the best popular culture related activities for the classroom are those that tie into course content, have an interactive component, and work at least as well as other pedagogical techniques would for the same material (Greenwald and Nestler 2004a, b). In this chapter I will discuss how this work developed as I highlight the unique challenges and rewards of working with popular culture in teaching, scholarship, and outreach. Through our educational website Simpsonsmath.com (Greenwald and Nestler 2001) I became involved with *Futurama*, another television show, and filmed a 25-minute feature for the *Futurama* DVD movie *Bender's Big Score* (Cohen et al. 2007). I will summarize the reception by the department of mathematical sciences at Appalachian State University (ASU), the institution, and the broader mathematical community. I will also reflect on how our work has affected students, general audiences, and my career.

24.2 *The Simpsons* and *Futurama* in the Classroom and Beyond

For years, Nestler and I analyzed the mathematical moments in *The Simpsons* as we developed and refined activities for our classes. We wanted to share these publicly with teachers and students at other schools, so we began to participate in informal outreach activities, including presentations at conferences and schools, as well as wider outreach through our educational website Simpsonsmath.com that we founded in 2001. We created classroom activity sheets for the website filled with questions designed to engage students with the mathematical moments in the show. These are aimed mainly at high school and college students and general audiences. Later my interest expanded into *Futurama* as well.

24.2.1 The Simpsons *and Simpsonsmath.com*

The Simpsons is an award-winning animated sitcom. It centers around the life of a nuclear family, Homer and Marge Simpson and their children Bart, Lisa and Maggie, as well as their neighbors and relatives. As of 2016 it is the longest running sitcom in television history and contains many references to scholars and academic subjects, including mathematics and its own mathematician character, Professor Frink. *The Simpsons* first aired in December of 1989 when I was an undergraduate student at Union College. My mother died the year before and I was the guardian of my younger brother, so time was precious; however, there was always a spare hour for friends. Our group gathered for dinner, television and the premiere of *The Simpsons*. The laughter was good for the soul and provided some much needed relaxation. There was even a quip about odds and multiplication, but we dismissed it as a fluke. The next episode aired in January 1990. It was "Bart the Genius" (Vitti and Silverman 1990) and we were surprised that it showcased an entire mathematics word problem, and even a separate calculus joke about derivatives. That same year Ernest Boyer's *Scholarship Reconsidered: Priorities of the Professoriate* (Boyer 1990) was published, and this was a pivotal moment in SoTL. Little did I know that these two events would later converge, to my great benefit. Group gatherings for dinner and *The Simpsons* continued in graduate school at the University of Pennsylvania. It is there that I met Andrew Nestler, a fellow graduate student, and we bonded over mathematics as well as *The Simpsons*.

The mathematical references in the show are diverse and range from basic arithmetic to advanced research topics in mathematics. For instance, *The Simpsons* showcases interesting numbers and equations, makes references to geometry and mathematical physics, and jokes about innumeracy and women in mathematics. Many of the mathematical moments appear briefly but prominently, e.g., in a close-up of a blackboard; however, in a few episodes mathematics has been key to the main plot. Second grader Lisa Simpson, the oldest daughter of the titular family, is featured in these episodes. For instance, in "Girls Just Want to Have Sums" (Selman and Kruse 2006) Lisa pretends to be a boy to do mathematics. Matt Selman wrote the episode in response to Lawrence Summers' controversial comments about the innate abilities of women in mathematics (Summers 2005). In "MoneyBart" (Long and Kruse 2010) Lisa uses sabermetrics, the statistical analysis of baseball. In yet another episode, "Mathlete's Feat" (Price and Polcino 2015), Lisa is a member of Springfield Elementary's Mathlympics Team. For a television show, the breadth and depth of the mathematical moments are quite remarkable.

Boyer (1990) noted that "Pedagogical procedures must be carefully planned, continuously examined, and relate directly to the subject taught," and we have taken these as guiding principles in our work. In developing Simpsonsmath.com, Nestler cataloged the mathematical references as I interviewed the writers and investigated their mathematical backgrounds. There are hundreds of items detailed in Nestler's online guide to mathematics and mathematicians on *The Simpsons*. We created separate webpages for the episodes where mathematics is fundamental to the main plot, and for each of the mathematician writers. We developed worksheets related to arithmetic, calculus, geometry, number theory, pre-calculus, probability and topology. We tested the worksheets in classes and refined them before we included them online. We chose a website format so we could share our work freely and update it regularly, rather than a book that would be outdated as soon as more references appeared on the show. Although we are on opposite sides of the country, we continue to improve the website and watch television together, by hitting play at the same time on our respective digital video recorders as we chat by phone.

24.2.2 Sample Activity on the Digits of π

Many references to π appear on *The Simpsons*, prompting us to develop activities related to the irrationality of π as well as unsolved questions about whether π is normal. A worksheet is available on Simpsonsmath.com for the following activity.

In the episode "Marge in Chains" (Oakley et al. 1993), Apu Nahasapeemapetilon, who runs a popular convenience store near the Simpsons, takes the stand during a court case. When the validity of his memory is challenged, he claims he can recite π to 40,000 decimal places, and notes that the 40,000th digit is one. My students laugh when Homer Simpson, the donut-loving patriarch of the Simpson family, responds: "Mmm, pie."

We ask students for the definition of π, how many digits they know by heart, the probability that a person would guess correctly if he had randomly guessed the 40,000th digit, and whether it is humanly possible to memorize that many digits. We have shared this activity with middle grade, high school, and college students on Pi Day, and it leads to interesting discussions about the world record for memorizing digits of π and why anyone would want to memorize or compute so many digits. We introduce these questions if the students do not.

Students are often amazed to learn that Hideaki Tomoyori, the world record holder during the making of and original airdate of this episode, knew exactly 40,000 digits. We share quotations that highlight Tomoyori's motivation and methodology, as well as a psychological study that investigated his cognitive abilities (Takahashi et al. 2006). Researchers compared Tomoyori to a control group and concluded that he was not superior. They attributed his achievement to extensive practice. We also discuss the motivation to calculate trillions of digits of π, noting that series algorithms are used to stress test computers, and number theory questions about the distributions of the digits are much more interesting than any specific digit.

Apu was correct in the episode—the 40,000th digit of π is indeed one, and students ask how this story ended up in an episode of *The Simpsons*. The show's writers wanted to honor Tomoyori's accomplishment. Mathematician Jon Borwein first shared the story with us of how the writers obtained the digit. The writers enlisted the help of mathematician and computer scientist David Bailey. We have a copy of the fax the writers sent to Bailey containing an image of Bart Simpson, the fourth grade son on the show. They asked Bailey for the 40,000th digit of π. At the time, Bailey was working at NASA, and he faxed all 40,000 digits. The writers told Nestler and me that they placed a huge pile of fax pages into another episode "22 Short Films About Springfield" in honor of all those pages Bailey sent. The pile elucidates the magnitude of that many digits of π and students laugh when I show them a picture of it. Students are also interested to learn that research on π continues. Bailey et al. (1997) published a series representation of π that we share with students who have taken calculus:

$$\pi = \sum_{i=0}^{\infty} \frac{1}{16^i}\left(\frac{4}{8i+1} - \frac{2}{8i+4} - \frac{1}{8i+5} - \frac{1}{8i+6}\right).$$

The Bailey-Borwein-Plouffe (BBP) formula can be used to compute binary digits of π in hexadecimal notation without the preceding digits, theoretically eliminating the need for many fax pages, at least in this setting.

24.2.3 Futurama, *DVD Feature, and Greenwaldian Theorem Activity*

In 2003, Art Benjamin, then co-editor of *Math Horizons*, a journal aimed at undergraduates, asked if I would write an article about *Futurama. Futurama* is a satirical science fiction cartoon that is often associated with *The Simpsons* because many of the same people developed and wrote for it. However, the focus of the show is quite different. In the pilot, Philip J. Fry accidentally falls into a cryogenic chamber and awakens 1000 years later, in the year 3000. Fry connects with his great-great-...-great nephew Professor Farnsworth, a scientist, inventor and owner of the Planet Express Delivery Company. Fry joins the delivery crew whose members include a one-eyed alien named Turanga Leela and a robot, Bender Bending Rodriguez, who is simply known as Bender in the show. *Futurama* revolves around their adventures.

Futurama makes many significant mathematical references, including the Goldbach conjecture, supersymmetric string theory, and taxicab numbers to name a few. Mathematical references abound in each show. I enlisted the help of chemical engineer Tom Georgoulias and computer scientist Marc Wichterich to write a *Math Horizons* article (Georgoulias et al. 2004). From this I developed my own educational website (Greenwald 2004). I subsequently met David X. Cohen, the head writer and an executive producer of *Futurama*, who holds a master's degree in computer science. He invited me to film a 25-minute feature presentation for the *Futurama* DVD movie *Bender's Big Score*. Live action is interspersed with animated content in this interactive mathematics lecture that is aimed at a general audience. I explore the mathematical moments in *Futurama* with help from the writers, executive producers, and animated characters. Cohen advertised it prominently on the back cover of the *Bender's Big Score* DVD as a special bonus feature.

As a surprise, Cohen included a "Greenwaldian theorem" on a blackboard in *Bender's Big Score* itself. In classes that explore non-Euclidean geometry or as a hands-on outreach activity, I can show Fig. 24.1, and we discuss why it is true. In the *Bender's Big Score* movie, Professor Farnsworth and Bubblegum Tate, a member of the Globetrotter's basketball and physics team, examine the blackboard. While I was certainly not the first to discover the spherical equation, I was thrilled to have my name up in the lights of the show. I created an activity sheet and posted it on my *Futurama* website. On a ball, as I demonstrate in the DVD feature and in the classroom, we mark three vertices of a spherical triangle from the north pole to the equator and over a bit. Next we use string to measure the lengths of a from the north pole to the equator and b along the equator. Now we create a Euclidean right triangle with a and b pulled tightly on the flat table. Students work in pairs because more than two hands are helpful to form the hypotenuse of that triangle, c_{flat}. Back on the sphere, we compare c_{flat} to the spherical c_{sphere}. As we place c_{flat} from the equator to the north pole, it is too long. Since c_{flat} satisfies the Pythagorean theorem,

$$a^2 + b^2 = c_{flat}^2 > c_{sphere}^2.$$

Fig. 24.1 Greenwaldian theorem (Image used with permission, courtesy of Twentieth Century Fox Home Entertainment, all rights reserved, *Futurama*™, © Twentieth Century Fox and its related companies)

While this is just an illustrative example, we next examine small and large right triangles in a dynamic geometry software program. We notice that the smaller the triangle, the flatter, and hence the Pythagorean theorem is closer to holding true. The opposite holds for large triangles.

24.3 Reception, Impact, and the Challenges and Rewards of Working with Popular Culture

SoTL, by definition (McKinney 2006), requires public sharing and review. McKinney advocates for SoTL in traditional formats, like peer-reviewed presentations and publications, as well as innovative formats to reach the public (McKinney 2012). One of the challenges of informal outreach is how to best judge its impact and success. My interactive mathematics lecture has been distributed on over one million *Futurama* DVDs worldwide, but I have no way of knowing how many people have actually watched it or how they have used it. In addition to the DVD sales, the Greenwaldian theorem has also aired on television and on streaming video sites, but it is impossible to gather data on how many people paid attention. A better measure is the number of visitors to Simpsonsmath.com, which lists over 800,000 page views. This count could be considered low because it does not include the first year of the site, Nestler's guide or my *Futurama* pages. On the other hand, since the count is by unique Internet Protocol (IP) address, the same viewer using different addresses will be counted multiple times.

Regardless of the difficulties in counting users, we do know the reach has been broad because we have received unsolicited feedback from all over the world, typically in the form of e-mail messages or from people who have approached us at conferences. We have other indications of success, such as those detailed below, even as we faced diverse challenges unique to the pedagogical use of popular culture.

24.3.1 Encouragement and Criticism from the Broader Mathematical Community

The broader mathematical community has consistently recognized and encouraged our outreach activities. They have also assisted us by sharing their critiques. In 2002 Brian Winkel, editor of the journal *Problems, Resources, and Issues in Mathematics Undergraduate Studies* (*PRIMUS*), asked Nestler and me to organize a special popular culture issue based on an upcoming conference session. At the same time, Winkel sent us long emails detailing what he referred to at the time as his severe bias against popular culture. We had many very interesting discussions about the value of studying and connecting to what the younger generation is in touch with, and how popular culture reflects cultural beliefs about mathematics and its value to society (Appelbaum 1995; Morrell 2002). These discussions were essential as they helped me formulate how I would justify my pedagogical use of popular culture to departmental colleagues and other teachers. Winkel (personal communication, February 29, 2016) recently noted: "It is pretty clear now that you and your colleagues were in the vanguard of appreciating mathematics and science in the popular culture and we all owe you our gratitude." Art Benjamin was also instrumental. I am not sure that I would have taken the time to work on the mathematics in *Futurama* had it not been for his encouragement to contribute an article on that topic to *Math Horizons*.

We have spoken to teachers at conferences for the Mathematical Association of America, the National Council of Teachers of Mathematics and the Ontario Association for Mathematics Education, to name a few, and the reception has been mostly positive. Huge audiences have attended our conference talks. The main area of critique has been that the activities are too high-level for elementary and middle grade students. Teachers have requested that we develop worksheets for their students. We did develop some activities with Carli Entin for *Scholastic Math* (Entin 2003) and I have used the Greenwaldian activity with middle grade students. However, even if elementary school children are watching *The Simpsons* or *Futurama*, I personally believe that the content is too mature for them. Both shows contain low-brow suggestive humor, especially *Futurama*, and *The Simpsons* movie was rated PG-13.

Teachers have suggested that we should distribute mathematical clips from the shows on our website or on DVD, but *The Simpsons* and *Futurama* are copyrighted by Twentieth Century Fox and this would violate educational fair use guidelines, as it is illegal to break DVD encoding to create clips. However, even without permission from the copyright holder or the purchase of performance rights, it is generally accepted that educators can still show a small portion played directly from a DVD owned by the library or an instructor in a face-to-face setting.

A few teachers told us that they were initially prepared to dislike the talk because it related to popular culture, but that they enjoyed the deep connections to mathematics and our thoughtfulness about effective use in the classroom. They say that they had not realized that popular culture could be a legitimate area of inquiry and investigation. On occasion, teachers send us their worksheets as well as feedback on the effectiveness of our worksheets. This has helped us improve our activities and we greatly appreciate the interaction.

Mathematical writers from *The Simpsons* and *Futurama* have been extremely positive and supportive, especially David X. Cohen, Ken Keeler and Jeff Westbrook. Westbrook has a PhD in computer science. In addition to their degrees, they have participated in the Joint Mathematics Meetings and a math club for adults. Popular science author Simon Singh interviewed us about our work and included it in a chapter of his book on the mathematics of *The Simpsons* (Singh 2013). His book has been well-received; for example the Joint Policy Board for Mathematics awarded Singh its 2016 Communications Award in part for this publication. The Mathematical Association of America honored me with a 2005 Alder teaching award based to some extent on the fact that Nestler's and my work had influenced others beyond my own classrooms, a criterion for the award.

24.3.2 Departmental and Institutional Challenges and Support

Overall, ASU has been very supportive of my work. Department and university documents recognize Boyer's model of scholarship (Boyer 1990), and they note that the scholarship of teaching is as significant and worthy a subject of inquiry as traditional research. The documents highlight public activities, forums, and presentations as valid forms of scholarship as long as there is indication of external validation. Hence, even though Nestler and I began sharing our work while I was an assistant professor, I felt encouraged to do so as a part of what would count as scholarship for my tenure portfolio.

One example of support from my colleagues in mathematical sciences occurred when my *Futurama* webpages indirectly overloaded our department server for an entire day in 2004. Slashdot, a website which advertises itself as "News for nerds, stuff that matters," linked to my *Futurama* pages as part of a discussion of the mathematics on the show. The sheer volume of users attempting to use the links resulted in too many page views per second, causing our server to come to a standstill, a phenomenon known as being "slashdotted." I stayed in communication with the department chair who helped diagnose the issue and explained the concept of the slashdot effect. The system administrator wrote to me: "No problem. At least I know what caused it. I'm going to look into what I can do to help… the next time it occurs," which seemed to me a very generous response. If there were any unhappy faculty, I never heard a word about it.

Another example of support from my department chair occurred when a member of the intellectual property department at Twentieth Century Fox called to ask Nestler and me to respond to their questions about our use of *The Simpsons* within and beyond our classrooms. Twentieth Century Fox was known for sending copyright infringement letters designed to shut down fan websites. My chair wrote a letter of support highlighting that this work was an integral part of my job at ASU and thus fell under educational fair use. Twentieth Century Fox never sent us a cease and desist letter.

I have been quite careful to explain to my colleagues that I only use *Futurama* or *The Simpsons* in the classroom when it relates to course goals and content I would have treated anyway, rather than using the shows themselves as the focus of investigation. For this reason when colleagues in other departments have suggested that I should teach a first-year seminar on *The Simpsons*, I have declined. I think that course would be hard to justify to mathematical sciences faculty. Instead, I created and regularly teach a first-year seminar on breakthroughs and controversies in science and mathematics, where I can still share my work as a small portion of the class. I am a faculty affiliate of the Gender, Women's and Sexuality Studies program and the director of the program requested that I design a new class on gender and popular culture. I am scheduled to teach this class in Fall 2016. Mathematical sciences chairs and colleagues have allowed me to teach outside of the department, partly because the university returns the credit hours to mathematical sciences and supports replacement costs.

That is not to say that there has always been universal support from the mathematical sciences department. In one of my departmental reviews, I was advised not to focus my scholarship on only one area. The faculty members were concerned that if *The Simpsons* ended, then my work would become obsolete long before the end of my career. None of us could have guessed at the time of the review that *The Simpsons* would still be on television today, and with the advent of streaming video sites, even cancelled shows like *Futurama* continue to be relevant to college students.

Outreach to the local community and beyond is a part of the university mission statement, so the university values my work on *The Simpsons* and *Futurama*. My department chair recently recommended that I be profiled as an example of a successful faculty member to prospective students. The university also referenced my outreach activities in a number of my teaching awards, giving some indication of its significance in this context.

24.3.3 Reception from Students and General Audiences

Ongoing reflection and analysis of student responses is fundamental for my SoTL work because it is the primary mode I use to assess and improve the activities. The best part of the work is that it helps me motivate students and better connect with them. I have seen shy students energized and students afraid of mathematics become willing to explore an activity just because the question is related to a cartoon. The sample 40,000th digit of π activity above is a good example of this. Discussion of the digits of π usually begins before I even ask any questions. It is also quite common for students who have been silent up to that point to approach me after I first use popular culture, to speak with me about their enjoyment of it, and this usually leads to increased participation overall. Of course, there is excitement from fans of *The Simpsons* or *Futurama*, but even among those who have never seen a show, many students are interested in anything connected to the entertainment world.

I have given expository talks all over the US and in Canada, and have been invited to speak at programs for specific groups of mathematics students, including career days at various high schools or colleges. Once in a while students at other venues I speak at are forced to attend as a part of their course grade. This is a more challenging setting, especially because some large lecture halls do not have desks, making certain activities impossible. The expository talks work best when audience members are willing and able to participate. Happily, that has been the case the vast majority of the time. A number of talks and interviews are designed for a general audience rather than groups of students, such as the National Museum of Mathematics (MoMath) and National Public Radio's (NPR) *Science Friday* program. Audiences have almost always responded extremely well in these settings, laughing, answering my questions and engaging in the mathematics.

Unsolicited comments or emails from students are not a great way to evaluate the success of activities, because I usually hear just positive comments in that context. Typical comments are that they never knew mathematics could be so fun and interesting, and that they are inspired to study further. Some faculty have done their own student evaluations in outreach programs at which I have spoken, and have provided me with summative feedback. I can judge the reception in my classes through comments in course evaluations. I have had a few negative comments from students who do not like the discussion of my work on *Futurama* in the context of a first-year seminar class. The same students indicated that they prefer classes whose teachers regularly let them out early. However, other students have noted that this was one of their favorite parts of class, and one of the course goals is for faculty members to introduce their research. It would be easy to dismiss the negative comments as those coming from non-serious students, but as part of SoTL I think it is imperative to ask what I can do better. The next time I teach the class I plan to have even more explicit discussions about the goals of the course and the point of sharing my work in that context. Humor is subjective and the more Nestler and I have to explain why a joke is funny, the less funny it is. For example, a non-native English speaker may not understand the reference to 'hardy har har' in the calculus joke from "Bart the Genius." The vast majority of the comments have been extremely positive, but only time will tell whether this continues to be true. Future students may connect less with these popular culture references than today's students do.

I evaluate my classes each semester to make improvements and modifications. Sometimes the improvement is to eliminate an activity from class or switch to a new technique. For example, early on some of the faculty teaching an introduction to mathematics for liberal arts asked students to fill out tax returns, so I tested Homer Simpson's taxes. Many of the students enjoyed it, but some did not. Upon SoTL assessment and reflection, it felt like I was drifting too close to making Homer the focus of investigation rather than the mathematics, so I removed the activity completely. I have learned that a cartoon does not actually have to be very funny for the majority of students to appreciate the reference. The key is that it relates to course content and will work at least as well as another pedagogical technique for that context. For instance, the *Futurama* episode "I, Roommate" (Horsted and Haaland 1999), brings M.C. Escher's 1953 *Relativity* lithograph to life in a humorous way as the robot Bender falls "down" staircases. The students label gluing instructions for the quotient space to identify the places on the edges of the room that are equivalent, by following Bender's path. The combination of dynamic movement and humor helps students visualize a finite universe in three dimensions better than other techniques. I could imagine a time in the future when I will need to reduce the activities in Introduction to Mathematics, which is the class where I utilize these references the most. I will continue to assess their use, but for now they work very well to engage the majority of my students. I actually deem the semester successful when I have some complaints from the student fans that we should have investigated more mathematical references from the shows, because I think it is good to underutilize, leaving them wanting more.

Overall, high school students and college students are usually much more excited about my work with *Futurama* than with *The Simpsons*, because *Futurama* has retained its cult status among that age group. Students I have never met before have come to my office or approached me around campus, even in the grocery store. One student was so excited that he interrupted a class he was not attending to ask for my autograph. It can be challenging to set boundaries with the students who are so excited about my work.

24.3.4 Impact on My Career

My contributions in this area are ongoing because *The Simpsons* is still airing new episodes, and these episodes sometimes contain new mathematical references for us to consider. As part of SoTL, I spend a lot of time thinking about how to best help students learn and understand course material, and in implementing diverse ways in order to help the class succeed, and so I regularly evaluate the effectiveness of pedagogical techniques to refine them. There is always something new and interesting to explore in this context.

When I try to reflect on direct and indirect impacts on my career, it is hard to measure the recognition and opportunities arising from our work. For instance, while it is easy to count the invited talks on this subject—currently there are 78—how do I interpret the teaching awards I have earned at least in part due to the outreach? Another example of how my career has been affected was being asked to co-edit the *PRIMUS* special issue on popular culture with Nestler. I co-edited a second issue with Chris Goff. We also wrote and published refereed articles focusing on SoTL (Goff and Greenwald 2007; Greenwald 2007; Greenwald and Nestler 2004a, b), which in turn led to an invitation to serve on the editorial board of *PRIMUS*. I have extended humor to linear algebra, where I have begun to develop and test comics I create myself. The content is not directly related to *The Simpsons* or *Futurama*, but it arose out of that work. My Erdős-Bacon number is a direct consequence of the activities. Mathematicians measure collaboration distance through an Erdős number, named for Paul Erdős. It is defined inductively using paper collaborations, with an Erdős number of one assigned to those who wrote a paper with Erdős and an Erdős number of $k+1$ assigned to those who wrote a paper with collaborators having an Erdős number k. Similarly, people measure connectivity in the Hollywood film industry using film roles and the actor Kevin Bacon. My work has possibly given me a Bacon number. A standard measure for Bacon numbers is inclusion on the Internet Movie Database (IMDb). By this measure, if you count documentaries, I have a Bacon number of three through David X. Cohen to Edward Asner to Bacon. My Erdős number is four. The Erdős-Bacon number is the sum of both, and is seven in my case, for those who count documentaries, and infinity otherwise. In turn, this has led to consulting work on a planned Erdős-Bacon number documentary.

Our mathematical outreach activities have allowed us to connect with many people, and in this way have enriched my own professional experiences. For instance, I participated in Raytheon's MathMovesU program for middle grade students at Upper Senate Park on Capitol Hill in Washington, DC, where I was delighted to speak alongside Senator Edward Kennedy, Representative Jo Ann Davis, mathematician Jonathan Farley and Olympic gold medalist Apolo Ohno. I am obviously a great fan of *The Simpsons* and *Futurama* and this work provides me with an outlet to enjoy the shows at a deeper level. My mother used to worry that all those hours I spent watching television were a waste of my time, so I find it especially rewarding to combine my interests in mathematics teaching and cartoons in this creative endeavor. It is a great intellectual challenge to use cartoons in outreach activities in a meaningful and effective manner. The rewards are well worth the substantial time and effort.

References

Appelbaum, P. (1995). *Popular culture, educational discourse, and mathematics*. Albany, NY: State University of New York Press.

Bailey, D., Borwein, P., & Plouffe, S. (1997). On the rapid computation of various polylogarithmic constants. *Mathematics of Computation, 66*(218), 903–913.

Boyer, E. (1990). *Scholarship reconsidered: Priorities of the professoriate*. Princeton, NJ: Carnegie Foundation.

Cohen, D. X., & Groening, M. (Executive Producers). (1999–2009). *Futurama* (television series). Boston, MA: Twentieth Century Fox.

Cohen, D. X., Groening, M., & Keeler, K. (Executive Producers). (2007). *Futurama: Bender's big score* (DVD). Boston, MA: Twentieth Century Fox Home Entertainment.

Entin, C. (2003). Math on *The Simpsons*. *Scholastic Math, 23*(9), cover, 4–5.

Georgoulias, T., Greenwald, S., & Wichterich, M. (2004). *Futurama* πk: Mathematics in the year 3000. *Math Horizons, 11*(4), 12–15.

Goff, C., & Greenwald, S. (2007). To boldly go: Current work and future directions in mathematics and popular culture. *PRIMUS, XVII*(1), 1–7.

Greenwald, S. (2004). *Futurama* math: Mathematics in the year 3000. Retrieved May 10, 2016 from http://cs.appstate.edu/~sjg/futurama/.

Greenwald, S. (2007). Klein's beer: *Futurama* comedy and writers in the classroom. *PRIMUS, XVII*(1), 52–66.

Greenwald, S., & Nestler, A. (2001). Simpsonsmath.com. Retrieved May 10, 2016 from http://simpsonsmath.com.

Greenwald, S., & Nestler, A. (2004a). r dr r: Engaging students with significant mathematical content from *The Simpsons*. *PRIMUS, XIV*(1), 29–39.

Greenwald, S., & Nestler, A. (2004b). Using popular culture in the mathematics and mathematics education classroom. *PRIMUS, XIV*(1), 1–4.

Groening, M., & Brooks, J. L. (Executive Producers). (1989-present). *The Simpsons* (television series). Boston, MA: Twentieth Century Fox.

Horsted, E. (Writer), & Haaland, B. (Director). (1999, April 6). I, Roommate (television series episode). In D. X. Cohen & M. Groening (Executive Producers), *Futurama*. Boston, MA: Twentieth Century Fox.

Long, T. (Writer), & Kruse, N. (Director). (2010, October 10). MoneyBart (television series episode). In M. Groening & J. L. Brooks (Executive Producers), *The Simpsons*. Boston, MA: Twentieth Century Fox.

McKinney, K. (2006). Attitudinal and structural factors contributing to challenges in the work of the scholarship of teaching and learning. *New Directions for Institutional Research, 129*, 37–50.

McKinney, K. (2012). Making a difference: Application of SoTL to enhance learning. *Journal of the Scholarship of Teaching and Learning, 12*(1), 1–7.

Morrell, E. (2002). Toward a critical pedagogy of popular culture: Literacy development among urban youth. *Journal of Adolescent & Adult Literacy, 46*(1), 72–77.

Oakley, B., (Writer), Weinstein, J. (Writer), & Reardon, J. (Director). (1993, May 6). Marge in chains (television series episode). In M. Groening & J. L. Brooks (Executive Producers), *The Simpsons*. Boston, MA: Twentieth Century Fox.

Price, M. (Writer), & Polcino, M. (Director). (2015, May 17). Mathlete's feat (television series episode). In M. Groening & J. L. Brooks (Executive Producers), *The Simpsons*. Boston, MA: Twentieth Century Fox.

Schacht, S., & Stewart, B. (1990). What's funny about statistics? A technique for reducing student anxiety. *Teaching Sociology, 18*(1), 52–56.

Selman, M. (Writer), & Kruse, N. (Director). (2006, April 30). Girls just want to have sums (television series episode). In M. Groening & J. L. Brooks (Executive Producers), *The Simpsons*. Boston, MA: Twentieth Century Fox.

Singh, S. (2013). *The Simpsons and their mathematical secrets*. New York: Bloomsbury.

Summers, L. H. (2005). Remarks at NBER conference on diversifying the science & engineering workforce. Retrieved May 11, 2016 from http://www.harvard.edu/president/speeches/summers2005/nber.php.

Takahashi, M., Shimizu, H., Saito, S., & Tomoyori, H. (2006). One percent ability and ninety-nine percent perspiration: A study of a Japanese memorist. *Journal of Experimental Psychology: Learning, Memory, and Cognition, 32*(5), 1195–1200.

Vitti, J. (Writer), & Silverman, D. (Director). (1990, January 14). Bart the genius (television series episode). In M. Groening & J. L. Brooks (Executive Producers), *The Simpsons*. Boston, MA: Twentieth Century Fox.

Chapter 25
Transforming Post-Secondary Education in Mathematics

Tara Holm

Abstract In this chapter I introduce and describe the work of mathematicians and mathematics educators in the group Transforming Post-Secondary Education in Mathematics (TPSE Math or TPSE, for short). TPSE aims to coordinate and drive constructive change in education in the mathematical sciences at 2- and 4-year colleges and universities across the nation. It seeks to build on the successes of the entire mathematical sciences community.

This chapter reviews the events that led to the founding of TPSE Math and articulates its vision and mission. In its first phase with national events, TPSE found broad consensus within the mathematical sciences community on the challenges facing the community. Learning from educational transformations in other scientific fields, and with the support of the Mathematical Advisory Group of 34 mathematical sciences department chairs and leaders, TPSE moves into a second phase, focused on action. This chapter is a snapshot in time; TPSE's continuing activities will be documented and disseminated. The chapter concludes with a reflection on the impact that my involvement in this work has had on my career.

Keywords Education policy • Higher education • Mathematics education • TPSE

25.1 Introduction

The education landscape has changed dramatically in the last half century. Higher education has become essential to economic mobility. At the same time, colleges, universities, and students are under severe financial pressure. And new pedagogies

---

MSC Codes
97B40
97B10
97A30

T. Holm (✉)
Department of Mathematics, Cornell University, Ithaca, NY 14853-4201, USA
e-mail: tara.holm@cornell.edu

J. Dewar et al. (eds.), *Mathematics Education*, Association for Women in Mathematics Series 7, DOI 10.1007/978-3-319-44950-0_25

and technologies make it possible to reach students in many more ways. These and other forces will change higher education (Bok 2013).

Mathematics departments play a central role in undergraduate education: few departments teach a larger percentage of the undergraduate student body. Mathematicians must respond to the challenges facing higher education. If we opt out, we risk losing the substantial role that mathematics departments currently play, and we endanger the health of the US mathematical sciences research enterprise.

A group of mathematicians and mathematics educators called Transforming Post-Secondary Education in Mathematics (TPSE Math or TPSE, pronounced "tipsy", for short) is working to support the mathematics community in this endeavor. Eric Friedlander, president of the American Mathematical Society (AMS) 2011–2012, asked me to join the TPSE leadership team when I was Chair of the AMS Committee on Education (CoE). I was invited by the editors to write this chapter relating my personal perspective on TPSE. Based on my involvement with TPSE, I chronicle the formation of TPSE and the foundation of work that it builds on, and report on TPSE's current partnerships and plans. I conclude with my personal history and experiences with TPSE.

25.2 Landscape Preceding TPSE

In the past 20 years, there have been many calls to improve mathematics instruction. With particular attention to research universities, the AMS Task Force on Excellence exhorted, "To ensure their institution's commitment to excellence in mathematics research, doctoral departments must pursue excellence in their instructional programs" (Ewing 1999, p. 3). Departments must maintain a relevant and broad curriculum. In addition to what they must teach, departments must also address questions of how to teach it (Ewing 1999). In 2003, Halpern and Hakel (2003, p. 38) noted, "it would be difficult to design an educational model that is more at odds with the findings of current research about human cognition than the one being used today at most colleges and universities."

The AMS Task Force on First-Year Mathematics (Lewis and Tucker 2009, p. 755) made three key suggestions towards the pursuit of excellent instruction:

- "Harness the power of technology to improve teaching and learning;
- "Leadership matters—success in this area depends upon the value assigned to it by a department's leadership;
- "Invest in teaching graduate students to be good teachers."

Preparing graduate students to teach, a particular role for research universities, is intertwined with any discussion of undergraduate education: graduate students represent the future of the professoriate. In their landmark Proceedings of the National Academy of Sciences study, Freeman et al. (2014) leave no doubt that active learning techniques improve student performance. Doctoral departments must adjust graduate teaching preparation accordingly, and offer support to all mathematics faculty members. Guidance for graduate students and current faculty must be broad, as no one technique will work in every classroom.

There are several aspects of the current post-secondary education landscape that particularly inform TPSE Math choices. Every 5 years, the Conference Board of Mathematical Sciences (CBMS) undertakes a statistical study of undergraduate programs in mathematics. One trend that their data make clear is that more and more students are enrolling in courses at 2-year colleges. In 2010, over 45 % of enrollments in mathematics, statistics and computer science courses taught in mathematics departments were taught in 2-year colleges (Blair et al. 2013, Table S.1). So TPSE made the commitment early on to include 2-year colleges and questions of credit transfer as an integral part of its work. Students in the twenty-first century are different from their mid-twentieth century counterparts. Babcock and Marks (2011) have analyzed a number of datasets to report that from 1961 to 2004, the amount of time a typical undergraduate spends on academic work has dropped by nearly a third, from around 40 h per week to 27. This is no doubt influenced by the dramatic increase in the cost of post-secondary education. The financial pressures that both students and universities face affect the types of programs that are feasible. Simultaneously, the increasing costs ratchet up the need to articulate the value mathematics adds to post-secondary education. Finally, there are now very good techniques available to assess the effectiveness of educational initiatives. TPSE Math is committed to evaluating the success of the projects it pursues and the innovations it endorses. It aims to cultivate collaborations between mathematicians, researchers in mathematics education and evaluators to strengthen assessment procedures.

There is renewed federal interest in higher education in general, and undergraduate science, technology, engineering and mathematics (STEM) education in particular. President Obama identified post-secondary education as key to a stronger economy and twenty-first century success of the nation. He asked the President's Council of Advisors on Science and Technology (PCAST) to prepare a report on producing one million more STEM graduates over the next decade. In that report, PCAST points to a US Department of Commerce report that projects a 17 % increase in the need for STEM-trained graduates over this time period (PCAST 2012). The mathematics community was taken aback when PCAST suggested that "faculty from mathematics-intensive disciplines other than mathematics" should develop and teach courses in college-level mathematics, and that there should be a "new pathway for producing K–12 mathematics teachers from … programs in mathematics-intensive fields other than mathematics" (PCAST 2012, p. 30). When writing its report, PCAST did not broadly consult the mathematics community. They were surprised not to find a journal focused on undergraduate mathematics education among AMS journals. They were not aware of the many successful innovations in post-secondary mathematics instruction (P. LePage, personal communication, October 22, 2015). This is an important lesson for the mathematics community: we must renew efforts to promulgate our successes beyond just our community.

In a more positive light, the National Research Council (2013) has described how mathematics has become essential to modern science, and recommends that undergraduate education in the mathematical sciences reflect this new stature. The good news is that all of these recommendations have spurred action within the mathematics community.

Calls for innovation and transformation of mathematics instruction are not new. Still the current spotlight on mathematics comes at a time when there is broad awareness of the challenges we face and an increased focus on educational outcomes. The mathematical sciences community has been and is increasingly involved in developing solutions. This is a rare opportunity to capitalize on the power of collective action and support the transformation of post-secondary education in the mathematical sciences.

25.3 Formation of TPSE Math

In February 2013, Carnegie Corporation of New York assembled a group of higher education leaders in the mathematical sciences to take stock and envision how to enhance the role of their field in post-secondary education. As a result of this meeting, Phillip Griffiths founded the group TPSE Math. TPSE has listened to and will continue to work with the mathematical sciences community to determine how best to achieve systemic change. It is now positioned to forge alliances among state and federal agencies, the policy community, university administrators, higher education associations, and professional organizations to secure the financial and structural support necessary to achieve these goals.

In May 2016, TPSE Math incorporated as an Educational Program affiliated with the University System of Maryland Foundation. The Board of Governors includes the following:

1. Phillip Griffiths (Board Chair) is Professor Emeritus of Mathematics and former Director of the Institute for Advanced Study. He was Provost of Duke University.
2. Eric Friedlander is the Dean's Professor of Mathematics at the University of Southern California and is a Past President of the AMS.
3. S. James Gates, Jr. is the John S. Toll Professor of Physics at the University of Maryland and a member of the President's Council of Advisors on Science and Technology (PCAST).
4. Mark Green is Professor Emeritus of Mathematics at the University of California, Los Angeles, and former Director of the Institute for Pure and Applied Mathematics (IPAM).
5. Tara Holm is Professor of Mathematics at Cornell University and former Chair of the AMS Committee on Education.
6. Karen Saxe is the DeWitt Wallace Professor of Mathematics at Macalester College, a past Vice President of the Mathematics Association of America (MAA), and a leader of MAA's Common Vision Project.

7. Uri Treisman is Professor of Mathematics and of Public Affairs at the University of Texas, Austin. He founded and directs the Charles A. Dana Center.

In 2015, William (Brit) Kirwan, Chancellor Emeritus of the University of Maryland, joined TPSE as a Senior Advisor. He chairs the Conference Board of the Mathematical Sciences (CBMS) and was appointed Executive Director of TPSE Math in May 2016.

TPSE Math envisions a future where postsecondary mathematics education will enable any student, regardless of his or her chosen program of study, to develop the mathematical knowledge and skills necessary for productive engagement in society and in the workplace. TPSE's mission statement articulates its vision (TPSE Math 2015):

> TPSE Math will facilitate an inclusive movement to strengthen postsecondary education in mathematics by working closely with—and mobilizing when necessary—faculty leaders, university administrations, membership associations, and relevant disciplinary societies in the pursuit of mathematically rich and relevant education for all students, whatever their chosen field of study. TPSE Math will identify innovative practices where they exist, advocate for innovation where they do not, and work with and through partners to implement and scale effective practices.

TPSE takes a national-level approach to transformation, seeking to leverage resources from non-profit foundations and federal agencies to increase the capacity of the profession to achieve change. The mathematical sciences community must proceed with coherence but not uniformity, ever heeding local needs of individual institutions to ensure appropriate changes will take root.

From 2013 to 2015, TPSE surveyed what is happening in the mathematics community through one national and four regional meetings. It engaged consultants from Parthenon-EY, a strategy consultancy, to help evaluate the state of post-secondary education in the mathematical sciences. Through meetings and research, TPSE has found a high level of consensus among faculty and administrators about the need for renewal of the post-secondary curriculum. Professional development opportunities can support interested faculty to develop and enhance their pedagogical practice. TPSE evaluated the capacity of the professional societies to support community-wide change. The following have been identified as areas where the TPSE leadership can have the biggest impact by working in concert with existing programs and societies to leverage existing capacity towards a common goal:

1. Curriculum pathways (lower and upper division, allowing students to reach the mathematics relevant to their field of study);
2. Graduate co-curricular training; and
3. Leadership development.

The TPSE Math leadership team has been successful at engaging large swaths of the mathematics community. In national and regional meetings, it has learned from and fostered conversations among departmental leaders from a broad range of institutions, college and university administrators, and representatives of federal and non-profit funding agencies. TPSE has brought together leaders of those professional societies that include post-secondary mathematics education as part of their primary mission:

1. American Mathematical Association of Two-Year Colleges (AMATYC).
2. American Mathematical Society (AMS).
3. American Statistical Association (ASA).
4. Association for Women in Mathematics (AWM).
5. Conference Board of the Mathematical Sciences (CBMS).
6. Mathematical Association of America (MAA).
7. Society for Industrial and Applied Mathematics (SIAM).

Starting in 2016, TPSE turned to the action phase of its work.

25.4 Foundation on Which TPSE Builds

25.4.1 History of Change in US Education in Mathematics

The last period of dramatic change in high school and college mathematics curricula began in the 1950s (Tucker 2015). The Cold War prompted unprecedented public support for science education and calculus became the ultimate goal of high school mathematics. Supported in part by the Ford Foundation, Advanced Placement (AP) calculus came into being. Since then, the AP calculus exam has shifted to being a test of calculus knowledge rather than more general problem solving (Bressoud et al. 2012). The variety of mathematics relevant to the world has expanded remarkably in the 60 years since then. We must open new pathways to offer students the mathematics they need. This is a particular challenge in mathematics, where theories do not become false or go out of fashion.

25.4.2 Work by the Professional Societies

One role that the AMS CoE plays is to cooperate with the other professional mathematics societies on matters concerning education. At the 2015 AMS CoE meeting, three other societies gave presentations about their current projects. In October 2015, David Bressoud (Macalester College) provided an analysis of the MAA's studies of first-semester calculus instruction across in the US (Bressoud 2015). Donna Lalonde (ASA) reported on the varied outreach efforts that ASA undertakes to engage students at all levels in the statistical sciences (LaLonde and Nichols 2015). Rachel Levy (Harvey Mudd College) is working with SIAM to create infrastructure to support internships in business, industry and government for mathematics students (Levy 2015). SIAM has also reported on the value of mathematical modeling in the K-12 and undergraduate curriculum (Turner et al. 2014), reflecting the increasing use and applications of mathematics across disciplines.

The MAA has been a leader of the mathematical community on the topic of undergraduate mathematics and a tireless supporter of educational initiatives.

Roughly every 10 years since 1953, the Committee on the Undergraduate Program in Mathematics (CUPM) has produced the benchmark curriculum guide for mathematics departments (MAA 2015). While the other professional societies are consulted during the preparation of each guide, the MAA provides the lion's share of support for the process. There is a new effort underway to create a companion instructional practices guide that will provide support for adopting new classroom practices. CUPM also has a subcommittee focused on Curriculum Renewal Across the First Two Years (CRAFTY) whose 2011 report (Ganter and Haver 2011) is particularly relevant to TPSE's work. One of the broadest impacts the MAA has had may be through the New Experiences in Teaching Project (Project NExT). Since 1994, this professional development program has helped over 1600 new faculty members become more effective teachers. After the yearlong program ends, NExT Fellows have access to the NExT listserv, providing continuous support and discussion about all matters related to teaching.

More recently, the MAA coordinated the *Common Vision* project, bringing together representatives from the five professional societies concerned with post-secondary mathematics education—AMATYC, AMS, ASA, MAA, and SIAM. Three members of the Common Vision leadership team are also affiliated with TPSE Math (the principal investigator Karen Saxe, Uri Treisman and myself). This project included a workshop and culminated in a report (Saxe and Braddy 2015) detailing the commonalities among the five societies' curricular recommendations and recording their shared opinions of the pressing need to transform post-secondary education in the mathematical sciences (Holm and Saxe 2016).

25.4.3 Innovations in Doctoral Departments

Each October, the AMS CoE meets in Washington DC and hosts a forum for discussion of issues in mathematics education. During my tenure as Chair of the CoE, the focus of the Committee's work shifted from K–12 education to post-secondary mathematics education, as this is the aspect of mathematics education most closely related to the daily work of AMS members. While we have come to expect a high level of commitment to undergraduate education in mathematics departments at primarily undergraduate teaching institutions, the CoE has been made aware of an impressive array of programs at research universities. A few examples from 2014 to 2015 meetings are described below, and more details are available at the AMS CoE website (AMS 2016).

Both the University of Michigan and its Mathematics Department are deeply committed to their educational mission. The 1999 AMS Task Force on Excellence found "a culture in the mathematics department that encourages and rewards innovation, one that is well-rounded, that strikes a balance between teaching and research, and that supports the work of students and colleagues at all levels" (Ewing 1999, p. 84). In his 2014 presentation, Stephen DeBacker of the Mathematics Department reported that innovations in active learning in first-year calculus have taken root and have now expanded to labs for second-year courses. Preparing graduate students and postdoctoral fellows to teach in the Michigan program ensures consistency and success. He cited as essential the full support of the university administration in this endeavor, and concluded, "A successful undergraduate program requires the efforts of nearly everyone in the department" (DeBacker 2014).

The Mathematics Department at the University of Illinois has developed several successful programs that reach a wide range of students studying mathematics (Ando 2014). Mathematicians have collaborated with engineering faculty to develop workshops where students must apply their calculus knowledge to solve problems inspired by real-world applications in engineering. The University uses Treisman's collaborative learning model, which it calls the Merit Program, to support students from traditionally underrepresented populations. Merit scholars are more successful than their peers in Illinois calculus. Finally, the Illinois Geometry Lab (http://www.math.illinois.edu/igl/) provides research experiences for 40 undergraduates each semester.

Through his San Francisco State University-Colombia Combinatorics initiative, Federico Ardila (2015) has forged connections between US students and their Colombian peers, each serving as role models for the other. Starting with the two principles that mathematical ability is uniformly distributed and that every student can have a meaningful mathematical experience, Ardila took an intentional approach to building a bridge between the two communities, using technology to offer courses and facilitate groups working together between the two countries. Of the 200 students who have participated in the program, 45 are currently working on their doctorates, including 26 women and 20 students from underrepresented groups.

The CoE also heard about the important role that the EDGE program and the National Alliance for Doctoral Studies in Mathematics have played in mentoring the next generation of mathematical scholars from groups traditionally underrepresented in the mathematical sciences, ensuring a diverse talent pool (Math Alliance 2013; Wilson 2015).

25.5 Building Systemic Change

In spite of the successes detailed here and the many more described on the AMS CoE web site, we still face significant challenges. Very few efforts are scaled or transferred; many rely on a charismatic individual for their continuation. Large research universities play a significant role in setting the standards for post-secondary education, but the mathematics research community has not been engaged in a coordinated way with undergraduate mathematics education. There is little more than informal cooperation among the mathematics professional societies, not through lack of will, but simply because each society is focused on its own mission. CBMS, the umbrella organization of all mathematical sciences societies, is understaffed and has too broad a mission to spearhead an initiative on undergraduate education. Despite the good will of its members, it is not in a position to generate change at the national scale.

By contrast, TPSE Math has a very precise focus, and strong connections to many college and university administrations, higher education associations, foundations, federal agencies, state government offices, and mathematics departments and societies. It aims to reinforce and augment the relationships that the professional societies have already built in the policy sphere. For these reasons, TPSE is especially well positioned to forge and sustain effective partnerships that will propel transformation in post-secondary education in mathematics.

Generating systemic change is a notoriously complex challenge. Fortunately, there are models in other disciplines that have been successful, aspects of which can be adapted for the mathematical sciences community

25.5.1 Life Sciences

The biological sciences changed dramatically in the second half of the twentieth century. The ultimate goal of understanding life remains constant, but the types of questions life scientists can ask and the tools available to answer them have developed rapidly. Beginning in 2008, the American Association for the Advancement of Science (AAAS) started a series of conversations with faculty, administrators, students, biological sciences professional societies, and funding agencies on the future of undergraduate biology education. First-year courses were completely overhauled, with careful attention to the desired outcomes for biology students and for general education students. Using the *Vision & Change in Undergraduate Biology Education* recommendations as a guide (Vision and Change 2010), the Partnership for Undergraduate Life Sciences Education (PULSE) has developed tools that promote department-wide implementation of new curricula and evidence-based pedagogies (PULSE 2014). A key feature of PULSE is its use of social connections, both in workshops and through a faculty ambassador network, to propagate change. This has been shown to lead to greater adoption of innovations than evidence presented via literature (Henderson and Dancy 2011). The landscape in the biological sciences is far more diverse than in the mathematical sciences. For example, there are dozens of professional societies of scientists in the life sciences. Nevertheless the essential aspects of PULSE could transfer to the mathematical sciences community.

25.5.2 Physical Sciences

Physicists and physics educators have received much attention for the development of a number of pedagogical techniques and assessment tools. The Force Concept Inventory (FCI) was the first test of its kind (Hestenes et al. 1992). It measures the change in students' conceptual understanding from the beginning to the end of a first physics course. Jerome Epstein has developed a Calculus Concept Inventory, which has not been adopted to nearly the same extent as the FCI (Epstein 2007). Eric Mazur and his colleagues have demonstrated the power of peer instruction to increase students' conceptual learning (Crouch and Mazur 2001). Carl Wieman was awarded the National Science Foundation (NSF) Director's Award for Distinguished Teaching Scholars in 2001, the same year that he won the Nobel prize. Wieman strongly advocates for activities that keep students actively engaged in exploring physics during class time. There are significant differences between first-year physics classrooms and first-year mathematics ones. In physics, students may be more homogeneous and there is not nearly as much variation in what a student's first post-secondary physics course could be. Still, we should analyze how and why it is that new instructional strategies are so much more quickly implemented by physical scientists, while being mindful that the reality of innovation may not be as widespread as self-reports lead us to believe (Dancy and Henderson 2010). It is also important to remember that active engagement in doing mathematics is a long-held value of many mathematicians. R. L. Moore and his followers have used a version of inquiry-based learning since the first half of the twentieth century (Wilder 1976). This technique is used in many advanced mathematics courses, and more recently is being adapted for appropriate use in first-year courses.

25.5.3 *First Steps in the Mathematical Sciences*

The mathematics community has achieved some transformation at the state level: mathematics departments at all post-secondary public institutions in certain states are beginning to work together. In 2013, the Ohio Mathematics Steering Committee was charged with "develop[ing] expectations and processes that result in each campus offering pathways in mathematics that yield

1. Increased success for students in the study of mathematics,
2. A higher percentage of students completing degree programs, and
3. Effective transferability of credits for students moving from one institution to another" (Ohio Mathematics Initiative 2014, p. 2).

This is indicative of how degree completion now dictates state policy in higher education. Ohio's college and university mathematics department chairs identified key challenges in achieving these goals and are now making progress towards addressing them. More and more states are following suit with their own mathematics task forces, many with the help of Treisman and the Dana Center. Going forward, transfer of credits from one higher education institution to another will become an increasing challenge, especially given financial pressures that are causing more students to divide their education among multiple institutions

25.6 Work by TPSE Math

From the beginning, TPSE has been intent on engaging the entire mathematics community. It has organized events at the Joint Mathematics Meetings (JMM), as well as a national and a series of regional TPSE meetings. The goal has been and is to work with the mathematics community to identify the most urgent issues, see how they are being addressed, and determine which early experiments and models can be scaled up and used by others. TPSE has also sought opinions from the "demand" side of the equation, including employers and disciplinary partners.

25.6.1 Preparation

TPSE's first public event was a standing-room-only panel discussion at the Baltimore JMM in 2014 (TPSE Math 2014a). Moderated by Philip Griffiths, panel members included Michelle Cahill (Carnegie Corporation of New York), Jo Handelsman (Yale University and now Associate Director for Science at the White House Office for Science and Technology Policy), Brit Kirwan and Joan Leitzel (Ohio Mathematics Initiative). Jerry McNerney, the only US Congressman with a PhD in Mathematics, also joined the conversation, providing valuable insights from his perspective on Capitol Hill. The discussions clarified how current governmental policy affects higher education in general and mathematics in particular. All agreed that we have a moment of opportunity for innovation and transformation of post-secondary mathematics education.

25.6.2 Information Gathering

In June 2014, TPSE organized a national meeting at the University of Texas, Austin, to bring together leaders from academia, business and government to discuss challenges and explore scalable solutions (TPSE Math 2014b). It was at this meeting that the following questions were articulated:

1. How can the undergraduate curriculum be reshaped to raise the level of numeracy among citizens and better align current teaching with the expanded role of mathematics?
2. How can pathways be opened to enable non-majors and developmental students to reach the level of numeracy needed for careers that demand analytical thinking and twenty-first century quantitative skills?
3. How will new technologies and teaching trends affect pedagogy and the economic model of mathematics departments?
4. How can a broader, more relevant undergraduate experience better prepare students for the workplace of the future, including interdisciplinary opportunities?

5. How can graduate students be equipped to teach more broadly about the uses of mathematics while maintaining depth in their own research?

Next came four regional meetings. Each meeting attracted participants from a variety of colleges, universities, and funding agencies. The meetings were open to all but TPSE began by using local contacts and the professional societies to invite a diverse group of stakeholders. The regional meetings were structured around panel discussions on a wide variety of topics. These are summarized as follows:

1. In November 2014, at the University of Maryland Baltimore County (UMBC), three panels considered these topics: (a) a description of TPSE Math and other initiatives with similar goals, (b) disparities in participation by various populations, and (c) issues facing non-R1 institutions. There were 50 participants.
2. In February 2015, at the University of California, Los Angeles (UCLA), three panels addressed these topics: (a) the role of mathematics in career preparation, (b) the role of (the field of) mathematics education in post-secondary education in the mathematical sciences, and (c) system-wide efforts to improve post-secondary education in the mathematical sciences. There were 49 participants.
3. In September 2015, at the University of Chicago, a 2-day meeting with six panels discussed: (a) the role of college and university administrations, (b) the role of mathematics departments, (c) preparation of graduate students as future faculty, (d) secondary school teacher training, (e) enhanced opportunities for highly motivated undergraduates, and (f) the role, relevance and reform of calculus. There were 63 participants.
4. In December 2015, at Duke University, a 2-day meeting offered five panels that considered: (a) multiple math pathways, (b) math and other disciplines, (c) math courses for non-STEM undergraduates, (c) adaptive learning, and (d) statistics and big data. Here, TPSE announced some of its next steps in the action phase (Sect. 25.6.4 below). There were 66 participants.

Further details and videos of some panels are available on the Meetings page of the TPSE Math website (TPSE Math 2015).

While each regional meeting had a slightly different focus, there emerged from all meetings a broad consensus among the mathematics community that all departments are facing pressure on these issues, some of the important work to address them has begun, and that we as a community must find a coherent way forward that allows for local variation. Moreover, transformation cannot rely on continuous additional resources. The declining state support of public universities is unlikely to reverse. Departments may be able to get one-time allocations to support a phase transition to more innovative curricula and pedagogies. To succeed in the long term, sustainability needs to be built in from the beginning.

25.6.3 Building Relationships

In addition to organizing meetings, the TPSE leadership has been building relationships with the professional societies, department leaders and mathematicians more broadly. It is committed to building a community of mathematics leaders who collectively drive the transformation process. At the San Antonio 2015 JMM, TPSE organized a discussion with the leadership of the professional societies and associations. It was an unusual gathering, possibly the first time that the leadership of AMATYC, AMS, the Association of Public and Land-grant Universities (APLU), CBMS, MAA, the National Council of Teachers of Mathematics (NCTM), and SIAM, as well as directors from NSF, all met at the same time and place. In March 2015, I met with mathematics department chairs from research universities at the Mathematical Sciences Research Institute's Sponsors' Day event. Green, Treisman and I gave an update on TPSE Math's activities to CBMS at their May 2015 meeting (Green et al. 2015). At all of these meetings, there was a strong consensus that now is the time to promote transformation.

25.6.4 Action Phase

Grounded in the consensus it found, TPSE Math now aims to advance several projects that are most likely to have a significant impact. These include partnering with associations and societies to achieve shared goals and seeking to provide a scaffolding to support the mathematics community efforts.

At a meeting in Washington DC in March 2015, TPSE initiated the Mathematics Advisory Group (MAG) (TPSE Math 2016). Funded by the Sloan Foundation, 34 department chairs and leaders have been identified who will gather to begin to develop an action plan to carry out, scale up and evaluate the effectiveness of major reforms. TPSE sought leaders who represent the gamut of higher education institutions, and who bring a diversity of views and experiences. This core group will be a key action and communication partner, advising TPSE on "grass roots" issues at the departmental level and helping identify successful and valuable models. The MAG will help TPSE convene a larger meeting of 100 to 200 department chairs and leaders to share information and establish partnerships of committed departments. Members of the MAG may become TPSE Math Ambassadors, willing to advise departments about transformation. Through the MAG and Ambassador network, TPSE also plans to strengthen the preparation of graduate students as educators and mentors.

TPSE Math has also embarked on several partnerships to enhance curriculum pathways. It will serve as an advisory partner to ITHAKA S+R (Ithaka 2004), a research group that studies the use of technology to improve teaching and learning, as well as the economic impact of such technologies. TPSE will also serve as an advisory partner to APLU, the American Association of State Colleges and Universities (AASCU), and the Dana Center in developing multiple pathways in lower division mathematics courses to improve completion rates and quality of instruction. TPSE also plans to promote renewal and creation of upper division curricula in response to the growing demands from other disciplines.

With the recent appointment of Kirwan as Executive Director, TPSE Math is preparing for the next stage of action, seeking substantial financial support from a number of non-profit donors and federal agencies to build expertise in data analytics to better assess needs and evaluate outcomes.

25.7 A Personal Journey to TPSE Math

As a faculty member in a research-intensive mathematics department, my principal job is mathematics research. I am also a practitioner in mathematics education, and I strive to use research findings in mathematics education to inform my classroom practice. Through my service in the AMS and now with TPSE Math, I work on transforming undergraduate mathematics education at the national level.

25.7.1 Path to Involvement

My personal interest in undergraduate mathematics education dates back to my days as a high school student taking my first college mathematics course. I am fortunate to have had outstanding teachers, and I am indebted to my mentors at each stage of my education.

As a graduate student at MIT, given only a couple days of teaching assistant training, my initial teaching assignments were recitations where I was forbidden to lecture, instead I was supposed to engage students in problem solving. I was one of three math graduate students to take an education course at MIT, discussing practical and theoretical aspects of general undergraduate education. That course certainly highlighted for me some of the ways in which mathematics is similar to and yet very different from other disciplines. I had the good fortune to spend a couple months teaching mathematics to computer science students in a non-traditional program at the short-lived ArsDigita University (ArsDigita 2002). This was an opportunity to teach incredibly motivated students using problem sets guiding them through parts of calculus, statistics and discrete mathematics. Together with Shai Simonson (Stonehill College), we engaged the students in a new-to-them research project that involved mathematics and computer experimentation to understand a card trick (Holm and Simonson 2003).

After graduate school, I spent 3 years at UC Berkeley, funded in part by an NSF postdoctoral fellowship. I taught three courses during my time there, and experienced first-hand the challenge of teaching upper division courses in large lecture format. After a year at the University of Connecticut, I started a tenure-track position at Cornell University. I was happy to find a position in a strong research department that also has a deep commitment to excellent undergraduate education. I was the first Cornell faculty member to participate in MAA's Project NExT. The sessions and materials from NExT workshops and the ongoing support through the electronic network have proved a tremendous resource in my teaching. One particular session, Joe Gallian's (University of Minnesota, Duluth) advice to "just say Yes" to opportunities to give back to the profession, has influenced my approach to service at my university and through the professional societies.

Soon after earning tenure, I was asked to run for election to the governing boards of the AMS and the AWM; I was elected to each in 2011 and 2012 respectively.

AMS Council members also serve on one of the five policy committees: Education, Meetings and Conferences, the Profession, Publications, or Science Policy. I was assigned to the Committee on Education (CoE), and chaired it from 2012 to 2016. During that time period, Eric Friedlander was President of AMS. One of his primary objectives as President was for the Society to increase its participation in the improvement of post-secondary education (E. M. Friedlander, personal communication, October 29, 2011). We worked together to shift the AMS CoE focus from K–12 to post-secondary education in the mathematical sciences. My fortuitous appointment to the CoE has opened opportunities to have an impact on undergraduate education in the mathematical sciences in ways that I could never have predicted. In particular, I was invited to join the leadership teams for the Common Vision project and TPSE Math as a direct result of my work with the AMS CoE.

The most pleasant surprise in my role as Chair of the AMS CoE was the opportunity to laud the work that mathematicians and mathematics educators are doing. In my AMS and more general mathematical travels, every mathematics department I have visited or heard about has some interesting project afoot. Faculty members want their students to engage deeply in thinking about mathematics. The professional societies seek to support their members' teaching mission. The mathematical sciences community must acknowledge and publicize these existing successes. Identifying the most promising innovations and determining the best way to adapt them and scale them for use at different institutions is no small task; this is at the heart of TPSE's mission.

A skeptic might point out that AMS CoE and TPSE Math meetings are assemblies of the willing, and ask whether the broader mathematics community is on board. Anecdotally, I have found general interest and support from mathematicians in the research community. For example, at the Spring 2016 Texas Geometry and Topology Conference, I was asked to give both a research talk and a second talk about my work with the AMS CoE and TPSE Math. Both talks were well attended, and the latter generated thoughtful discussions among all participants, from graduate students to the most senior topologists in the room (Holm 2016). Moving forward, when TPSE identifies departments to serve as lodestars, it will be important to select those that have high levels of faculty commitment to transformation. Their early successes will serve as models to promote change at all institutions.

25.7.2 Reflections

I conclude with a personal perspective. As indicated earlier, I am not a mathematics education researcher, but I believe strongly that the mathematical sciences community must maintain the bridges between researchers in mathematics education and practitioners of mathematics education, particularly at the post-secondary level. Moving forward we need to improve our communication and collaboration. Working with the AMS CoE and with TPSE Math, I have had the tremendous opportunity to engage with the senior leadership in mathematics and policy leaders in academia and government.

I count myself lucky to be a member of a supportive research department where faculty members are encouraged to contribute to all aspects of the profession. I have no illusions: my work with the AMS and TPSE Math did not get me tenure or promotion to full professor. It was considered a favorable part of my dossier, but my research is the *sine qua non*. These service opportunities did arise at a good time in my career. I had young children at home. Particularly while my second child was an infant and I was on parental leave, I appreciated the opportunity to be engaged with the mathematics community in this way. TPSE was in an early phase when most of the work consisted of phone meetings and email correspondence. This all fit into to the spare time I might find at odd times of the day. TPSE now involves more travel, but it is work that I continue to be able to fit in with the rest of my research and teaching. I welcome it as a chance to think at the community-wide and national level about the future of our profession.

Cornell University does provide strong support for its faculty. The Office of Faculty Development and Diversity offers a number of professional development opportunities and mentoring programs. For example, they have encouraged faculty to raise their voices beyond the walls of academe by offering a Public Voices fellowships through the Op-Ed project. The Center for Teaching Excellence offers workshops, lunches and logistical support for faculty who want to innovate in their classrooms (Cornell University 2012). They supported the Mathematics Department in bringing the Discovering the Art of Mathematics leaders (Fleron et al. 2008) to Cornell to sponsor a workshop for Cornell and Ithaca College faculty members introducing their teaching materials for introductory general education courses and their inquiry-based approach. Implementing innovative teaching practices has also been an attractive cause for university fund raising. By the end of 2015, donations had funded nearly $1 million in grants to faculty members to support curricular renewal and innovation (Cornell University 2016).

Through my work with TPSE Math, I have come to understand better the political and financial forces that are reshaping the way the public and the mathematics community perceive the role of mathematics in today's society and for the future. All mathematics departments are under pressure: the federal and state governments are curtailing their contributions to universities; university administrations are slashing resources; and everyone is demanding more from higher education institutions. The alignment of these forces creates an opportunity to work for systemic change. It is my hope that all levels of the mathematics community—from department colleagues and administrators to the leadership of professional societies—will come together and work to ensure that our students are prepared for a future we cannot yet imagine.

Acknowledgements The author thanks the anonymous referees and the editors for detailed reviews that greatly improved the structure of this chapter. TPSE Math is deeply grateful for the continuing support of the Arthur P. Sloan Foundation and Carnegie Corporation of New York.

References

AMS. (2016). *AMS Committee on Education*. Retrieved March 9, 2016 from http://www.ams.org/about-us/governance/committees/coe-home.

Ando, M. (2014). *Helping students do mathematics: A field report from one large public university*. Retrieved March 9, 2016 from http://www.ams.org/about-us/governance/committees/Ando.COE2014.pdf.

Ardila, F. (2015). *Building bridges to broaden and deepen representation*. Retrieved March 9, 2016 from http://www.ams.org/about-us/governance/committees/Ardila.COE2015.pdf.

ArsDigita. (2002). *ArsDigita University*. Retrieved March 1, 2016 from http://aduni.org.

Babcock, P. S., & Marks, M. (2011). The falling time cost of college: Evidence from half a century of time use data. *The Review of Economics and Statistics, 93*(2), 468–478.

Blair, R., Kirkman, E., & Maxwell, J. (2013). *Statistical abstract of undergraduate programs in the mathematical sciences in the United States*. Providence, RI: American Mathematical Society.

Bok, D. (2013). *Higher education in America*. Princeton, NJ: Princeton University Press.

Bressoud, D. (2015). Update on MAA's studies of calculus. Retrieved March 1, 2016 from http://www.ams.org/about-us/governance/committees/Bressoud.COE2015.ppt.

Bressoud, D., Camp, D., & Teague, D. (2012). *Background to the MAA/NCTM statement on calculus*. Retrieved March 1, 2016 from http://www.nctm.org/uploadedFiles/Standards_and_Positions/Position_Statements/MAA_NCTM_background.pdf.

Cornell University. (2012). *Center for teaching excellence*. Retrieved March 1, 2016 from http://www.cte.cornell.edu.

Cornell University. (2016). *Engaged Cornell*. Retrieved March 1, 2016 from http://engaged.cornell.edu.

Crouch, C. H., & Mazur, E. (2001). Peer instruction: Ten years of experience and results. *American Journal of Physics, 69*(9), 970–977.

Dancy, M., & Henderson, C. (2010). Pedagogical practices and instructional change of physics faculty. *American Journal of Physics, 78*(10), 1056–1063.

DeBacker, S. (2014). It takes a math department. Retrieved March 1, 2016 from http://www.ams.org/about-us/governance/committees/DeBacker.COE2014.pdf.

Epstein, J. (2007). Development and validation of the Calculus Concept Inventory. *Proceedings of the ninth international conference on mathematics education in a global community, 9*, 165–170.

Ewing, J. E. (Ed.). (1999). *Towards excellence: Leading a mathematics department in the 21st century*. Providence, RI: American Mathematical Society.

Fleron, J., Hotchkiss, P., Ecke, V., & von Renesse, C. (2008). Discovering the art of mathematics. Retrieved June 1, 2016 from https://www.artofmathematics.org.

Freeman, S., Eddy, S., McDonough, M., Smith, M., Okoroafor, N., Jordt, H., & Wenderoth, M. (2014). Active learning increases student performance in science, engineering, and mathematics. *Proceedings of the National Academy of Sciences, 111*(23), 8410–8415.

Ganter, S., & Haver, W. (Eds.). (2011). *Partner discipline recommendations for introductory college mathematics and the implications for college algebra*. Washington, DC: Mathematical Association of America.

Green, M., Holm, T., & Treisman, U. (2015). *TPSE math blog*. Retrieved March 1, 2016 from https://d3n8a8pro7vhmx.cloudfront.net/math/pages/155/attachments/original/1430745188/Green-Holm-Treisman_CBMS_5_1_15.pdf?1430745188.

Halpern, D., & Hakel, M. (2003, July/August). Applying the science of learning to the university and beyond: Teaching for long-term retention and transfer. *Change*, 36–41.

Henderson, C., & Dancy, M. (2011). *Increasing the impact and diffusion of STEM education innovations: A white paper commissioned for the characterizing the impact and diffusion of engineering education innovations forum, New Orleans, LA, February 7–8, 2011*. Retrieved March 16, 2016 from http://homepages.wmich.edu/~chenders/Publications/Henderson2011Diffusion%20of%20Engineering%20Education%20Inovations.pdf.

Hestenes, D., Wells, M., & Swackhammer, G. (1992). Force concept inventory. *Physics Teacher, 30*, 141–158.

Holm, T. (2016). *Reflections on the AMS Committee on Education*. Retrieved March 14, 2016 from https://d3n8a8pro7vhmx.cloudfront.net/math/pages/6/attachments/original/1457961569/Holm_CoE.pdf?1457961569.

Holm, T., & Saxe, K. (2016). A common vision for undergraduate mathematical sciences programs in 2025. *Notices of the American Mathematical Society, 63*(6), 630–634.

Holm, T., & Simonson, S. (2003). Using a card trick to teach discrete mathematics. *PRIMUS, 13*(3), 248–269.

Ithaka. (2004). Retrieved June 1, 2016 from http://www.sr.ithaka.org.

LaLonde, D., & Nichols, R. (2015). *ASA education and outreach programs*. Retrieved March 1, 2016 from http://www.ams.org/about-us/governance/committees/LaLondeNichols.COE2015.pptx.

Levy, R. (2015). *Industrial mathematics opportunities and career pathways for undergraduate and graduate students*. Retrieved March 1, 2016 from http://www.ams.org/about-us/governance/committees/Levy.COE2015.pptx.

Lewis, J., & Tucker, A. (2009). Report of the AMS first-year task force. *Notices of the American Mathematical Society, 56*(6), 754–760.

MAA. (2015). *Committee on the undergraduate program in mathematics*. Retrieved March 1, 2016 from http://www.maa.org/programs/faculty-and-departments/curriculum-department-guidelines-recommendations/cupm.

MAA. (2016). *MAA national studies of college calculus*. Washington DC: MAA. Retrieved March 9, 2016 from http://www.maa.org/programs/faculty-and-departments/curriculum-development-resources/national-studies-college-calculus.

Math Alliance. (2013). The National Alliance for Doctoral Studies in the Mathematical Sciences. Retrieved March 1, 2016 from http://mathalliance.org/.

National Research Council. (2013). *The mathematical sciences in 2025*. Washington DC: National Academies Press.

Ohio Mathematics Initiative. (2014). *Rethinking postsecondary mathematics*. Columbus, OH: Ohio Mathematics Initiative. Retrieved March 1, 2016 from https://ohiohighered.org/sites/ohiohighered.org/files/uploads/math/MATH-REPORT_FINAL_4.22.14.pdf.

PCAST. (2012). *Engage to excel: Producing one million additional college graduates with degrees in science, technology, engineering and mathematics*. Office of Science and Technology Policy, President's Council of Advisors on Science and Technology. Washington DC: PCAST. Retrieved June 1, 2016 from https://www.whitehouse.gov/sites/default/files/microsites/ostp/pcast-engage-to-excel-final_feb.pdf.

PULSE. (2014). *PULSE Home page*. Retrieved March 1, 2016 from http://www.pulsecommunity.org.

Saxe, K., & Braddy, L. (2015). A common vision for undergraduate mathematical sciences programs in 2025. Retrieved 2 July, 2016 from http://www.maa.org/sites/default/files/pdf/CommonVisionFinal.pdf.

TPSE Math. (2014a). *JMM panel discussion 1/17/14: Summary*. Retrieved March 1, 2016 from http://www.tpsemath.org/jmm_panel_summary.

TPSE Math. (2014b). *Transforming post-secondary education in mathematics*. Princeton, NJ: TPSE Math. Retrieved March 1, 2016 from https://d3n8a8pro7vhmx.cloudfront.net/math/pages/47/attachments/original/1415904260/TPSE_Report_pages_web.pdf?1415904260.

TPSE Math. (2015). Transforming post-secondary education in mathematics. Retrieved March 10, 2016 from http://www.tpsemath.org/.

TPSE Math. (2016). *Mathematics advisory group*. Retrieved March 1, 2016 from http://www.tpsemath.org/mag.

Tucker, A. (2015). The history of the undergraduate program in mathematics in the United States. In S. Kennedy et al. (Eds.), *A century of advancing mathematics* (pp. 219–238). Washington DC: Mathematical Association of America.

Turner, P., Crowley, J., Humpherys, J., Levy, R., Socha, K., Wasserstein, R. (2014). *SIAM reports: Modeling across the curriculum.* Philadelphia, PA: Society for Industrial and Applied Mathematics. Retrieved March 1, 2016 from http://www.siam.org/reports/ModelingAcross%20Curr_2014.pdf.

Vision & Change. (2010). *Vision and change home page*. Retrieved March 1, 2016 from http://visionandchange.org/.

Wilder, R. (1976). Robert Lee Moore 1882-1974. *Bulletin of the American Mathematical Society, 82*, 417–427.

Wilson, U. (2015). Lessons learned in building diversity. Retrieved March 1, 2016 from http://www.ams.org/about-us/governance/committees/about-us/governance/committees/Wilson.COE2015.pdf.

Index

J. Dewar et al. (eds.), *Mathematics Education*, Association for Women in Mathematics Series 7, DOI 10.1007/978-3-319-44950-0

T

Printed by Printforce, the Netherlands